METHODS AND TOOLS FOR DROUGHT ANALYSIS AND MANAGEMENT

Water Science and Technology Library

VOLUME 62

The titles published in this series are listed at the end of this volume.

METHODS AND TOOLS FOR DROUGHT ANALYSIS AND MANAGEMENT

edited by

GIUSEPPE ROSSI
Department of Civil and Environmental Engineering, University of Catania, Italy

TEODORO VEGA
Sicilian Regional Agency for Waste and Water, Water Observatory, Palermo, Italy

and

BRUNELLA BONACCORSO
Department of Civil and Environmental Engineering, University of Catania, Italy

Springer

A C.I.P. Catalogue record for this book is available from the Library of Congress.

ISBN 978-1-4020-5923-0 (HB)
ISBN 978-1-4020-5924-7 (e-book)

Published by Springer,
P.O. Box 17, 3300 AA Dordrecht, The Netherlands.

www.springer.com

Cover Illustration:
Drought management planning process. Figure 5 in Chapter 16 by Rossi, G., Castiglione, L. and Bonaccorso, B.: *Guidelines for planning and implementing drought mitigation measures.*

Printed on acid-free paper

TABLE OF CONTENTS

PART V DROUGHT IMPACTS AND MITIGATION MEASURES

LIST OF CONTRIBUTORS

Joaquín ANDREU ÁLVAREZ
Júcar Basin Agency, Environmental Ministry, Avenida Blasco Ibáñez 48.3, 46021 Valencia, Spain.
joaquin.andreu@chj.mma.es

Secondo BARBERO
Operative Centre, Piedmont Regional Agency for Environmental Protection, C.so Unione Sovietica 216, 10134 Torino, Italy.
secondo.barbero@regione.piemonte.it

Barbara BEGLIUTTI
Agency of Flumendosa Basin, Viale Elmas 116, 09123 Cagliari, Italy. c/o
paola.buscarinu@enteflumendosa.it

Brunella BONACCORSO
Department of Civil and Environmental Engineering, University of Catania, Viale A. Doria 6, 95125 Catania, Italy.
bbonacco@dica.unict.it

Isabella BORDI
Department of Physics, University of Rome "La Sapienza", Piazzale Aldo Moro 2, 00185 Roma, Italy.
Isabella.Bordi@roma1.infn.it

Lucio BOTARELLI
Hydrometeorological Service, Emilia Romagna Regional Agency for Environmental Protection, V.le Silvani 6, 40122 Bologna, Italy.
lbotarelli@arpa.emr.it

Cristiana BRAGALLI
Centre for Studies on Urban Hydraulics, c/o Department of Structural, Transport, Water Engineering and Territory Survey, University of Bologna, Viale del Risorgimento 2, 40136 Bologna, Italy.
cristiana.bragalli@mail.ing.unibo.it

Paola BUSCARINU
Agency of Flumendosa Basin, Via Mameli 88, 09122 Cagliari, Italy.
paola.buscarinu@enteflumendosa.it

Carlo CACCIAMANI
Hydrometeorological Service, Emilia Romagna Regional Agency for Environmental Protection, V.le Silvani 6, 40122 Bologna, Italy.
ccacciamani@arpa.emr.it

Giovanni CALCAGNO
Department of Soil Conservation, University of Calabria, Via P. Bucci, 41b, 87036 Arcavacata di Rende, Cosenza, Italy.
calcagno@dds.unical.it

Antonino CANCELLIERE
Department of Civil and Environmental Engineering, University of Catania, Viale A. Doria 6, 95125 Catania, Italy.
acance@dica.unict.it

Patrizia CAPIZZI
Department of Chemistry and Physics of the Earth and Applications to Georesources and Natural Hazards, University of Palermo, V. Archirafi 26, 90123 Palermo, Italy.
patrizia.capizzi@unipa.it

Leila CASTIGLIONE
Department of Civil and Environmental Engineering, University of Catania, Viale A. Doria 6, 95125 Catania, Italy.
lcasti@dica.unict.it

Pietro COSENTINO
Department of Chemistry and Physics of the Earth and Applications to Georesources and Natural Hazards, University of Palermo, V. Archirafi 26, 90123 Palermo, Italy.
pietro.cosentino@unipa.it

Giuseppe DI MAURO
Department of Civil and Environmental Engineering, University of Catania, Viale A. Doria 6, 95125 Catania, Italy.
gdimau@dica.unict.it

Vincenzo FERRARA
Department of Geological Sciences, University of Catania, C.so Italia 57, 95129 Catania, Italy.
vferrara@unict.it

Javier FERRER POLO
Water Resources Planning Office, Júcar Basin Agency, Environmental Ministry, Avenida Blasco Ibáñez, 48.3, 46021 Valencia, Spain.
jferrer@chj.mma.es

Gianluca FIANDACA
Department of Chemistry and Physics of the Earth and Applications to Georesources and Natural Hazards, University of Palermo, V. Archirafi, 26, 90123 Palermo, Italy.
gianluca.fiandaca@unipa.it

Giovanna FONTANA
Hydrometeorological Service, Emilia Romagna Regional Agency for Environmental Protection, V.le Silvani 6, 40122 Bologna, Italy.
gfontana@arpa.emr.it

Gabriele FRENI
Centre for Studies on Urban Hydraulics c/o Department of Hydraulic Engineering and Environmental Applications, University of Palermo, Viale delle Scienze Ed. 8, 90128 Palermo, Italy.
freni@idra.unipa.it

Goffredo LA LOGGIA
Centre for Studies on Urban Hydraulics c/o Department of Hydraulic Engineering and Environmental Applications, University of Palermo, Viale delle Scienze Ed. 8, 90128 Palermo, Italy.
glal@idra.unipa.it

Stefano MARCHESI
Hydrometeorological Service, Emilia Romagna Regional Agency for Environmental Protection, V.le Silvani 6, 40122 Bologna, Italy.
smarchesi@arpa.emr.it

Vittorio MARLETTO
Hydrometeorological Service, Emilia Romagna Regional Agency for Environmental Protection, V.le Silvani 6, 40122 Bologna, Italy.
vmarletto@arpa.emr.it

Gabriele MARRAS
Agency of Flumendosa Basin, Via Mameli 88, 09122 Cagliari, Italy.
gabriele.marras@enteflumendosa.it

Raffaele MARTORANA
Department of Chemistry and Physics of the Earth and Applications to Georesources and Natural Hazards, University of Palermo, V. Archirafi 26, 90123
Palermo, Italy.
r.martorana@unipa.it

Arcangela MATERA
Hydrometeorological Service, Emilia Romagna Regional Agency for Environmental Protection, V.le Silvani 6, 40122 Bologna, Italy.
amatera@arpa.emr.it

Giuseppe MENDICINO
Department of Soil Conservation, University of Calabria, Via P. Bucci, 41b, 87036 Arcavacata di Rende, Cosenza, Italy.
menjoe@dds.unical.it

Paolo MESSINA
Department of Chemistry and Physics of the Earth and Applications to Georesources and Natural Hazards, University of Palermo, V. Archirafi 26, 90123 Palermo, Italy.
paolo.messina@unipa.it

Giuseppina MONACELLI
Agency for Environmental Protection and Technical Services (APAT), V. Curtatone 3, 00185 Roma, Italy.
giuseppina.monacelli@apat.it

Elsa E. MOREIRA
Department of Mathematics, Faculty of Sciences and Technology, New University of Lisbon, Quinta da Torre, 2829-516 Caparica, Portugal.
efnm@fct.unl.pt

Antonella MORGILLO
Hydrometeorological Service, Emilia Romagna Regional Agency for Environmental Protection, V.le Silvani 6, 40122 Bologna, Italy.
amorgillo@arpa.emr.it

Dialecti PANGALOU
Laboratory of Reclamation Works and Water Resources Management, National Technical University of Athens, 9, Iroon Polytechniou str., 157 80, Athens, Greece.
pangalou@survey.ntua.gr

Javier PAREDES ARQUIOLA
Department of Hydraulic and Environmental Engineering, Polytechnic University of Valencia, Camino de Vera S/n, 46021 Valencia, Spain.
jparedea@hma.upv.es

Ana A. PAULO
Agricultural Engineering Research Center, Institute of Agronomy, Technical University of Lisbon, Tapada da Ajuda, 1349-017, Lisbon, Portugal.
apaulo@isa.utl.pt

Valentina PAVAN
Hydrometeorological Service, Emilia Romagna Regional Agency for Environmental Protection, V.le Silvani 6, 40122 Bologna, Italy.
vpavan@arpa.emr.it

Santino PELLERITO
Department of Chemistry and Physics of the Earth and Applications to Georesources and Natural Hazards, University of Palermo, V. Archirafi 26, 90123 Palermo, Italy.
sanpell@yahoo.it

Renata PELOSINI
Environmental Monitoring and Forecasting Area, Piedmont Regional Agency for Environmental Protection, C.so Unione Sovietica 216, 10134 Torino, Italy.
r.pelosini@arpa.piemonte.it

Luis S. PEREIRA
Agricultural Engineering Research Center, Institute of Agronomy, Technical University of Lisbon, Tapada da Ajuda, 1349-017 Lisbon, Portugal.
lspereira@isa.utl.pt

Miguel A. PÉREZ MARTÍN
Department of Hydraulic and Environmental Engineering, Polytechnic University of Valencia, Camino de Vera S/n, 46021 Valencia, Spain.
mperezm@hma.upv.es

Davide RABUFFETTI
Environmental Monitoring and Forecasting Area, Piedmont Regional Agency for Environmental Protection, C.so Unione Sovietica 216, 10134 Torino, Italy.
d.rabuffetti@arpa.piemonte.it

Christian RONCHI
Environmental Monitoring and Forecasting Area, Piedmont Regional Agency for Environmental Protection, C.so Unione Sovietica 216, 10134 Torino, Italy.
c.ronchi@arpa.piemonte.it

Ricardo D. ROSA,
Agricultural Engineering Research Center, Institute of Agronomy, Technical University of Lisbon, Tapada da Ajuda, 1349-017, Lisbon, Portugal.
ricrosa@isa.utl.pt

Giuseppe ROSSI
Department of Civil and Environmental Engineering, University of Catania, Viale A. Doria 6, 95125 Catania, Italy.
grossi@dica.unict.it

Alessio SALANDIN
Environmental Monitoring and Forecasting Area, Piedmont Regional Agency for Environmental Protection, C.so Unione Sovietica 216, 10134 Torino, Italy.
a.salandin@arpa.piemonte.it

Giovanni M. SECHI
Hydraulics Section – Department of Territorial Engineering, University of Cagliari, Piazza d'Armi, 09123 Cagliari, Italy.
sechi@unica.it

Alfonso SENATORE
Department of Soil Conservation, University of Calabria, Via P. Bucci, 41b, 87036 Arcavacata di Rende, Cosenza, Italy.
senatore@dds.unical.it

Andrea SULIS
Hydraulics Section – Department of Territorial Engineering, University of Cagliari, Piazza d'Armi, 09123 Cagliari, Italy.
andrea.sulis@unica.it

Alfonso SUTERA
Department of Physics, University of Rome “La Sapienza”, Piazzale Aldo Moro 2, 00185 Roma, Italy.
alfonso.sutera@roma1.infn.it

Dimitris TIGKAS
Laboratory of Reclamation Works and Water Resources Management, National Technical University of Athens, 9, Iroon Polytechniou str., 157 80, Athens, Greece.
dtigkas@mail.ntua.gr

Fausto TOMEI
Hydrometeorological Service, Emilia Romagna Regional Agency for Environmental Protection, V.le Silvani 6, 40122 Bologna, Italy.
ftomei@arpa.emr.it

George TSAKIRIS
Laboratory of Reclamation Works and Water Resources Management, National Technical University of Athens, 9, Iroon Polytechniou str., 157 80, Athens, Greece.
gtsakir@central.ntua.gr

Haris VANGELIS
Laboratory of Reclamation Works and Water Resources Management, National Technical University of Athens, 9, Iroon Polytechniou str., 157 80, Athens, Greece.
harrivag@central.ntua.gr

Antioco VARGIU
Environmental Monitoring and Forecasting Area, Piedmont Regional Agency for Environmental Protection, C.so Unione Sovietica 216, 10134 Torino, Italy.
antyvar@libero.it

Teodoro VEGA
Region of Sicily – Presidency, Via G. Bonsignore 1, 90135 Palermo, Italy.
teodoro.vega@tin.it

Pasquale VERSACE
Department of Soil Conservation, University of Calabria, Via P. Bucci, 41b, 87036 Arcavacata di Rende, Cosenza, Italy.
versace@dds.unical.it

Angel VILLALOBOS
Department of Hydraulic and Environmental Engineering, Polytechnic University of Valencia, Camino de Vera S/n, 46021 Valencia, Spain.
anvildea@doctor.upv.es

Franco ZINONI
Hydrometeorological Service, Emilia Romagna Regional Agency for Environmental Protection, V.le Silvani 6, 40122 Bologna, Italy.
fzinoni@arpa.emr.it

PREFACE

Frequent drought events that occurred recently in different Mediterranean regions have highlighted a general inadequacy of the current strategies for mitigating drought impacts on the different socio-economic sectors related to water use. In particular, lack of effective drought monitoring and forecasting systems, difficulties in transferring advanced methodologies for drought risk assessment to water managers, as well as the complexity in defining simple and objective criteria to properly select and implement mitigation measures, represent the main limits for an appropriate drought management policy. Such limits arise from the difficulty to shift from a reactive approach in facing drought, to a pro-active one, based on planning in advance suitable measures, including long-term actions to reduce drought vulnerability and short-term actions to mitigate the most harmful impacts.

These key issues have been tackled by universities and public agencies of four countries (Greece, Italy, Portugal and Spain) involved in the projects Sedemed (2003–2004) and Sedemed II (2004–2006), funded by the European Commission within the Programme Interreg IIIB – MEDOCC (Axis 4, Measure 4), coordinated by the Sicilian Regional Hydrographic Office (now Regional Agency for Waste and Water – Water Observatory). The main objectives of the projects included the definition of an integrated network for real time monitoring of drought processes, the development of a common framework of methodologies for drought analysis and forecasting, as well as the definition of appropriate mitigation strategies, to be shared among the Mediterranean countries.

Methods and Tools for Drought Analysis and Management is the result of the investigations carried out by a group of researchers and experts in the fields of hydro-meteorologic monitoring and water supply systems analysis and management. During the Sedemed projects they have had the opportunity to exchange their findings and points of view through several coordination meetings and workshops. Thus, the book reflects the skills and experiences of people from different organizations with different missions, but it is also the outcome of the interactions among the partners working in a common project.

In preparing the book a specific focus has been given to the methods and tools for collecting and processing hydrometeorological data for drought monitoring and forecasting that can be considered valid both from a scientific and practical point of view and on the modelling tools to improve water resources management under shortage conditions due to drought.

The book consists of 19 chapters divided in five parts. In the first part, methods and tools for drought monitoring and forecasting are presented, with special reference to the SPI index, analysed at different spatial scales. The second part is devoted to drought identification and characterization through the application of new or modified agro-meteorological indices and by remote sensing techniques. The third part refers to water resources management under drought conditions based on Decision Support Systems (DSS), including simulation and optimization models, also considering water quality aspects. The fourth part illustrates methods and tools for monitoring groundwater on the basis of hydrogeological and hydrodynamic characteristics of aquifers, with particular reference to degradation conditions associated to a reduction of precipitation input and sea-water intrusion. Finally, the fifth part presents guidelines and general criteria to select and implement drought mitigation measures, with special reference to the agricultural sector and to the urban areas.

We would like to thank all the contributors for the valuable work carried out during the development of the projects and for the further effort made in the preparation of the manuscript. Particular gratitude is expressed to Eng. Giuseppe Geraci for his encouragement to the initiative of a larger dissemination of Sedemed results, to Prof. Antonino Cancelliere for his precious scientific support in the review process and to Engs. Maria Teresa Noto and Guido Sciuto for the help in editing this book.

December 2006 The editors

PART I

DROUGHT MONITORING AND FORECASTING

CHAPTER 1

DROUGHT MONITORING AND FORECASTING AT LARGE SCALE

I. BORDI AND A. SUTERA

Department of Physics, University of Rome "La Sapienza", Rome, Italy

Abstract: In dealing with drought, a proactive approach requires a drought monitoring, a drought forecasting and an estimation of extreme dry event return times. For this purpose a useful tool is the drought bulletin where all the available information and analyses are collected and updated. In this chapter the main aspects of such approach are described and some sample applications for the Mediterranean area are shown. Results suggest that for a drought risk assessment there is not a unique analysis procedure; however, an ensemble of different kinds of analyses can provide useful information in preventing and minimizing the negative effects of drought events

Keywords: Drought monitoring, drought forecasting, extreme events, drought bulletin

1. INTRODUCTION

Drought is a normal, recurrent feature of climate. It occurs in all climatic zones, but its characteristics vary significantly from region to region. It differs from aridity that is restricted to low rainfall regions and is a permanent feature of climate. Defining drought is therefore difficult (Redmond, 2002). The Glossary of Meteorology (1959) defines a drought as "a period of abnormally dry weather sufficiently prolonged for the lack of water to cause serious hydrological imbalance in the affected area. Drought is a relative term, therefore any discussion in terms of precipitation deficit must refer to the particular precipitation-related activity that is under discussion". This means that whatever the definition, drought cannot be viewed solely as a physical phenomenon but it should be considered in relation to its impacts on society. The American Meteorological Society (1997) groups drought definitions and types into four categories: meteorological, agricultural, hydrological and socioeconomic. Meteorological drought is usually an expression of precipitation departure from normal conditions over a period of time, while agricultural drought occurs when there is not enough soil moisture to meet the needs of a particular crop at a particular time. This kind of drought happens after a meteorological one, this is because agriculture is usually the first economic sector to be affected by a drought. Hydrological drought refers to deficiencies in surface

G. Rossi et al. (eds.), Methods and Tools for Drought Analysis and Management, 3–27.

and subsurface water supplies due to precipitation reduction over an extended period of time. Finally, socioeconomic drought associates the supply and demand of some economic good with elements of meteorological, agricultural and hydrological drought. It occurs when water shortage starts to affect people and the water demand exceeds supply. On the other hand, drought produces several impacts that affect many sectors of the economy. This is because water is integral to our ability to produce goods and provide services. The impacts of drought are usually categorized as economic, environmental and social.

An objective evaluation of drought condition in a particular area is the first step for planning water resources in order to prevent and mitigate the negative impacts of future occurrences. For this purpose, along the years, several indexes have been developed to evaluate the water supply deficit in relation to the time duration of precipitation shortage (see Keyantash and Dracup, 2002, Heim, 2002 and references therein). Among them we mention the Percent of Normal, the Standardized Precipitation Index (SPI), the Palmer Drought Severity Index (PDSI), Deciles and the Crop Moisture Index (CMI). Usually these indexes are based on precipitation amount and measure the deviation of actual precipitation from a historically established norm. Some of them, instead, take into account also other climatological variables such as temperature, evapotranspiration or soil moisture. However, if we wish to compare drought conditions of different areas, which often have different hydrological balances, the most important characteristic of an index is its standardization. For this purpose, the SPI seems to be the most powerful index. It is a standardized index and can be computed on different time scales, so as to allow to monitor all the aforementioned kind of drought.

However, the reliability of drought analyses carried out by applying these indexes strongly depends on the quality of the primary data. In particular, in assessing dry (wet) periods, it is highly desirable to have at hand a data set that: (i) is easy to access, (ii) uniformly covers the globe, (iii) has a time-duration sufficiently long to be trustworthy in a statistical sense and (iv) is optimal in the sense of capturing consistently dry and wet events. Most of the available records may meet one or more of these requirements, but hardly they meet all of them (especially the first). Thus, in meteorological studies it has become a popular practise to disregard raw observations in favour of "analysed data", i.e. a set of observations which have been processed through several quality checks, including the ones of their consistency with atmospheric models of great complexity. In meteorology, the analysed fields are the result of complex interactions between available observations and model results. The final products of this procedure are uniformly gridded fields on a global scale of wind, temperature, specific humidity and mass that are released for further applications. In the latest decade or so, two of such re-analyses became easily available: one produced by the National Centers for Environmental Prediction/National Center for Atmospheric Research (NCEP/NCAR) and the other one by the European Centre for Medium-Range Weather Forecasts (ECMWF), the ERA-40 (see Kalnay *et al.*, 1996, Simmons and Gibson, 2000). In principle, these data satisfy the criteria above described and, therefore, they may be used to assess dry (wet) periods over the globe for approximately the last fifty years. A comprehensive check of such re-analyses and their reliability against observations is not yet available, especially when the purpose is the evaluation of the long-term aspects of climatic features, such as dry periods. Notice that this has a great deal of relevance for predicting the future behaviour of drought episodes. In the present chapter, an intercomparison between the performances of the two reanalyses in capturing the trend unveiled in some regions using observations (such as Sicily and China, Bonaccorso *et al.*, 2003; Bordi *et al.*, 2004) is presented, pointing out the positive outcomes and the negative ones.

Besides the limitations of the data, in recent years drought monitoring has been greatly improved: several indexes have been tested and also a synthesis of multiple indexes is operationally used by the most famous Drought Monitoring Centres in the world (as an example, see the URL http://www.drought.unl.edu/dm). Predicting future dry events remains, instead, a complex task, because of the random character of precipitation field, which is the basic variable commonly used for drought assessment. On the other hand, forecasting future dry events in a region is very important for finding sustainable solutions to water management and risk assessment of drought occurrences. A technique, commonly used in Statistics for predicting the future behavior of a given time series is the Auto Regressive model (AR, Coles, 2001). So that, if a long time series of precipitation is at hand, we may use AR model to estimate its future behavior. It is intuitive, however, that an AR method would extract the seasonal cycle of precipitation as the leading prediction. This forecast may have little part in predicting drought, since the latter depends strongly on the departure from the seasonal behavior of precipitation. An alternative approach, here proposed, may be the estimation of the probability function of monthly precipitation for forecasting (for a fixed probability) future precipitation values and, therefore, the corresponding drought index, such as for example the SPI.

Furthermore, for assessing risk of highly unusual events, such as extreme droughts, extreme value statistics need to be applied. The analysis of extremes is concerned with probabilistic and statistical questions related to very low/high values in sequences of random variables. It is usually based on the estimation of extreme event return times. The return time or return period of an event can be defined as the average number of observations to be made to obtain one observation equalling or exceeding its magnitude. Usually, two methods are applied to sample the original data: Annual Maximum (AM) and Partial Duration (PD) series, also known as Peaks Over Threshold series (Beguería, 2005). In the present chapter we apply these two methods for assessing the return times of dry/wet periods in a given area. Then, we analyse the SPI extremes for Sicily (which can be considered representative of the climatic condition characterizing the Mediterranean basin), illustrating how the index better describes dry/wet spells than precipitation does.

On these grounds, it is clear that an approach based on the assessment of drought risk needs all the available information derived from the analyses above. For this purpose a drought bulletin, where all the results are collected, seems to be a useful tool since it can provide the basic information for helping water resources managers in planning preventing and mitigation actions. In the present chapter we show a prototype drought bulletin for the Western Mediterranean area (hereafter MEDOCC area) where all the analyses carried out for the large scale are shown and updated every month.

The chapter is organized as follows. In Section 2 applications of some drought indexes are shown for the MEDOCC area and in Section 3 an analysis of climatic variability is shown using different re-analysis data. Section 4 is focused on the comparison between the results obtained from the AR and a new forecast method. In Section 5 the extreme values analysis of SPI time series for Sicily is described, while in Section 6 the drought bulletin is presented. In the final section, conclusions and suggestions for future investigations are discussed.

2. MONITORING DROUGHT

As mentioned in the Introduction, there are several indexes that measure how much precipitation for a given period of time has deviated from historically established norms. In the following paragraph we show some applications for the MEDOCC area using the NCEP/NACR re-analysis dataset.

2.1 Data and Methods

The input data needed for the computation of drought indexes are monthly precipitation and temperature fields that have been retrieved from the NCEP/NCAR re-analysis data set for the period spanning from January 1948 to June 2006. These fields are $1.9° \times 1.9°$ or $2.5° \times 2.5°$ grid spacing in longitude and latitude gridded precipitation rates or temperature respectively. They have been derived from the primary meteorological fields by means of a re-analysis procedure. Details about the assimilation model and the re-analysis project can be found in Kalnay *et al.* (1996). We use NCEP/NCAR re-analysis both for its time extension, which allows a large-scale assessment of drought, and because the data are easily accessible through the net.

Among the indexes available in the international literature, the most commonly used for drought monitoring are the Percent of Normal, the SPI and the PDSI. The index Percent of Normal is computed by dividing actual precipitation by normal precipitation, typically considered to be a 30-year mean, and multiplying by 100%. This can be computed for a variety of time scales. Usually these time scales range from a single month to a group of months representing a particular season, to an annual or water year. Normal precipitation for a specific location is considered to be 100%. One of the disadvantages of using the percent of normal precipitation is that the mean, or average, precipitation is often not the same as the median precipitation, which is the value exceeded by 50% of the precipitation occurrences in a long-term climate record. The reason for this is that precipitation on monthly or seasonal scales does not have a normal distribution. This is a great limitation when we wish to compare climatic conditions of different areas. This shortcoming has been overcame by the SPI because it is standardized. The SPI has been introduced by McKee (1993) and is based on precipitation field alone. It was designed to quantify the precipitation deficit for multiple time scales, which reflect the impact of drought on the availability of the different water resources. For example, groundwater, streamflow and reservoir storage reflect the longer-term precipitation anomalies. The index computation for any location is based on the long-term precipitation record cumulated over the selected time scale. This long-term record is fitted to a probability distribution (usually a Gamma distribution, Guttman, 1999), which is then transformed through an equal-probability transformation into a normal distribution. Positive SPI values indicate greater than median precipitation, and negative values indicate less than median precipitation (Bordi and Sutera, 2001). Being the SPI normalized, wetter and drier climates can be represented in the same way, and wet periods can also be monitored using the SPI.

Before the introduction of the SPI, which has several positive characteristics, the PDSI was widely applied for drought monitoring purposes. The index was introduced by Palmer (1965) and it is based on the supply-and-demand concept of the water balance equation for a two-layer soil model. Initially, several local coefficients are calculated which define local hydrological norms related to temperature and precipitation averaged over some calibration period (at least 30-yr period, according to the World Meteorological Organization recommendation). The basis of the index is the difference between the amount of precipitation required to retain a normal water balance level and the actual precipitation. The calculation of the coefficients above depends heavily on the soil water capacity of the underlying layer (i.e. AWC, Available Water Capacity). We have computed the PDSI for the Mediterranean region setting the AWC to the mean value of 150 mm (Bordi and Sutera, 2001).

2.2 Applications

In this subsection maps of Percent of Normal, SPI and PDSI for the MEDOCC area are shown (see Figure 1). They are updated to June 2006. The Percent of Normal on 3-month time scale shows dry conditions in almost the whole area with the exception of the central Italy. On longer time scale (24-months) the central and southern Italy is under normal conditions while the remaining regions are affected by drought. The SPI partially confirms this analysis. Severely or extremely dry conditions characterize the Alps and the southern France, while moderately dry conditions affect the Mediterranean basin. Most of western Italy is under normal conditions. On longer time scale, Italy is still characterized by wet conditions with the exception of the Alps, and the remaining Mediterranean

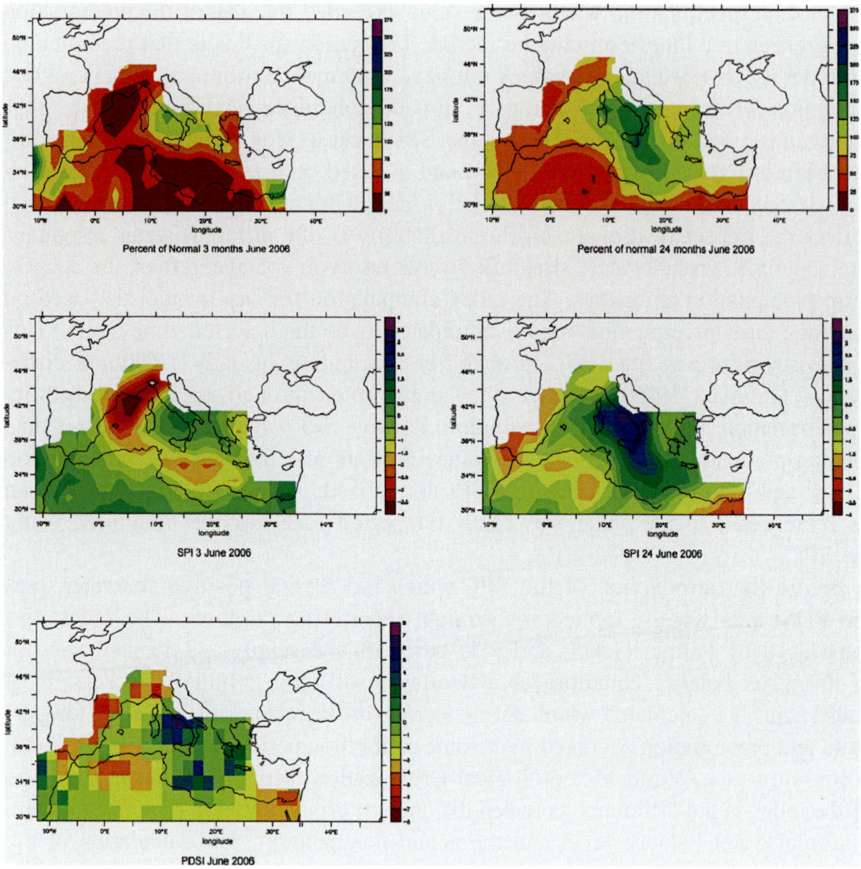

Figure 1. Percent of Normal on 3 and 24-month time scales, SPI on 3 and 24-month time scales and PDSI. Maps refer to June 2006

regions are affected by dry events. Finally, the PDSI shows a climatic condition very close to that derived by the computation of the SPI on 24-month time scale.

3. DROUGHT AND LONG-TERM CLIMATIC VARIABILITY

The long-term aspects of drought over the globe during the last decades have been evaluated by computing the Standardized Precipitation Index on 24-month time scale using the NCEP/NCAR and ERA-40 data sets. The SPI, in fact, seems to be a useful tool for monitoring dry and wet periods on multiple time scales and comparing climatic conditions of areas governed by different hydrological regimes. To reveal possible discrepancies between the analyses carried out with the two data sets, we studied the leading space-time variability of drought by applying the Principal Component Analysis (PCA) to the SPI time series.

3.1 Data and Methods

The two data sets used, the NCEP/NCAR and the ERA-40, have different spatial resolutions and different assimilation schemes have been applied in their re-analysis procedures. The ERA-40 precipitation data are available at 2.5° × 2.5° regular latitude/longitude grid, while, as mentioned before, those from the NCEP/NCAR have 1.9° × 1.9° horizontal resolution. The ECMWF data are provided every six hours and it is possible to download total precipitation or its different components (i.e. convective and stratiform precipitation), while daily or monthly precipitation rates are available for the NCEP/NCAR data set. The NCEP/NCAR re-analysis, which is available back to 1948, comprises different data sources such as observations from land stations and ships, upper air rawinsondes, satellite and numerical weather forecasts, which are assimilated in an AGCM (Atmospheric Global Circulation Model) and re-analysed by means of a "frozen" state of an AGCM (for more information see Kalnay *et al.*, 1996). The ECMWF 40-year re-analysis project (ERA-40, Simmons and Gibson, 2000) has been recently finalised and the data are available for climatic studies. This analysis of the state of the atmosphere, which covers the period from September 1957 to August 2002, complements the hitherto available NCEP/NCAR and ERA-15 re-analyses. The ERA-40 project applies a modern Variational Data Assimilation technique (used in daily operational numerical forecasting at ECMWF) to the past conventional and satellite observations. Because the two data sets cover different periods, we choose the common one ranging from January 1958 to December 2001. We carry out the analysis for the globe and the Euro-Mediterranean region (the area selected is 27.5°-70°N, 12.5°-62.5°E).

The leading time and spatial variability of drought has been investigated by applying the Principal Component Analysis to the SPI on 24-month time scale (SPI-24). The PCA is a classical statistical method widely used in data analysis, for identifying patterns and compression, reducing the number of dimensions. The method consists in computing the covariance matrix of the data with the corresponding eigenvalues and eigenvectors (see for details Rencher, 1998, Peixoto and Oort, 1992, Bordi et al., 2006). In guiding a proper interpretation of the results shown in the next subsection, we remark that the spatial patterns (eigenvectors), properly normalised (divided by their Euclidean norm and multiplied by the square root of the corresponding eigenvalues), are called "loadings"; they represent the correlation between the original data (in our case, the SPI-24 time series at single grid points) and the corresponding principal component time series.

3.2 Results

Let us consider, as shown in Figure 2a and b, the first loadings obtained by decomposing the total variance of the SPI-24 by means of a PCA for both ERA-40 and NCEP/NCAR data. They explain respectively 28.2% and 18.2% of the total variance. The associated principal component scores are shown in Figure 2c. The

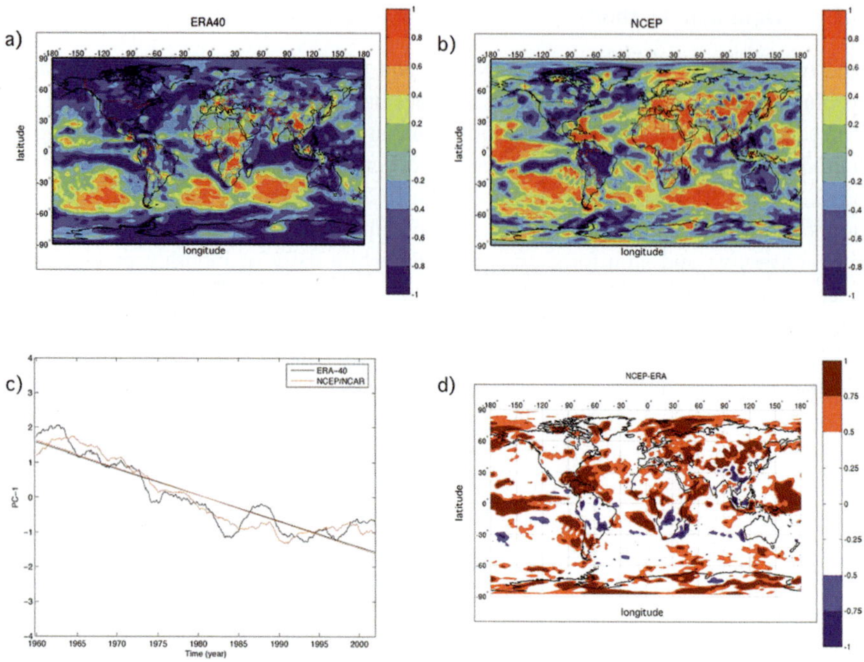

Figure 2. (a) First loading pattern of the SPI-24 for the globe obtained using the ERA-40 data set, contour interval is 0.2; (b) as in a) but for NCEP/NCAR data set; (c) first PC scores for ERA-40 (black line) and NCEP/NCAR (red line), straight lines denote the fitted linear trends; (d) differences between the first loading of NCEP/NCAR and that of ERA-40, absolute values less than 0.5 are denoted by white colour

two signals have a high correlation coefficient (0.96) and denote a long-term linear trend (see the straight lines in the figure) crossing the time axis around the eighties.

It can be noted that the two re-analyses reveal a common linear trend in the first principal component of the SPI-24 field, which explains more than 80% of the total variation of the signal. The presence of this long-term trend means that, looking Figure 2a and b, the red (blue) areas have been switched from prevalent wet (dry) conditions to prevalent dry (wet) conditions. In particular, the first loading for ERA–40 has positive correlation with the corresponding score greater than 0.5 in about 5.4% of the total grid points and negative correlation less than −0.5 in about 44.9% of points. For NCEP/NCAR the percentage of grid points showing values greater than 0.5 is about 15.0%, while that with values less than −0.5 is about 13.5%. This means that in most of grid points the SPI-24 time series for ERA-40 have a high anti-correlation with the PC score shown in Figure 2c; such behaviour is not confirmed by the NCEP/NCAR data set. On the other hand, the integrals in spherical coordinates of the first loadings provide values of −0.16 for ERA-40 and 0.03 for NCEP/NCAR, denoting the presence of a weak "global" trend towards wet conditions for the ECMWF re-analysis and the absence of a "global" linear trend for the other data set, i.e. the areas in the world characterised by positive/negative trends balance themselves.

In illustrating more quantitatively the differences we first interpolate the two loadings on a common grid of $1° \times 1°$ degree and then compute their difference (NCEP minus ERA-40, see Figure 2d). It can be seen that the two loadings have a good agreement (absolute differences less than 0.5) in about 65.3% of points. Thus, we may conclude that in the last forty years or so, a linear trend in the SPI-24 is detected by the two independent data sets, although the locations where this trend should be observed most likely remain not uniquely identified by the two re-analyses.

To be sure that these differences are not due to the coarse spatial resolution of the ERA-40 precipitation field, we employed the 'kriging' technique (Cressie, 1991) to this data set and repeated the analysis (i.e., we compute the SPI-24 and apply the PCA). The analysis (here not shown) provides results in agreement with those obtained by using the original precipitation data, suggesting that the origin of the detected differences cannot be uniquely attributed to the spatial resolution.

In Figure 3a–c the first loading patterns and scores of the SPI-24 computed for the European area using ERA-40 and NCEP/NCAR precipitation data are shown. They explain, respectively, 22.9% and 21.7% of the total variance. Both PC scores are characterised by a linear trend superimposed to short-term fluctuations, which provide a low degree of co-variability between the two signals. The correlation coefficient of the two PC scores is, in fact, 0.53. The trend statistics suggests that the linear fits explain different percentages of the scores variability (ERA-40 about 33% and NCEP about 72%) even if the NCEP/NCAR trend is within the error band of that for ERA-40. On the other hand, the leading spatial pattern for ERA-40, shows positive correlation between the SPI-24 time series and the corresponding score in the Balkans, Italy, central Europe and Spain, and negative correlations elsewhere. The first loading for NCEP/NCAR, instead, reaches maximum values in North Africa, central Spain, north-eastern Europe, part of Italy, Balkans, Greece and Middle East. For ERA-40 the first loading has no grid points with values greater than 0.5, while about 39% of grid points have values less than –0.5. The NCEP/NCAR loading, instead, has positive values greater than 0.5 in about 37% of points, while values less than –0.5 in only 2% of points. This means that the loading for ERA-40 shows prevalent negative correlations between the SPI-24 time series and the corresponding PC score, with opposite occurrences in the case of NCEP/NCAR. Although the spatial patterns seem to preserve the main features shown by the global scale analysis, some discrepancies are now more noticeable.

From a comparison of the loadings (see Figure 3d) no discernable pattern may be recognised, since the two maps differ virtually everywhere. The two loadings, in fact, have a good agreement (absolute differences less than 0.5) only in about 22% of the points, and in the remaining points the loading differences are strictly positive and greater than 0.5.

To shed light on these discrepancies we directly compare the index time series, instead of loadings and scores, and proceed as follows. We consider a location over

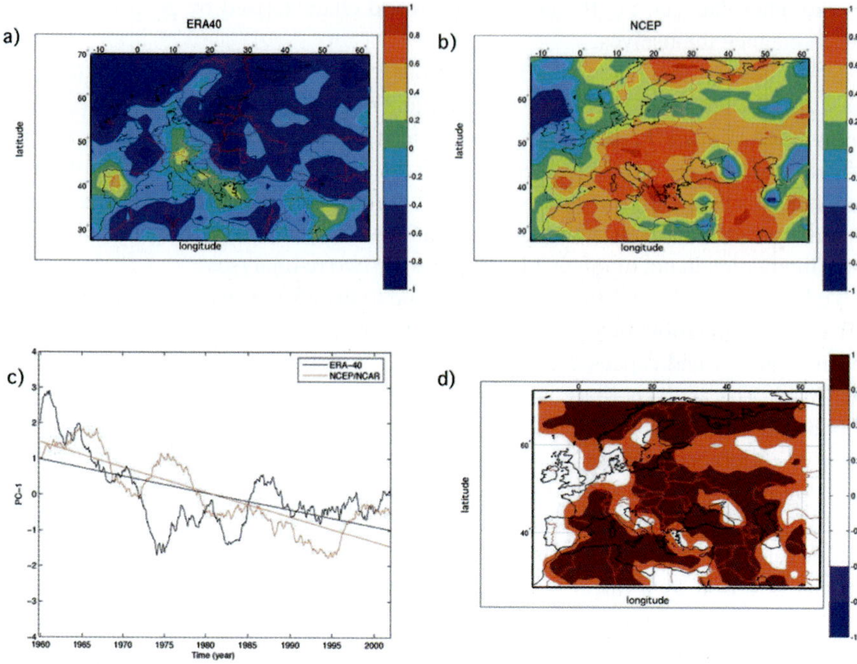

Figure 3. (a) First loading pattern of the SPI-24 for the Euro-Mediterranean sector obtained using the ERA-40 data set, contour interval is 0.2; (b) as in (a) but for NCEP/NCAR data set; (c) first PC scores for ERA-40 (black line) and NCEP/NCAR (red line), straight lines denote the fitted linear trends; (d) differences between the first loading of NCEP/NCAR and that of ERA-40, absolute values less than 0.5 are denoted by white colour

Europe where a remarkable difference between the first loadings have been detected, i.e. where the SPI-24 time series for ERA-40 has a negative correlation with the PC-1 (i.e. the trend), while the index time series for NCEP/NCAR has positive correlation with the corresponding score. For instance, let us consider the ERA-40 grid point at 50.0°N-25.0°E and the nearest four points in the NCEP/NCAR grid (say 50.53°N–24.38°E, 50.53°N–26.25°E, 48.63°N–24.38°E, 48.63°N–26.25°E). Next, we compute the SPI-24 time series averaged over the latter points. The time behaviours of these SPI series are shown in Figure 4.

The signals have a correlation coefficient of about 0.1, justifying the lack of correlation when the first principal component is considered. It must be noted, however, that, up to the seventies, the two re-analyses show a remarkable different SPI behaviour, while in the remaining part of the record the two series seems to be more alike. Thus, we suspect that the low correlation between the two SPI time series is mainly related to the differences occurring at the beginning of the record, which are crucial in determining the first principal component linear trend. Since in this early decade the data considered had a lower resolution (satellite data were virtually absent), we may suggest that the observed difference may be attributed

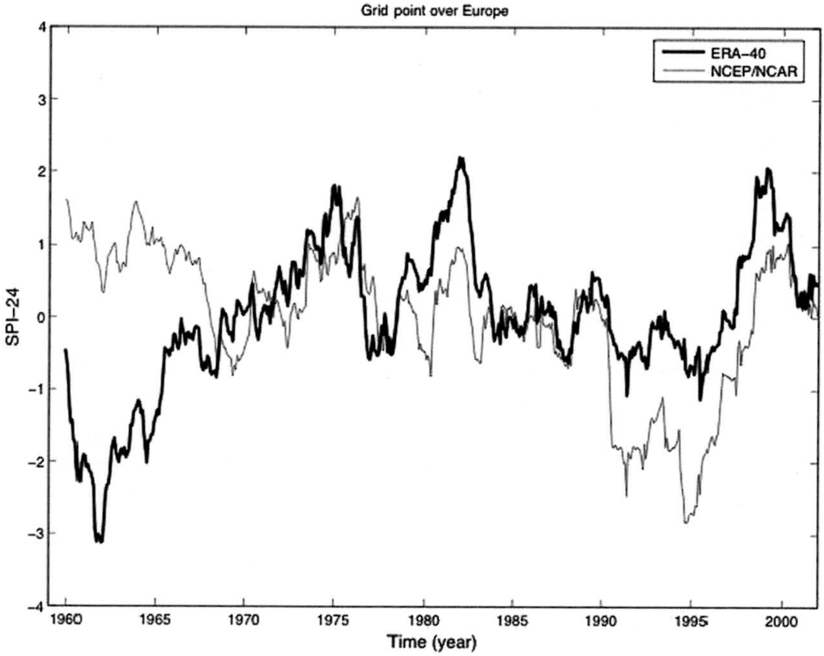

Figure 4. SPI-24 time series at a selected location over Europe (see the text): thick line refers to ERA-40 data set, thin line to NCEP/NCAR

to the two assimilating models. We feel that a careful comparison with the SPI computed using rain gauge observations would be useful to clarify the nature of the problem.

In summary, results suggest that on the global scale, the two re-analyses agree in their first principal component score, but not in the associated loading: both re-analyses capture a linear trend, though the areas where this feature should be most likely observed are not uniquely identified by the two data sets. Moreover, while the ERA-40 reveals the presence of a weak net "global" trend towards wet conditions, the NCEP/NCAR re-analysis suggests that the areas in the world characterised by positive/negative trends balance to zero. For the European sector the two re-analyses show remarkable differences both in the first loading and in representing the timing of the wet and dry periods. Also for these areas a linear trend, superposed on other short-term fluctuations, is detectable in the first principal component of the SPI field.

4. FORECASTING DROUGHT

Assuming that the SPI describes all the facets of a dry condition, it may be possible to exploit long time series of this index to forecast future occurrences of drought. However, there are at least two main hindrances that do not allow a readily solution to the forecast by standard time series methods (Box and Jenkins, 1970):

i) Precipitation is usually not correlated on the time scale on which drought manifests itself;
ii) The time correlations found in the SPI on long time scales are simply an effect of cumulating precipitation on the selected time scale, unless SPI's multiyear periodicities are revealed.

Thus, the application of forecast methods to SPI time series appears to be a wrong approach from the out start. We must consider precipitation and exploit the knowledge acquired by sampling it for a long time.

4.1 Data and Methods

To evaluate the ability in predicting future values of the SPI, we use monthly precipitation time series from 36 stations in Sicily covering the period 1926-2000. The stations have been extracted from a larger set according to Alecci et al. (2000) criterions, which are mainly the record length, data quality and homogeneous spatial distribution. For illustrative purposes, we decided to consider the monthly precipitation averaged over the 36 stations, neglecting local variations since they add very little to the understanding of the problem. In computing the SPI on 1-month time scale (this is to avoid any spurious correlations) we employ the algorithm presented in Bordi and Sutera (2001).

To forecast the SPI-1 values we use two different approaches: the Auto Regressive (AR) and the GAmma Highest Probability (GAHP) method. A common approach for modelling multivariate time series is the autoregressive model of order p (AR(p) model):

$$x_i = w + \sum_{i=1}^{p} A_l x_{t-l} + \varepsilon_t \tag{1}$$

where x_t is the m-dimensional state vector that have been observed at equally spaced instants t. The matrices $A_1, \ldots A_p \in \Re^{m\times m}$ are the coefficient matrices of the AR model and the m-dimensional vector $\varepsilon_t = noise(C)$ are uncorrelated random vectors with mean zero and covariance matrix $C \in \Re^{m\times m}$. The m-dimensional vector w is a vector of intercept terms, which allows for an eventually nonzero mean of the time series. First, the approach needs the estimation of the following parameters: the order p of the AR model, the intercept vector w, the coefficient matrices $\mathrm{A}_1 \ldots \mathrm{A}_p$ and the noise covariance matrix C. For the selection of these parameters, the stepwise least squares algorithm is usually implemented (Neumaier and Schneider, 2001).

Thus, we start our analysis by computing the SPI on 1-month time scale (see figure 5a), trying to forecast its future values. However, it must be noted that the SPI-1 time series cannot be predicted using AR method because, as expected, it has a white noise power spectrum as clearly shown in Figure 5b.

The inability of the AR method to forecast any future value of the SPI-1 could be overcome by applying the AR method to precipitation time series. Using AR method

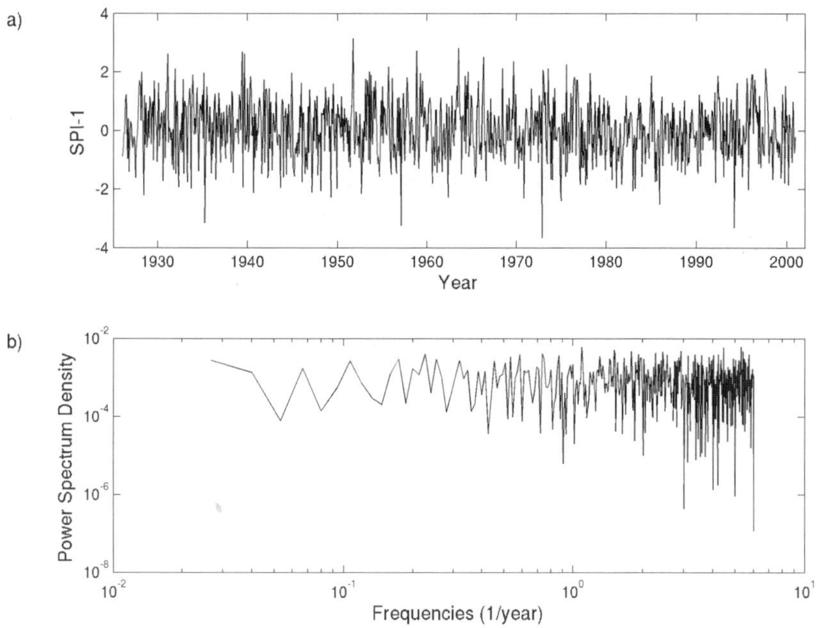

Figure 5. (a) SPI-1 for Sicily from 1926 to 2000 computed using the precipitation averaged over 36 stations; (b) Power spectrum density of the SPI-1

to evaluate future values of precipitation we find, however, more implications. The AR method gives, in fact, a regression of 12th or 13th order. Comparing the weights of the coefficients, it is possible to note as the first, the eleventh, the twelfth and the thirteenth are the highest ones. The AR method finds correlation in two main components: the seasonal cycle, mixing among the 11th, 12th and the 13th coefficients, and a simple regression with the previous event (the first coefficient), typical of the linear AR method. This result is clearly illustrated in Figure 6 where the precipitation time series (6a) and the associated power spectrum (6b) are shown. The power spectrum has a peak corresponding to the annual component, though mixed with the nearest frequencies. It follows that AR is just extracting the seasonal cycle, thus preventing to have a good knowledge of large deviations from it. Recalling that the latter ones lead to extreme conditions, it appears that the skill of this method should be low, especially for extremely dry (or wet) occurrences.

In avoiding these shortcomings, we propose a method that we denote as GAHP (GAmma Highest Probability) method. This approach forecasts the precipitation for a future month as the most probable value described by the probability density distribution of the precipitation for that month. Thus, the method needs the estimation of the parameters of the Gamma distribution function that best fits the frequency histogram of the observed precipitation for a given month of the year. Then, the

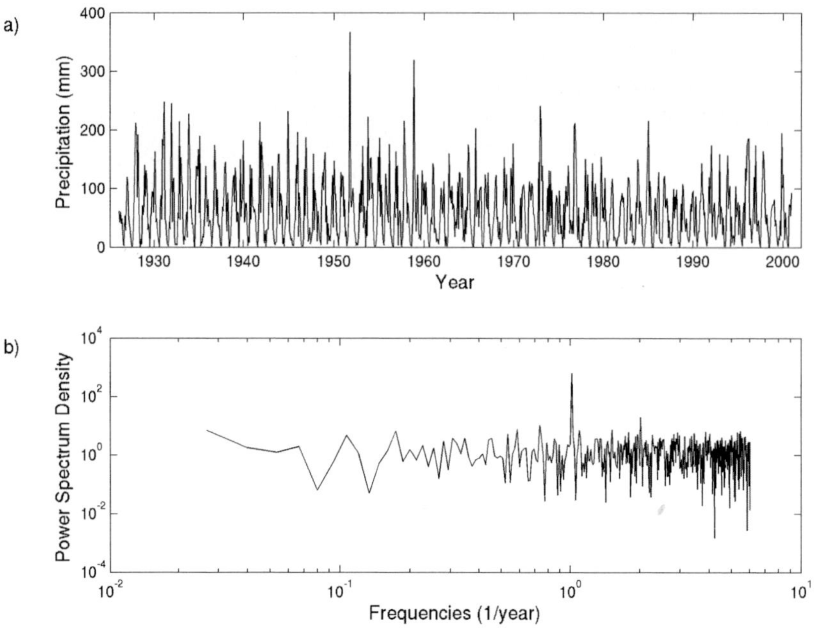

Figure 6. (a) Monthly precipitation averaged over the 36 stations in Sicily from 1926 to 2000; (b) Power spectrum density of the precipitation time series. Note the peak at 1-year period

precipitation predicted for the next month is the *mode* of the fitted distribution. This approach is based on three hypotheses:

1. Two consecutive precipitation events are not correlated;
2. The relation between precipitation values are only related to the seasonality;
3. Future events will be the most probable ones.

The first two assumptions are both related to the phenomenology of the precipitation for a particular month in the Mediterranean regions (see Bordi et al., 2005). The third has an empirical nature and may be easily substituted by other location measures of a given distribution.

4.2 Results

In this section we compare the forecast obtained with the AR and the GAHP methods. The procedure used to perform the forecast of the SPI-1 for the next year can be summarized as follows:

- Extraction of a subset from the entire precipitation dataset (for example, from January 1926 to December 1999);
- Computation of the 12 predicted values for monthly precipitation (one year) with the two methods;
- Evaluation of the SPI-1 with the new precipitation time series;

- Comparison between the predicted SPI-1 and the one obtained with the entire observed precipitation time series 1926- 2000.

To estimate the goodness of the prediction we use the mean squared prediction error (MSE) given by:

$$MSE(t_i) = [SPI_o(t_i) - SPI_F(t_i)]^2 \tag{2}$$

where SPI_o is the observed value, SPI_F is the forecast value and i indicates the month of the year. In order to have a larger statistics and to better validate our claimed skill, we apply the two methods to different time segments of precipitation time series and compute the mean MSE of the resulting SPI-1. Thus, we form a set of 75 random permutations of the precipitation data that preserve the right sequence of the 12 precipitation values registered in a year, i.e. the seasonal variability. For each permutation we evaluate the coefficients of the AR method and we compute the forecast for the last year. It must be noted that for our method it is not necessary to do any permutation because the estimation of the most probable value does not depend on the sequence of the value registered in a certain month. In this case, to obtain a test statistically significant we evaluate the MSE of the SPI-1 taking the squared difference between the predicted value of the index for a particular month and all the values for that month in the previous years.

Results are shown in Figure 7. In this case the GAHP method gives consistently better results than the AR method, especially for spring and summer. In fall and

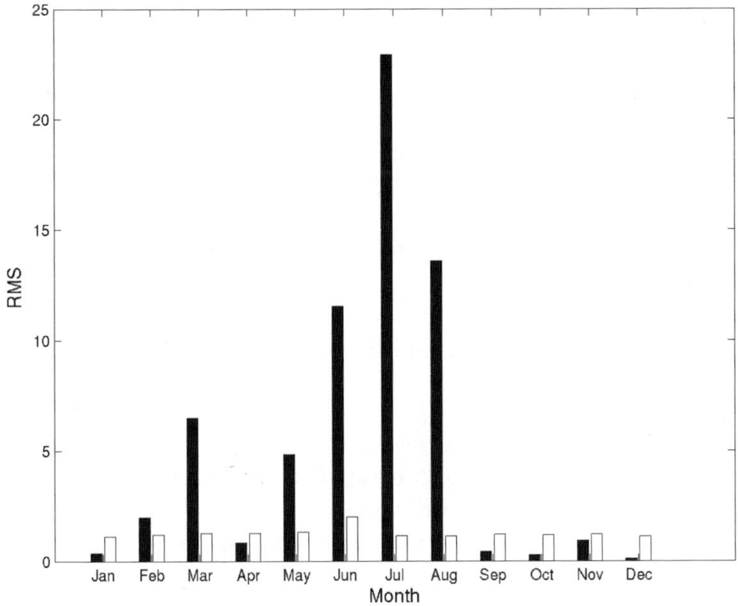

Figure 7. MSE of the SPI-1 computed forecasting precipitation with the AR (black bars) and GAHP method (white bars) averaged over the 75 realizations

winter, when the precipitation has highest values, the AR estimation seems as accurate as the GAHP one, although for these cases the difference between the observed and the forecasted values is too small to reach any firm conclusion.

5. EXTREME DRY/WET EVENTS

For assessing risk of highly unusual events, extreme value statistics needs to be applied, it plays an important role in engineering practices for water resources design and management. Here we analyze extreme wet and dry periods in a sample area (Sicily). First, we studied monthly precipitation extremes both using the annual maximum and partial duration methods, next, we studied the extremes of the SPI on 1-month time scale. Furthermore, we tried to identify the large-scale atmospheric conditions associated to the SPI extremes.

5.1 DATA AND METHODS

Used data are monthly precipitation time series from 36 stations in Sicily covering the period 1926-2000 as in the previous section. In order to illustrate the phenomenology of large-scale atmospheric conditions associated to dry/wet extremes in Sicily we consider the 500hPa geopotential height extracted from the NCEP/NCAR data set.

To quantify wet and dry conditions the SPI is applied. We limit the study to the SPI on 1-month time scale because the correlation introduced by longer time scales is an unwished property for applying standard extreme value technique.

The analysis of extremes is usually based on the estimation of extreme event return times. This implies the assumption that the extremes are independent random variables described by a probability distribution, which should not change from sample to sample, i.e. the data should be homogeneous. As anticipated in the introduction, two methods are applied to sample the original data: Annual maximum (AM) and partial duration (PD) series (also known as Peaks Over Threshold series). The AM series consists of the greatest events of each year in a given period of time. According to the Fisher-Tippett theorem (1928), the asymptotic distribution of a series of sample maxima (AM series) belongs to one of three basic distributions, regardless of the original distribution of the observed data. These three families were combined into a single distribution, which is now known as the Generalised Extreme Value (GEV) distribution (see Jenkinson, 1955). In PD analysis, on the other hand, we are interested in the behaviour of observations that exceed a high threshold. A theorem by Balkema and de Haan (1974) and Pickands (1975) shows that for sufficiently high threshold, the distribution function of the excess may be approximated by the Generalised Pareto (GP) distribution such that, as the threshold gets large, the excess distribution converges to the GP distribution.

5.2 Applications

First, the Sicily monthly precipitation time series averaged over the 36 stations has been computed and the annual maxima series has been extracted from the data for

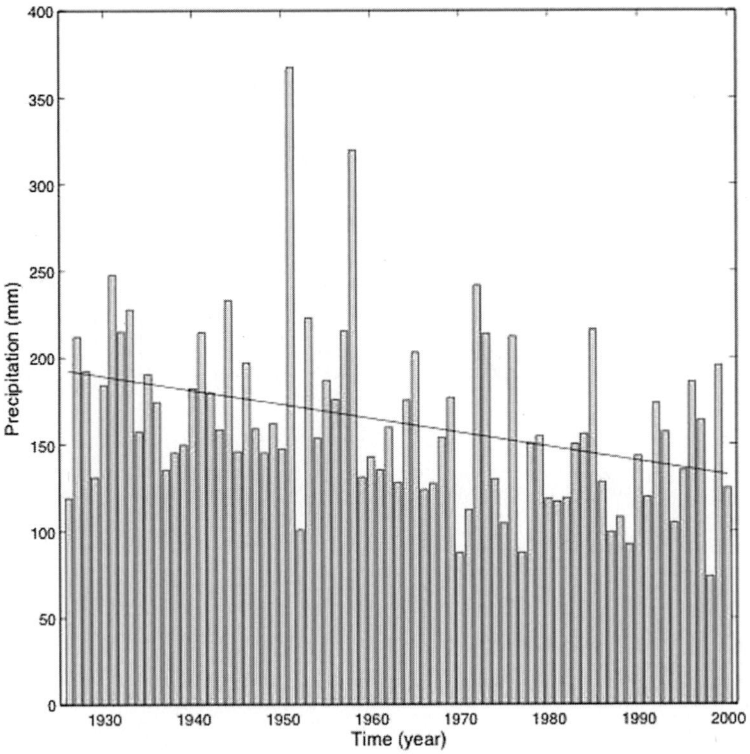

Figure 8. AM series of monthly precipitation averaged over the 36 stations in Sicily. Straight line is the computed linear trend; unit is mm

the 75 years considered starting from 1926. The resulting AM series is shown in Figure 8. Precipitation annual maxima show values ranging from about 73 mm to about 250 mm with two peaks higher than 300 mm occurring in 1951 and 1958. A trend towards lower values of annual extremes can be noted and it characterises the time series: a loss of about 50 mm has occurred in the last 75 years, while the time average of annual maxima is about 160 mm. Whether this trend is related to global change is uncertain, because of the local nature of our data, but we like to stress that its statistical significance is very high to warrant further analysis in the future. We remark also that trends may have impacts on extremes, but we have decided not to take into account these two problems in the present chapter and to postpone a proper analysis to another occasion.

Also the PD method with two different precipitation thresholds, 76 and 200 mm, is applied to the averaged precipitation time series.

Return times are then estimated following the two approaches. A comparison between results obtained for AM and PD methods (see Figure 9) suggests that return times for different precipitation amounts are comparable, especially when a low threshold is considered. Precipitation extremes between 200 and 300 mm have return

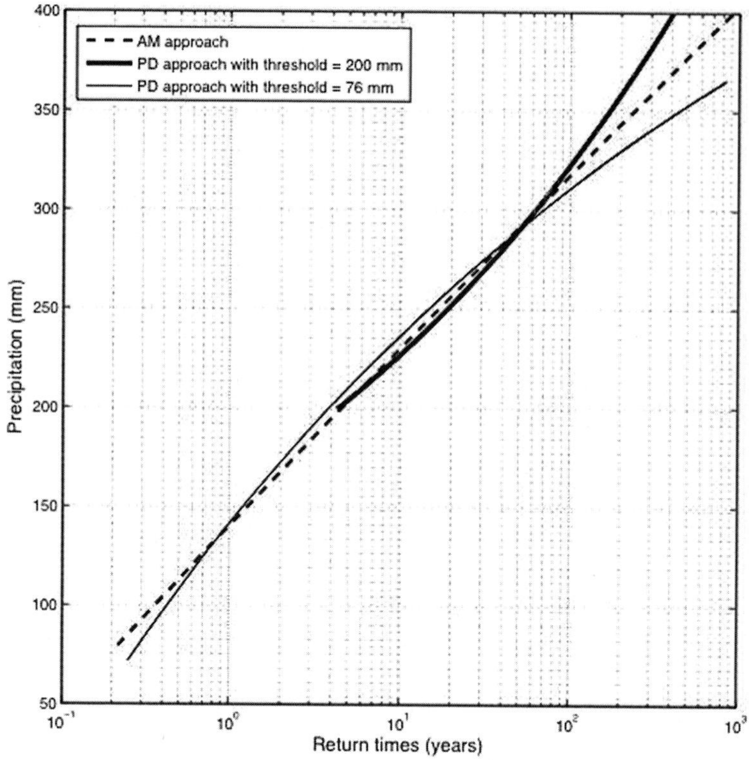

Figure 9. Return times curves for precipitation extremes in Sicily computed applying the AM (dashed line) and the PD method with the two thresholds (76 mm thin line, 200 mm thick line). Unit is year

periods ranging from 4 year to 60 years. Within this range of precipitation the return times estimated using the two thresholds show a good agreement, while for 300 mm they differ remarkably. However, as expected, for high precipitation values, we get a large error band, thus leaving the corresponding return times not clearly defined.

At this stage of our analysis, we can raise the question whether an AM extreme, occurring at a particular month of the year, really represents an abnormal departure from some location parameter of the empirical probability distribution of precipitation for that month, so that we can consider it as a wet extreme. For this purpose, we show in Figure 10 the probability distribution (histogram) of observed precipitation for December.

The solid line denotes the Gamma function fitting the distribution and filled circles, on the bottom of the figure, are the annual maxima precipitation occurring in December. As can be seen, only four values greater than 200 mm (corresponding to December 1927, 1933, 1944, 1972 with precipitation amount of 212, 228, 233 and 242 mm respectively) lie on the tail of the distribution, suggesting that only these annual maxima can be considered as extremes.

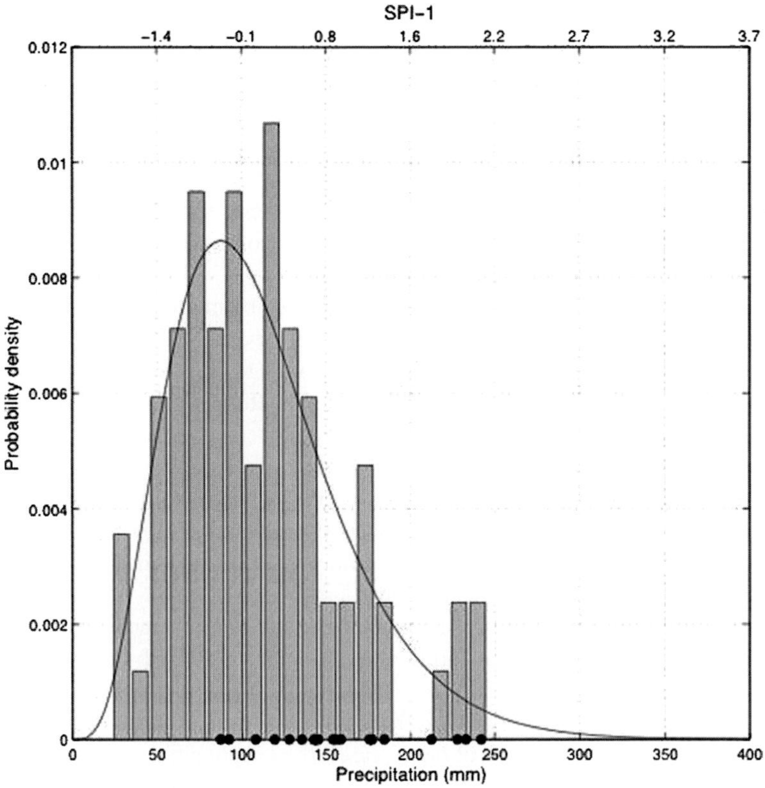

Figure 10. Probability density function of observed precipitation for December. Solid line denotes the Gamma function fitting the distribution and filled circles, at the bottom of the figure, are precipitation annual maxima occurring in December. Unit at the (bottom) x-axis is mm. On the top x-axis there are values of the SPI-1 for different precipitation amounts occurring in December

Let us consider, instead, the SPI-1 computed using the monthly precipitation occurring in December. In these cases SPI-1 values (see the x-axis on the top of the figure) show that only two of the four events (December 1944 and 1972), selected with the strategy above described, are extremely wet events since their SPI-1 are 2.0 and 2.1.

As a consequence of this result, we may suppose to move to PD approach selecting extremes as the precipitation values above the threshold of 233 mm. Unfortunately, this will lead also to a misinterpretation, since the distributions of precipitation from month to month change and so their tails. Thus, the threshold should be a function of the particular month when the extreme occurs. This, of course, will bear practical consequences since the return times will differ greatly as well. The same conclusions can be drawn for dry events.

In order to avoid these shortcomings, a suitable solution appears to be the use of the SPI-1 for extreme analysis. In this case, in fact, the distribution of the SPI is Gaussian, so that we definitively remove the ambiguity above illustrated. Thus, we compute the SPI-1 time series using monthly precipitation recorded at 36 stations and estimate return times for SPI-1=1.7 and SPI-1=-1.7 (see Figure 11a, b).

We have assumed that there is no spatial correlation among the stations, which has been tested to be very poor, so that return times may be assumed to be independently distributed.

Maps show that return times for extreme wet and dry events have different features. Extreme wet events are more frequent in the eastern part of Sicily (return times between 2 and 3 years), while in the remaining parts of the island such events are less frequent (return times between 3 and 6 years).

This is probably related to the orographic effects induced by mountain Etna on the precipitation field. On the other hand, extreme dry events occur more frequently in the Northern part of Sicily with return times ranging from 2 to 4 years. Furthermore, longer return times, from 8 to 12 years, characterise the Eastern part of the island.

It must be noted that this analysis refers to the return times of the standardized monthly precipitation, which has some interest only when the meteorological dry/wet extremes are considered. On the other hand, the estimation of return periods for the SPI on longer time scales requires the analysis of correlated time series. This should be the topic of future investigations.

In order to illustrate the large-scale atmospheric conditions characterizing the SPI-1 extremes in Sicily we proceed as follows. We select from the SPI-1 time series (based on averaged precipitation over stations) the annual maxima (minima) above (below) the threshold 2 (-2) from 1948 to 2000. This leads to 13 extreme events for wet and dry conditions. Then, for each of the 13 cases, the geopotential anomaly map is computed (that is, we consider the difference between the geopotential field for the month with extreme SPI-1 and the long-term mean for that month). The anomalies associated to few extreme cases are displayed as an example in Figure 12a-c. They correspond to the following events: November 1958 (SPI-1=2.7), August 1997 (SPI-1=2.1) and January 1983 (SPI-1=-2.1). As can be seen by the figure, these

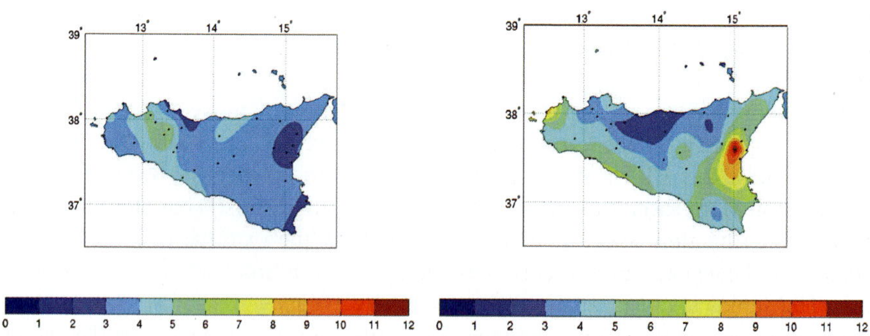

Figure 11. Spatial distribution of estimated return times for SPI-1=±1.7: extreme wet (left) and extreme dry (right). Units are years

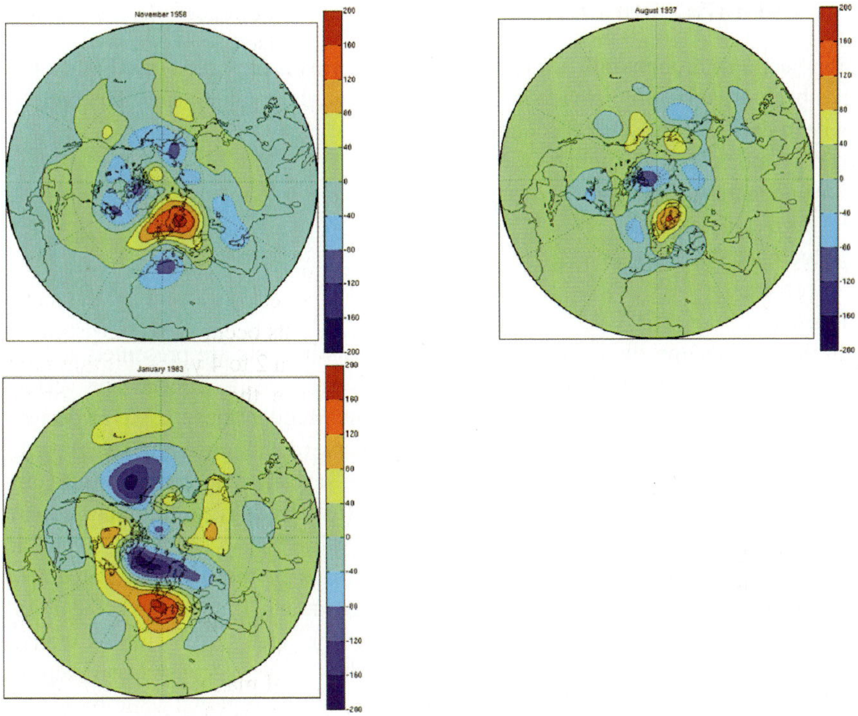

Figure 12. Anomaly maps of the 500 mb geopotential height corresponding to extreme events in November 1958 (SPI-1=2.7), August 1997 (SPI-1=2.1) and January 1983 (SPI-1=-2.1). Anomalies are deviations from the long-time monthly mean. Units are m, contour interval is 40 m

wet events are characterized by a dipole structure over the European area. Such a feature seems to be independent of the season when the extreme occurs; the negative anomaly of geopotential is localized in the Mediterranean basin extending to about 50N, while the positive anomaly characterizes Northern Europe. On the other hand, the extreme dry event shows a positive geopotential anomaly in the Mediterranean basin extending to Western Europe and a negative anomaly over North-Eastern Europe.

The magnitudes of the interannual standard deviations for each month suggest that these features are statistically significant. These patterns, in fact, may be considered typical of all extremely wet and dry events identified in the historical record. This is confirmed by the mean anomaly maps (here not shown) averaged over the 13 extreme wet and dry cases. These results suggest that, although the limitation in using only precipitation averaged over Sicily, extreme wet and dry events in the Mediterranean basin occur when dipole-like geopotential anomalies characterise the large-scale circulation. If these findings were corroborated by a deeper analysis, we would be able to predict in advance the meteorological extreme events.

6. DROUGHT BULLETIN

In this section a drought bulletin for the MEDOCC area is presented (see Figure 13). It has been organized to provide information on drought at large-scale. The bulletin has been divided into several sections as follows:

i) Why monitoring drought?
ii) Atmospheric conditions
iii) Large-scale analysis
iv) Regional-scale analysis
v) Forecast
vi) Archive

In the first section the concept of drought and its impacts are presented and the main indexes usually applied for monitoring drought are described. The second section provides the atmospheric conditions for the Mediterranean basin as derived from the NCEP/NCAR re-analysis. Anomaly maps (deviation from the climatological mean 1968-1996) of the sea level pressure, surface temperature and precipitation rate are shown for the month considered (usually it is the previous month with respect to the actual one for which the NCEP/NCAR data are available through the net). In the third section there are maps of the SPI and Percent of Normal on different time scales (3, 6, 12 and 24-month time scales), the PDSI and Deciles for a given grid point (i.e. Sicily). The analysis of climatic conditions for the month in question is provided according to the indexes classifications. The forth section is devoted to the analysis of drought at regional scale by means of rain gauges data. Thus, this section consists of several links to the regional institutions that provide such analysis, such as ARPA/SIM (Emilia-Romagna),

Figure 13. Drought bulletin home page

Hydrographic Office of Palermo (Sicily) or Civil Protection of Calabria. Next, there is a section containing the forecast maps of the SPI on different time scales. The forecast here proposed is based on GAHP method and it is 1, 2 and 3-months forecast. Finally, in the archive all the analyses are collected and movies of the SPI are provided.

The bulletin is updated every month and has the aim to support the water resources managers in planning proactive actions. The organization of the web pages is such that also the public with no particular expertise can benefit the information about drought.

7. CONCLUSIONS

The aim of the chapter was to provide the state of the art about drought monitoring and forecasting at large scale such as the Mediterranean area. Results suggest that we are able to monitor the climatic conditions in a satisfactory manner, especially when the SPI is considered. The SPI, in fact, seems to be a useful tool for monitoring dry and wet periods on multiple time scales and comparing climatic conditions of areas governed by different hydrological regimes.

Forecasting drought remains, instead, a complex task, because of the random character of the precipitation field. Some improvements have been done, by introducing a new statistical method based on the probability density distribution of monthly precipitation, or by analyzing the long-term climatic variability (i.e. climatic trend). Also the estimation of extreme events return times can provide useful information for predicting drought. In particular, the study of the relationship between extreme droughts and large-scale atmospheric circulation characterizing such conditions can help in identifying those atmospheric conditions that may lead to drought events (Bordi et al., 2006). We feel that further efforts should be done along this line, since in the Mediterranean area no particular teleconnection has been found between drought events and global patterns of climatic variability, such as ENSO (El Nino Southern Oscillation) or NAO (North Atlantic Oscillation).

Finally, the drought bulletin, here presented in a prototype form, seems to be a useful tool for collecting all the available information on drought and for supporting the water managers in planning proactive measures.

ACKNOWLEDGEMENTS

The financial support has been provided by the EU Project INTERREG IIIB - SEDEMED II (code 2003-03-4.4-I-010).

NCEP/NCAR data have been provided by the NOAA-CIRES Climate Diagnostics Center, Boulder, Colorado, USA, from their Web site at http://www.cdc.noaa.gov, while the ERA-40 precipitation data set by the European Centre for Medium-Range Weather Forecasts from their web site http://www.ecmwf.int. Rain gauges data have been provided by the Hydrographic Office of Palermo, for this the authors kindly thank Eng. G. Geraci and Eng. T. Vega for their support.

The authors like also to thank sincerely Prof. G. Rossi for his useful suggestions and discussions.

REFERENCES

Alecci, S., Arcidiacono, F., Bonaccorso, B. and Rossi, G. (2000) Identificazione delle siccità regionali e sistemi di monitoraggio, *Atti del Convegno "Siccità: monitoraggio, mitigazione, effetti"*. Villasimius, 21–23 Settembre 2000.

American Meteorological Society. (1997) Meteorological drought - Policy statement. *Bull. Amer. Meteor. Soc.* 78:847–849.

Balkema, A.A. and de Haan, L. (1974) Residual lifetime at great age. *Annals of Probability*. 2: 792–804.

Beguería, S. (2005) Uncertainties in partial duration series modelling of extremes related to the choice of the threshold value. *J. Hydrol.* 303: 215–230.

Bonaccorso, B., Bordi, I., Cancelliere, A., Rossi, G. and Sutera, A. (2003) Spatial variability of drought: an analysis of the SPI in Sicily. *Water Resour. Manage.* 17: 273–296.

Bordi, I. and Sutera, A. (2001) Fifty years of precipitation: some spatially remote teleconnections. *Water Resour. Manage.* 15: 247–280.

Bordi, I., Fraedrich, K., Jiang, J.-M. and Sutera, A. (2004) Spatio-temporal variability of dry and wet periods in eastern China. *Theor. Appl. Climatol.* 79: 81–91.

Bordi, I., Fraedrich, K., Petitta, M. and Sutera, A. (2005) Methods for predicting drought occurrences. *Proceedings of 6th International Conference of EWRA "Sharing a common vision for our water resources"*, September 7–10, 2005, Menton, France, p. 57.

Bordi, I., Fraedrich, K., Petitta, M. and Sutera, A. (2006) Large-scale assessment of drought variability based on NCEP/NCAR and ERA-40 Re-analyses. *Water Resour. Manage.* In press.

Box, G.E. and Jenkins, G.M. (1970) *Time series analysis: forecasting and control*. Holden-Day, Inc., San Francisco, CA.

Coles, S. (2001) *An introduction to statistical modelling of extreme values*. Springer-Verlag London, 200 pp.

Cressie, N.A.C. (1991) *Statistics for spatial data*. John Wiley & Sons, New York, 900 pp.

Fisher, R.A. and Tippett, L.H.C. (1928) Limiting forms of the frequency distributions of the largest or smallest member of a sample. *Proceedings of the Cambridge Philosophical Society* 24: 180–190.

Guttman, N.B. (1999) Accepting the Standardised Precipitation Index: a calculation algorithm. *J. Amer. Water Resour. Assn.* 35: 311–322.

Heim, R.R.Jr. (2002) A review of twentieth-century drought indices used in United States. *Bull. Amer. Meteor. Soc.* 83: 1149–1165.

Jenkinson, A.F. (1955) The frequency distribution of the annual maximum (or minimum) of meteorological elements. *Quart. J. R. Met. Soc.* 81: 158–171.

Kalnay, E. and co-authors. (1996) The NCEP-NCAR 40 year reanalysis project. *Bull. Amer. Meteor. Soc.* 77: 437–471.

Keyantash, J., Dracup, J.A. (2002) The quantification of drought: an evaluation of drought indices. *Bull. Amer. Meteor. Soc.* 83: 1167–1180.

McKee, T.B., Doesken, N.J., Kleist, J. (1993) The relationship of drought frequency and duration to time scales, Preprints. *8th Conference on Applied Climatology*. 17–22 January, Anaheim, CA, pp. 179–184.

Neumaier, A., Schneider, T. (2001) Estimation of parameters and eigenmodes of multivariate autoregressive models. *ACM Trans. Math. Softw.* 27, 1 (Mar.), 27–57.

Palmer, W. (1965) *Meteorological drought*. Tech. Rep., Vol.45, U.S. Weather Bureau, Washington, D.C., p.58.

Peixoto, J.P. and Oort, A.H. (1992) *Physics of climate*. Springer-Verlag, New York, Inc., 520 pp.

Pickands, J. (1975) Statistical inference using extreme order statistics. *Ann. Stat.* 3: 119–131.

Redmond, K.T. (2002) The depiction of drought: a commentary. *Bull. Amer. Meteor. Soc.* 83: 1143–1147.

Rencher, A.C. (1998) *Multivariate statistical inference and applications*. John Wiley & Sons, New York, Inc., 559 pp.

Simmons, A.J. and Gibson, J.K. (2000) The ERA-40 project plan. ERA-40 Project Report Series 1, ECMWF, Reading, U.K., 62 pp.

CHAPTER 2

MONITORING AND FORECASTING DROUGHT ON A REGIONAL SCALE: EMILIA-ROMAGNA REGION

C. CACCIAMANI, A. MORGILLO, S. MARCHESI AND V. PAVAN

Hydrometeorological Service of ARPA Emilia Romagna-Bologna, Italy

Abstract: This chapter presents an overview of the occurrences and effects of droughts through a study of the Standard Precipitation Index (SPI) in the Emilia-Romagna region, which is located in the northern-central part of the Italian peninsula. The link between this index and large-scale atmospheric circulation was investigated and the SPI index was also used to predict drought. The study describes the development of a method of forecasting SPI index based on an earlier Interregional project (SEDEMED), involving a statistical downscaling scheme model using as input the large-scale seasonal forecasts obtained from Atmospheric Global Circulation Models. The downscaling scheme, which has already been used with relatively good results to predict surface parameters of temperature and precipitation, is applied to the SPI index, providing a statistical regionalization of this indicator

Keywords: Drought, NAO, EB, SPI , Z500

1. METEOROLOGICAL DROUGHTS: INTRODUCTION

Drought is a natural phenomenon that occurs when precipitation is significantly lower than normal. Low precipitation can lead to severe hydrological deficit and cause serious problems for agriculture, the hydroelectric sector and industry, as well as deficit in the drinking water supply, with heavy consequences for the local population. In the long run, if a drought lasts many months or even years and involves a large area, it can permanently damage the environment and cause significant economic losses.

In the Italian climate, droughts are not only possible, but also relatively frequent. Europe and the entire Mediterranean area have suffered major droughts in recent years. In the '70s and '80s north-western Europe was often subject to drought conditions (in 1972 and 1976, and from 1988 to 1992) and in the last few years drought conditions have also been experienced in large areas of central and southern Europe and in countries like France, Italy, Portugal and Spain. The more and more frequent occurrence of sequences of heavy precipitation and flooding followed by periods characterized by low precipitation and drought has fueled fears that hydrological cycles could be changing, as a result of global warming.

G. Rossi et al. (eds.), Methods and Tools for Drought Analysis and Management, 29–48.

In the current literature, there are different definitions of drought depending on the duration of the phenomenon, on its spatial extension and on its effects or impacts on human activities. Different approaches and different choices of indicators may be used to describe the problem in relation to the different definitions. In this chapter, attention is focused only on meteorological and agricultural droughts and on the indicators correspondingly used to describe them.

"Meteorological drought" is defined as the lack of precipitation (expressed relative to a climatic value) for a sufficiently long period (a number of consecutive days of dry weather) to cause severe hydrologic imbalance in the area affected. Defined in this way "meteorological drought" depends on the area under examination and, more specifically, on what are the "normal" climatic conditions of that region. Therefore, in order to characterize and study the problem of meteorological drought, it is necessary to be familiar with the typical meteorological configurations on the synoptic and sub-synoptic scale responsible for precipitation in the studied region.

"Agricultural drought" links the characteristics of meteorological drought to agricultural needs and is defined as a condition in which the amount of moisture in the soil is insufficient for the needs of a particular crop. The deficit derives from the difference between the amount of moisture in the soil and the loss of moisture due to various physical processes (evaporation, runoff, surface infiltration, etc.). This type of drought depends on the type of crop and of soil, and will be discussed in greater detail in later chapters.

2. SYNOPTIC FLOWS AND LARGE-SCALE CIRCULATION: CONNECTION WITH DROUGHT

As said above, statistical analysis of any drought indicator is not sufficient to ensure a complete climatic characterization of "meteorological drought", but must be supported by an adequate characterization of the spatial-temporal variability of the atmospheric circulation, from large-scale to meso-scale, taking also into account the connections existing between the various spatial scales. It is this analysis of the variability of atmospheric circulation that enables to understand the causes of precipitation.

Moving down from the hemispheric scale, it is possible to identify several important modes of variability (Wallace and Gutzler, 1981), such as the North Atlantic Oscillation (NAO), the so-called Blocking phenomena, the cyclonic or anticyclones structures of the pressure field at the ground level, the Atlantic jet stream characterizing the upper troposphere in middle latitudes over the North Atlantic sector, all of which have a major impact on the weather and also on the climate of the European region and of our National territory. The interactions taking place between these structures of large-scale variability and the complex orography and morphology of the area are responsible for more local characterizations of the weather (and of the climate) even in very small areas such as Emilia-Romagna region in Italy, located in the south-eastern sector of the Po Valley and bordered on the south-west by the Apennine Mountains and on the east by the Adriatic Sea.

The NAO is known to be the structure of variability that most affects the climate in the area of interest. The NAO is defined as a sea-saw in atmospheric sea level pressure over the North Atlantic Ocean between the Azores and Iceland. The time evolution of this mode of variability can be described by means of several possible NAO indices. Although this mode characterize the variability over the North Atlantic all the year around, its amplitude is greatest in the winter and in this season the impacts of NAO on European climate are more intense.

The index is positive when pressure over the Azores exceeds the climatology while a sea level pressure deficit is present over Iceland, and it is negative when the mode is in its opposite phase. The negative or positive phase of the NAO has significant repercussions on the weather on the European continent, and exercises there a great influence on surface temperature and precipitation.

With regards to its influence on precipitation, many studies have been carried out on the continental scale as well as on the local scales of the Alpine area of Northern Italy, generally with reference to the winter season (Cacciamani et al., 1994, Hurrel et al., 1995, Quadrelli et al., 2001). In the Alpine region, a negative NAO phase causes an excess of precipitation in the region (Quadrelli et al., 2001). This can be explained in terms of a southern shift and a decrease in the intensity of the Atlantic jet-stream and of the associated Atlantic storm-track, leading to more favorable conditions for the passage of storms and for the establishment of cyclogenesis on Southern Europe and the Mediterranean (Hurrel et al., 1995).

During a positive NAO phase, the jet stream is more intense and moves towards northern Europe. The Atlantic storms are more likely to shift towards the higher latitudes in Europe bringing heavier precipitations to those areas. If this phase remains active for a protracted period of time, the Mediterranean basin and the Italian Peninsula may receive over such period low or no precipitation , this leading to the establishment drought conditions (Quadrelli et al., 2001). Northern Italy lies south of the jet stream end, and Emilia-Romagna, climatologically under the influence of masses of warmer air, can be at times subject to droughts. A typical example of positive and negative NAO phases is shown in Figure 1.

Another extremely important structure of variability characterizing the weather in the western Mediterranean regions is the Blocking (or meteorological blocking patterns) particularly when it is localized on the European area (European Blocking pattern, EB, Tibaldi et al., 1994). The structure of the blocking pattern (Rex, 1950) is characterized by the presence of a high pressure zone persisting for many days in the upper tropospheric geopotential height field (and often by a relative high pressure field at sea level), with a correspondent easterly flow occurring south of the geopotential height anomaly.

This pattern of circulation anomaly can be located in correspondence of the center of the Atlantic jet-stream over the North Atlantic (Atlantic Block, negative phase of the NAO) and/or at the end of the Atlantic jet-stream over continental Europe (European Blocking).

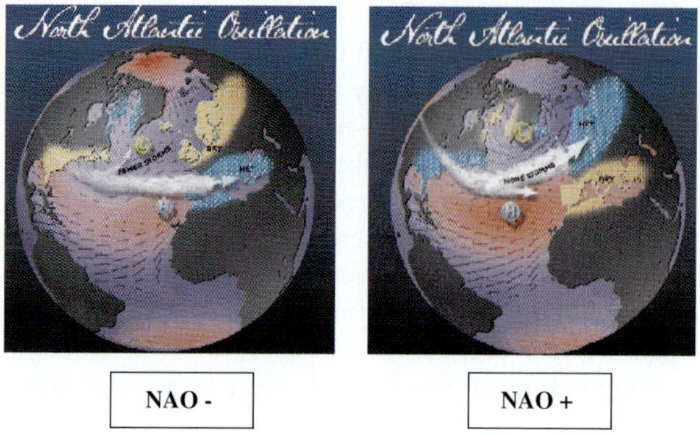

Figure 1. NAO in the negative and positive phases (from: http://www.ldeo.columbia.edu/NAO/main.html by Martin Visbeck)

During blocking situations, sometimes lasting for several weeks, central European air masses tend to flow mainly along the meridians. During these events, the entire Mediterranean area may be characterized by extended period of highly unusual weather: for example, when the axis of the block is located over the Iberian peninsula

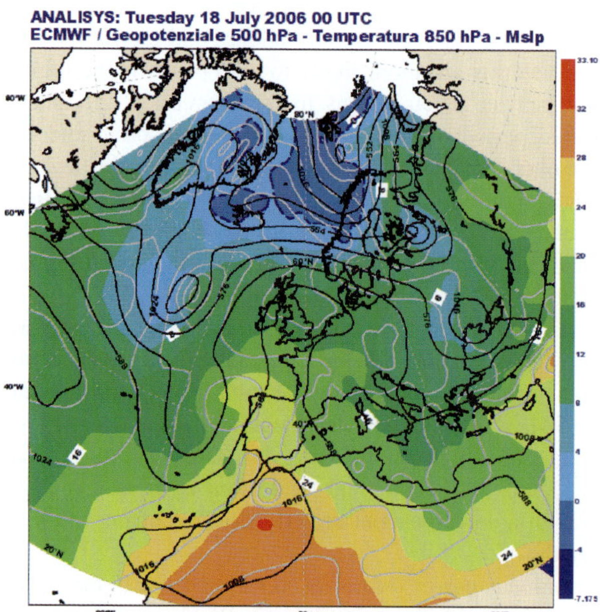

Figure 2. European Block on the Mediterranean area, responsible for the current drought conditions in the western Mediterranean area

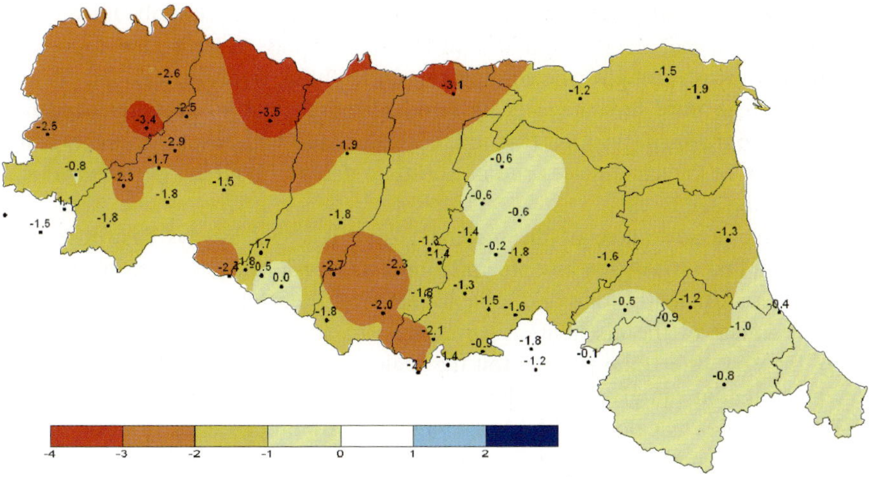

Figure 3. SPI index for July 2006 linked to the Block Pattern: the red contour indicates very drought conditions. The number of stations used here is the full historical stations available for Emilia Romagna (about 110 stations)

and France, Italy may experience a wave of northerly cold flows with persistent stormy weather for several days.

On the contrary, if the high pressure structure is located further east (for example, on the Italian peninsula), all of Central Europe and most of our country will remain under the influence of an anticyclonic pressure system characterized by good weather and stability.

The persistence of a situation of this kind can certainly lead to a lack of precipitation for very extended periods and thus to drought. For example, Figure 2 describes a typical situation of European Blocking that, in this case, lasted through June and July 2006. The promontory of the high pressure system remains over the central Mediterranean basin. There has been no significant precipitation and the temperature has exceeded 35 degrees in many areas. This situation is favorable to a continued absence of precipitation leading to a deepening in drought conditions.

Figure 3 illustrates the SPI drought index for the month of July 2006 over Emilia-Romagna as the consequence of the blocked pattern situation.

3. DROUGHT INDICATORS: SPI AND PRECIPITATION

One of the simplest indicators used to characterize drought is the Standardized Precipitation Index (SPI) introduced by McKee in an article published in 1993 (McKee et al., 1993) which made it possible to calculate the precipitation deficit on different time scales.

A number of advantages arise from the use of the SPI index. First of all, the index is simple and is based only on the amount of precipitation so that its evaluation is rather easy. In the second place, the index makes it possible to describe the drought on time scales that, while definitely arbitrary, typically describe the four most frequent types of drought (meteorological, agricultural, hydrological and socioeconomic). The third advantage is its standardization, that ensures independence from the geographical position (the index is calculated with respect to the average precipitation in the same place) and its ability to describe both dry periods and wet periods in the same way.

The main disadvantage of the method is that it is not always easy to find a distribution that models the raw precipitation data, nor it is always possible to have access to a sufficiently long and reliable temporal time series so as to produce a robust estimate of the distribution parameters. In addition, the application of the index for very brief time periods (on the scale of 1 or 2 months) in regions with low precipitation can give misleading both positive and negative SPI values.

The SPI index is calculated by modeling a density probabilities function on the distribution frequency of the total precipitation cumulated over different time scale of interest (3 months, 6 months, 12 months, 24 months). The probability density function is then transformed into a normal standardized distribution. The result is a standardized index whose values classify the category of drought characterizing each place and time scale. Table 1 shows the classification of the SPI index following the categories described in the original article.

The SPI index can be computed only when sufficiently long and possibly continuous time-series of monthly precipitation data are available (at least 30 years). The number of months on which the precipitation is accumulated is defined on the basis of the time window on which the index is calculated. This means that, in order to compute the SPI at 3 months, it is necessary to produce time series of cumulated precipitation over the current and the two previous months. A detailed mathematical description of the method of SPI index calculation is given in the Appendix.

At ARPA-SIM the values of the SPI index over the region is constantly updated for the four mentioned time windows, as one of a set of drought indicators that make up the Drought Monitor for Emilia-Romagna. In Figure 3, the map of the 3-month SPI index is evaluated for the month of July 2006.

From Figure 4 to Figure 6, three scenarios (calculated in the previous month, in this case in June 2006) are presented for the SPI3 index for the month of July 2006, obtained using three relevant scenarios of precipitation amount for July (the 25th, 50th and 75th percentile-probability density function of precipitation – pdf – for

Table 1. Classification of drought based on the SPI index

SPI value	Category
$0 < \text{SPI} < -0.99$	Mild drought
$-1.00 < \text{SPI} < -1.49$	Moderate drought
$-1.50 < \text{SPI} < -1.99$	Severe drought
$\text{SPI} < -2$	Extreme drought

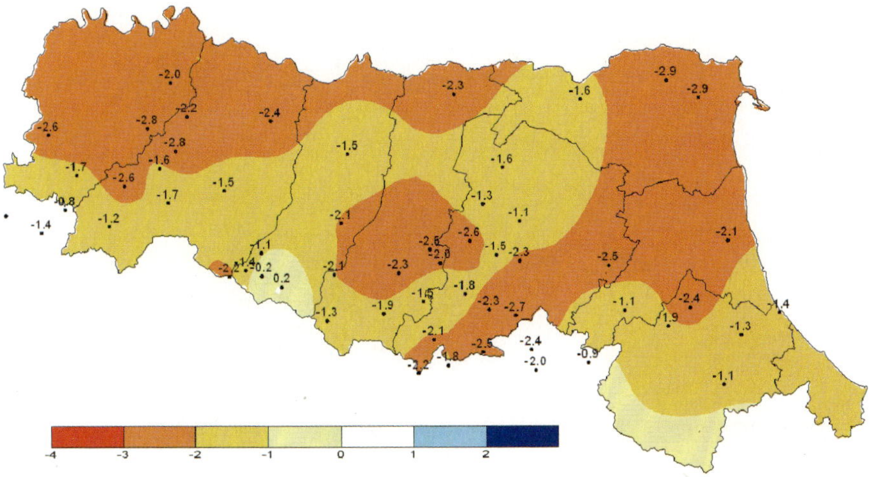

Figure 4. SPI 3-month index for the month of July 2006 calculated using as forecast the 25th percentile of the rainfall distribution function of the month of July

July). As mentioned earlier, the year 2006 is very similar to the year 2003, that is to say, very dry. From these figures it can been seen that even for the "wettest" (and thus more "optimistic") possible scenario, hypothesizing conspicuous rainfall (values on the order of the 75th percentile of the pdf for July), the SPI map calculated for the month of July still presents large areas with negative index values over the whole region.

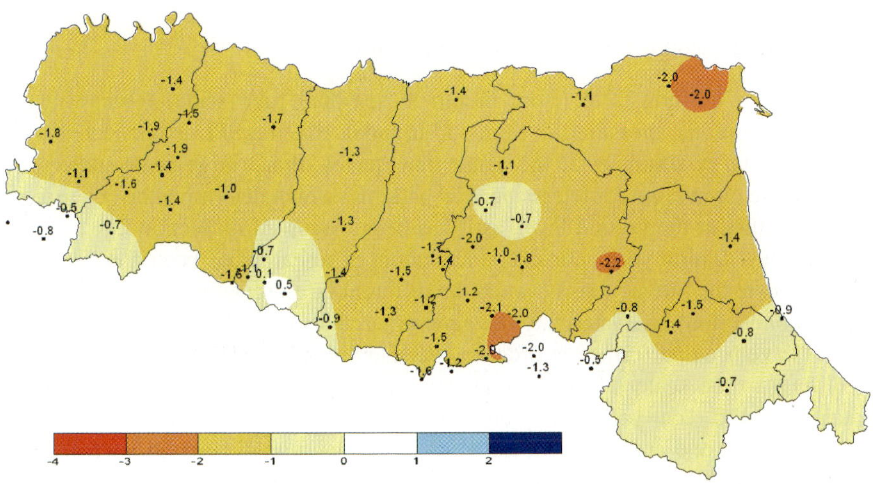

Figure 5. SPI 3-month index for the month of July 2006 calculated using as forecast the 50th percentile of the rainfall distribution function of the month of July

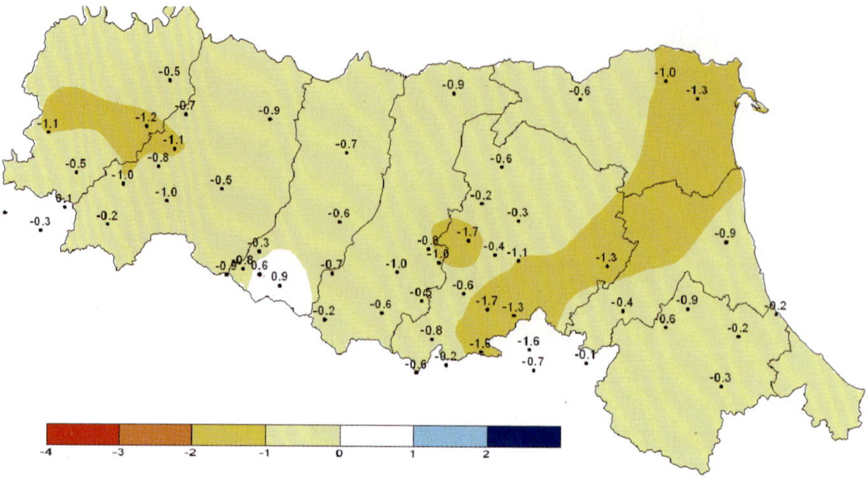

Figure 6. SPI 3-month index for the month of July 2006 calculated using as forecast the 75th percentile of the rainfall distribution function of the month of July

The SPI map in Figure 3 shows the observed situation (that realizes the "pessimistic" scenario of Figure 4). The use of SPI scenarios can be very useful in the stage of drought control as it enables us to plan the mitigating actions and/or contrasting actions on the basis of objective evidences, define the area of the region in which more serious problems exist and where it appears more necessary to intervene to prevent situations of water shortage.

4. DROUGHT FORECASTING MODELS

4.1 The Seasonal Forecast of the Spi Index

The problem of long term (from month to seasonal time scale) forecasting of drought conditions, measured in terms of an index like the SPI, is connected with the capability to simulate, at the same time scales, those large scale indices of atmospheric circulation (or teleconnection patterns) which determine the occurrence of the weather at the ground level, as it has been discussed in Section 2.

In the past several years, considerable amount of research has been done in order to investigate the interactions between the different large-scale modes of variability (or teleconnection) characterizing the general circulation of the atmosphere. Many studies have been performed in order to evaluate the predictability of these patterns at different time scales (from several days to month and seasonal scale). These patterns tend to recur periodically and with similar characteristics over rather long period of time, and are often associated with unusual climatic conditions that arise in areas of the oceans characterized by time scales of variability longer than those of the atmosphere. If the teleconnection between oceanic anomalies in a given zone of the terrestrial globe (for example the El Niño anomaly) and the atmospheric variability of other areas is strong enough, the conditions may favor an extension of the period over which there is sufficient meteorological predictability even up to rather long time scales (months or seasons) starting from that typical for meteorological forecasts (7–10 days). In particular, the tropical areas, dominated as already said by long time scales of variability (connected with the occurrence of anomalies of sea surface temperature over the tropical ocean), are much more predictable than the extra-tropical areas where, instead, the teleconnection links seem to be less strong. The project Tropical Ocean Global Atmosphere (TOGA) has produced results that suggest that it may be possible to predict certain climatic conditions associated with El Niño events more than a year ahead of time. For those regions whose climate is heavily affected by El Niño, what was learned from the TOGA project may help to produce meteorological forecasts capable of limiting the risks in the economic sectors most sensitive to climatic variability and extreme events like droughts.

In the extra-tropical regions, particularly on the European continent, long-term forecasts are not nearly as reliable. The margin of predictability is primarily linked to empirical-statistical relationships more than to clear dynamic causes. This is due to the fact that the climate in these regions tends to be affected more by typical atmospheric phenomena (which are less predictable than those of the ocean) characterized by atmospheric processes that are strongly non-linear, such as the aforementioned North Atlantic Oscillation (NAO) or the Euro-Atlantic Blocking (Pavan and Doblas-Reyes 2000). The success of long-term forecasts on the European continent depends to a large extent on the ability to reproduce, on a monthly to seasonal basis, these atmospheric processes. Unfortunately, until now, most of the Atmospheric General Circulation Models have been unable to reliably reproduce these patterns of atmospheric variability dominating the weather on the European continent, presenting systematical problems both in the reproduction of the climatology of these patterns and in their interannual variability.

In spite of these limitations, it has been shown that the paired use of more than one modeling chain (the "multi-model ensemble") can substantially improve the results of long-term forecasts (months to seasons) especially when significant large-scale circulation anomalies occur (Krishnaumurti et al., 1999, Pavan and Doblas-Reyes, 2000).

By using the technique of the multi-model ensemble it is possible to obtain an increase of the predictability of those large-scale patterns, defined above, resulting relevant for the forecasting, on long time scales (months to seasons), of climatic anomalies like, for example, drought, defined through the value of the SPI index.

The approach used to forecast SPI index consists of a series of large-scale fields, simulated by the General Circulation Models in ensemble mode (the "predictors") as inputs of statistical downscaling scheme, suitable to define the connection between the large-scale structures and the meteorological weather in the area and/or some indicators of it (the "predictants"), such as, for example, drought indices, in the specific application at hand.

The statistical downscaling scheme used here is a multiple linear regression (Wilks, 1995) in the context of a "Perfect Prognostic" approach (Klein et al., 1959).

The causal relationship between "predictor" and "predictand" is assessed using historical observational data time-series. Then, operationally, the model simulations are used as predictors.

According to this approach, in the operational stage, when the model simulation is used instead of observational data, it is necessary to take into account the error associated with the "imperfection" of the predictor in addition to the intrinsic error of the statistical method measurable by the fraction of explained variance not taken into account by the regression. However, it is clear that this type of approach is susceptible to natural and continuous improvements as the errors associated with the predictors tend to become smaller and smaller. For models which are theoretically "error-free" (or perfect, as the name Perfect Prognostic implies) this problem is entirely eliminated.

The following data-sets are used to develop this technique:

1. the ERA40 re-analysis of the European Center of Medium-range Weather Forecasting (ECMWF);
2. the analysis of precipitation over Northern Italy and the Alps produced by the Mesoscale Alpine Project (MAP);
3. the large-scale seasonal forecasts produced within the DEMETER project (Development of a European Multimodel Ensemble system for seasonal to inTERannual prediction).

The ERA40 re-analysis (which replaces the previous ERA15) was produced by the ECMWF using the T159L60 version of the forecasting model currently in use at the ECMWF. Analyses are produced every 6 hours for the period 1958–2002. A detailed description of the project is found in Uppala et al., 2002. The monthly averages of geopotential height at 500 hPa (Z500) were extracted and interpolated from the original grid to a spatial resolution of $2.5° \times 2.5°$.

The MAP data-set, used for precipitation analysis, is described by Frei and Schär 1998. This analysis was produced from surface observations provided by the various national networks belonging to the Alpine area and northern Italy that participated to the project. The analysis is available over a regular grid of $0.3° \times 0.22°$ for the period 1966–1999.

Seasonal forecasts for the indices of large scale circulation were obtained using the data produced within the DEMETER project (Palmer et al., 2004), a 6th Framework Program project of the European Community. The project produced a data-set of seasonal forecasts obtained using seven different models. Every model provided 9 different members for every season for the period 1958–2002.

4.2 The Downscaling Technique

The group of possible large-scale predictors of the downscaling scheme consists of a group of indices describing the time variability of the major patterns of large scale variability, which are highly correlated to the NAO and Blocking indices, as has been shown in the previous chapters. These indices of large scale variability can be defined making use of multivariate statistical analysis. In particular, the description of the variability of the large-scale circulation was obtained applying a standard principal components analysis (PC) to the monthly anomalies of Z500 for the area from 90°W to 60°E and from 20°N to 90°N and to air temperature at 850hPa (T850) for the area 10°W–60°E and 30°N–70°N.

Figures from 7 to 10 show, as an example, the patterns for every season of the Empirical Orthogonal functions (EOFs) associated with the first four PCs of Z500 anomalies for winter, spring, summer and autumn.

In EOF1 of Z500 for the winter season it is possible to recognize the typical bipolar configuration of the NAO, while in the EOF2 the typical configuration of European Blocking (compare with Figure 2).

After the EOFs decomposition of Z500 and T850, the associated PCs represent the possible predictors of the SPI index, that is to say "the predictands". Figure 11 outlines the logical flow of the method. The regression coefficients between predictors (PC of Z500 and T850) and predictants (PC of SPI3) were assessed using the "cross–validation" technique on all the available data observed. The large-scale predictors were obtained, by combining all the forecasts of the DEMETER models using the BLUE multi-model ensemble technique, which was applied with success in Pavan and Doblas-Reyes, 2000 and Pavan et al., 2005. Different weights were used for every model and pattern. Weights are determined using the BLUE technique (Best Linear Unbiased Estimate) suggested by Thompson (1977) and Sarda et al. (1996) using all the data available in cross-validation.

The quality of the forecast of the predictors was estimated considering the correlation between the SPI indices simulated with the downscaling technique and those

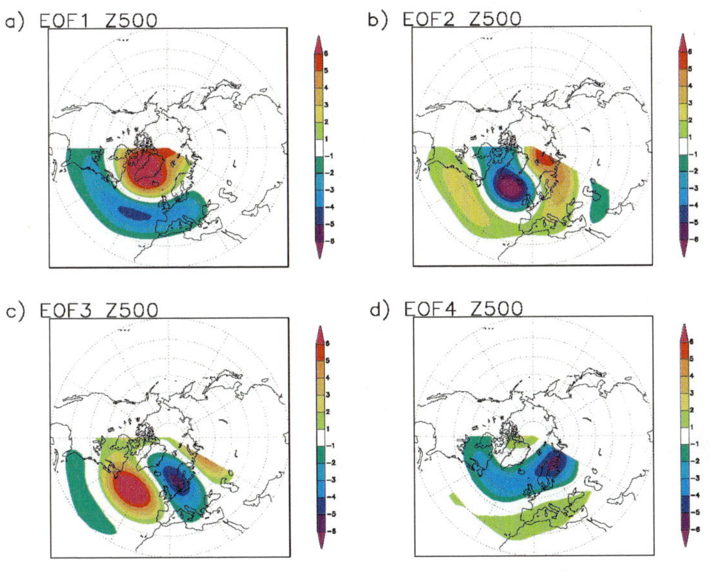

Figure 7. Empirical Orthogonal Functions for the winter season. The PCs associated are used as "predictors" for the seasonal drought forecast

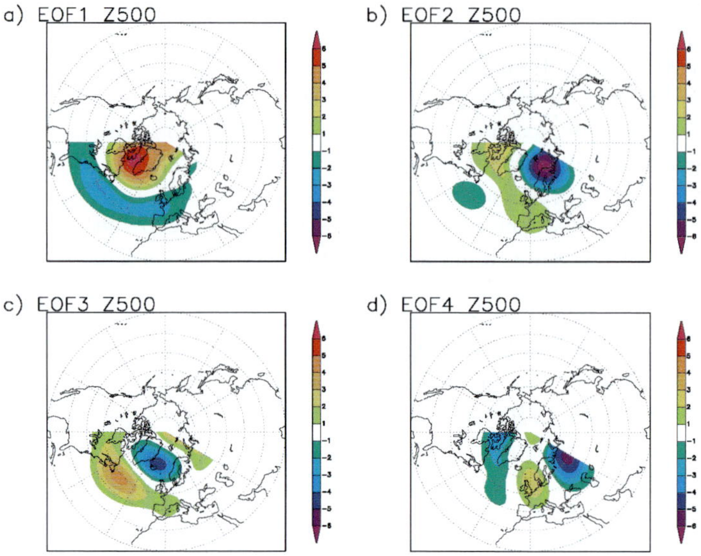

Figure 8. Empirical Orthogonal function for the spring season. The PCs associated are used as "predictors" for the seasonal drought forecast

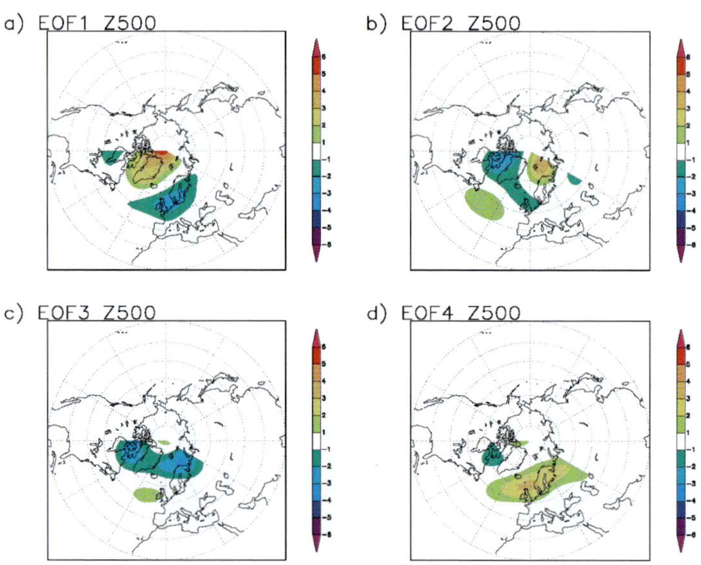

Figure 9. Empirical Orthogonal function for the summer season. The PCs associated are used as "predictors" for the seasonal drought forecast

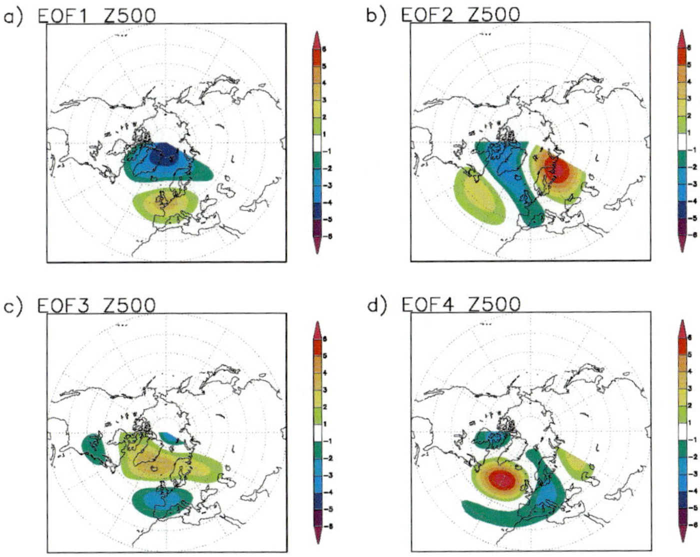

Figure 10. Empirical Orthogonal function for the autumn season. The PCs associated are used as "predictors" for the seasonal drought forecast

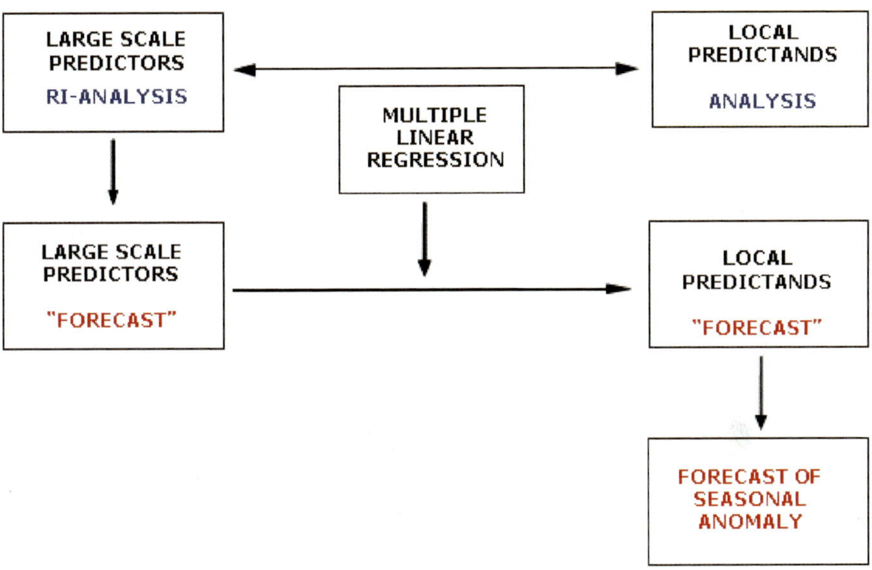

Figure 11. Diagram of downscaling method

observed (that is, calculated with observed data) both as regards performance in time of the PC (Principal Component) and as a map of correlation.

The first phase of the method is the assessment of the statistical link between the SPI index and the variability of the atmospheric circulation. This analysis has been performed by evaluating the correlation between the first 4 PCs of the SPI index calculated at 3 months for the Emilia-Romagna region and the NAO and EB indices, which are correlated with the PCs of Z500 and T850. In order to obtain an index as clean as possible from any "interferences", the temporal series of SPI relative to the last month of the season was taken into consideration (February for the winter season, May for spring, August for summer, November for autumn), so as to be sure of considering only the precipitation relative to the season under study.

The coefficients of correlation were calculated on the overall period of study, 1958-2002. Table 2 shows the coefficients of correlation between the first PC of the SPI index, in general linked with the mean field variability and representing the greatest variability of the index, and the NAO and EB indices calculated using ERA40 for the season under study.

The NAO index for winter is anticorrelated with PC1 of the SPI index as can be expected of the synoptic situation described in the preceding section (if the NAO index is negative there is more precipitation in northern Italy, thus an increase of the wet conditions, i.e. positive SPI values). In spring and summer, the highest correlations occur only with the blocking index (spring) or with both indices (summer), that is with regimes of high pressure or flows moving northward that cause a lack of precipitation. In fall, both the NAO and the EB are not significantly correlated with the first PC of the SPI index. The forecast, as we shall see, is therefore still satisfactory and this is explained by the fact that the other PCs (not shown here) that participate in the downscaling model are in any case significantly correlated with the NAO and EB indices.

In view of the results of high correlation observed, it was decided to use the first four PCs of the Z500 and T850 anomalies, drawn from the seasonal simulations of the different models used in the DEMETER project as potential predictors. The predictants were, instead, the 4 PCs of the SPI index at 3 months. The forecast is made using the multiple regression technique described above. The values of the coefficients of multiple regression used in this calculation were based on observed data.

Table 2. Coefficients of correlation between the NAO and EB indices and the first PC of the SPI index at three months for the different seasons

1958–2002	NAO	EB
SPI (Feb)	−0.4	0.1
SPI (May)	0.07	0.8
SPI (Aug)	0.6	0.7
SPI (Nov)	0.3	0.1

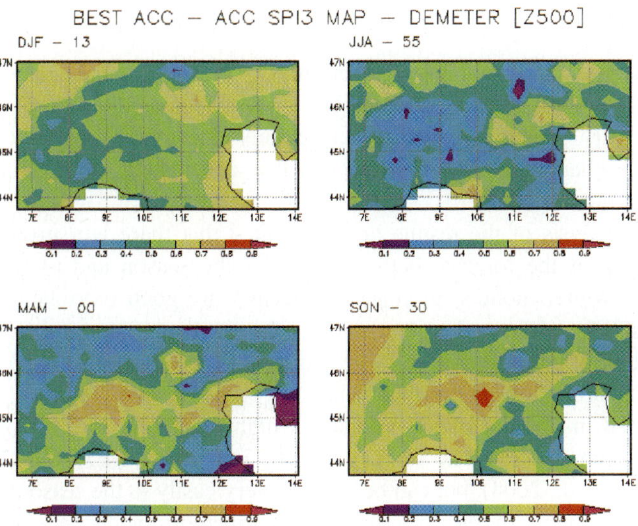

Figure 12. Map of correlation between the SPI index at 3 months obtained with the downscaling model and the SPI index calculated from observed data

On the basis of the link between the large scale and the SPI3 index, comparisons were made between a set of results obtained by modifying the areas (domains) of definition of the EOFs of the potential predictors in order to emphasize the importance of the different structures that make up their pattern of variability. The result of the comparison among these domains led to the map shown in Figure 12 in which, for each season, the area was selected in such a way as to maximize the area average of the correlation between the result of downscaling and the observed time series.

Table 3 shows the correlations values between the first PC of the SPI index at 3 months simulated by the downscaling model and PC1 of the same index calculated from the observed data.

Table 3. Table of correlation between the PC1s of the SPI3 index simulated by the DS model and PC1 of the SPI3 index calculated on observed data for the seasons DJF (Dec-Jan-Feb), MAM (Mar-Apr-May), JJA (Jun-Jul-Aug) and SON (Sept-Oct-Nov)

PC1-SPI3	Correlation
DJF	0.58
MAM	0.48
JJA	0.46
SON	0.65

5. CONCLUSIONS

Seasonal forecasts are among the main research activities of meteorological centers throughout the world. The current state of forecasts on a few regions, like Europe, is, unfortunately, not very satisfactory and this is mainly due to the negligible influence of large-scale phenomena like El Niño that can affect predictability for time scales longer than one month.

It is clear from the analysis of the results here presented that there is a strong variability in the quality of the forecast both in terms of the season and of the area considered. In the winter months, in general, forecasts are good (correlation everywhere better than 0.5), while in the other seasons the results are more mixed. It should be noted, however, that there are large areas in which the forecast has considerable accuracy (correlations better than 0.7).

The method described in this document produces a high resolution forecast of the SPI index starting from the seasonal forecasts of the European "DEMETER" research project (Palmer et al., 2004), but can be adapted very easily to the seasonal forecasts produced operationally by the large international meteorological services (such as ECMWF) by linking the forecast of this index with that of the average field of precipitation so as to optimize the result while preserving the consistency between the two measurements. At the present time, the technique has been applied only to Northern Italy as what is fundamental for the development of the model is the availability of a reliable observational analysis, currently only available for northern Italy (MAP analysis, Frei and Schär, 1998). Once a reliable analysis will be available on a national scale, the model will be easily implemented throughout the country.

ACKNOWLEDGEMENTS

This work was funded by the European Union under the INTERREG III B area MEDOCC project SEDEMED II. The authors wish to thank the entire Scientific Committee of the SEDEMED II projects for their useful suggestions during the project, and the editors for their precious comments during the preparation of this chapter.

REFERENCES

Abramowitz M, Stegun A, (1965) *Handbook of Mathematical formulas, Graphs, and Mathematical Tables*. Dover Publications, inc.: New York.

Cacciamani, C., Nanni, S., Tibaldi, S. (1994) Mesoclimatology of winter temperature and precipitation in the Po Valley of northern Italy. *International Journal of Climatology*, 14, no. 7, 777–814.

Edwards DC, McKee TB, (1997) *Characteristic of 20th century drought in the United States at multiple timescales*. Colorado State University: Fort Collins, Climatology Report No 92–2.

Frei, C. Scär, C. (1998) A precipitation climatology of the Alps from high resolution rain-gauge observations. *International Journal of Climatology* 18, 873–900.

Hurrel W. (1995) Decadal Trends in the North Atlantic Oscillation: Regional Temperatures and Precipitation. *Science* 269, Issue 5224, 676–679.

Klein, W.H., B.M. Lewis, Enger, I. (1959) Objective prediction of five-day mean temperature during winter. *Journal of Meteorology* 16, 672–682.

Krishnamurti, T.N., Kishtawal, C.M, Timoty, E.L., Bachiochi, D., Zhan Zhang, and co-authors (1999) Improved weather and seasonal climate forecasts from multi-model superensemble. *Science* 285, 1548–1550.

Lloyd-Hughes B, Saunders M.A. (2002) A drought climatology for Europe. International Journal of Climatology 22, 1571–1592.

McKee, T.B., Doesken, N.J., Kleist, J. (1993) The relationship of drought frequency and duration to time scale. *8th Conference on Applied Climatology*, Anaheim, California, 179–184.

Palmer, T.N., Alessandri A., Andersen U., Cantelaube P., Davey M., and co-authors (2004) DEMETER, Development of a European multi-model ensemble system for seasonal to interannual prediction. *Bull. Am. Meteorol. Soc.* 85, 853–872.

Pavan, V., Doblas-Reyes, F. (2000) Multi-model seasonal hindcasts over the Euro-Atlantic: skill scores and dynamic features. *Clim. Dyn*, 16, 611–625.

Pavan, V. Marchesi, S, Morgillo, A., Cacciamani, C. (2005) Downscaling of DEMETER winter seasonal hindcasts over Northern Italy. *Tellus* 57A, 424–434.

Quadrelli, R., Lazzeri, M., Cacciamani, C., Tibaldi, S. (2001) Observed winter Alpine precipitation variability and links with large-scale circulation patterns. *Climate Research* 17: 275–284.

Rex, D. F. (1950) Blocking action in the middle troposphere and its effect on regional climate. I. An aerological study of blocking action. *Tellus* 2, 196–211.

Sarda, J., Plaut, G., Pires, C., Vautard, R. (1996) Statistical and dynamical long-range atmospheric forecasts: experimental comparison and hybridization. *Tellus* 48A, 518–537.

Tibaldi, S., Tosi, E., Navarra, A., Peduli, L. (1994) Northern and Southern Hemispere seasonal variability of blocking frequency and predictability. *Monthly Weather Review* 122, No. 9, pp. 1971–2003.

Thom H.C.S. (1958) A note on the Gamma distribution. *Monthly Water Review* 86 vol. 4 117–122.

Thompson, P.D. (1977) How to improve accuracy by combining independent forecasts. *Monthly Weather Review* 105, 228–229.

Uppala, S.M., Kållberg, P.W., Simmons, A.J., Andrae, U., da Costa Bechtold, V., Fiorino, M., Gibson, J.K., Haseler, J., Hernandez, A., Kelly, G.A., Li, X., Onogi, K., Saarinen, S., Sokka, N., Allan, R.P., Andersson, E., Arpe, K., Balmaseda, M.A., Beljaars, A.C.M., van de Berg, L., Bidlot, J., Bormann, N., Caires, S., Chevallier, F., Dethof, A., Dragosavac, M., Fisher, M., Fuentes, M., Hagemann, S., Hólm, E., Hoskins, B.J., Isaksen, L., Janssen, P.A.E.M., Jenne, R., McNally, A.P., Mahfouf, J.-F., Morcrette, J.-J., Rayner, N.A., Saunders, R.W., Simon, P., Sterl, A., Trenberth, K.E., Untch, A., Vasiljevic, D., Viterbo, P., Woollen, J. (2005) The ERA-40 re-analysis. *Quart. J. R. Meteorol. Soc.* 131, 2961–3012.

Wallace J. M., Gutzler D. S. (1981) Teleconnections in the geopotential height field during the Northern Hemisphere winter. *Monthly Weather Review* 109, 784–812.

Wilks, D.S. (1995) *Statistical Methods in the Atmospheric Sciences. An introduction.* ACADEMIC PRESS.

APPENDIX: CALCULATING THE SPI INDEX

In most cases, the distribution that best models observational precipitation data is the Gamma distribution. The density probability function for the Gamma distribution is given by the expression:

$$g(x) = \frac{1}{\beta^{\alpha}\Gamma(\alpha)} x^{\alpha-1} e^{\frac{-x}{\beta}} \qquad \text{for } x > 0 \tag{1}$$

where $\alpha > 0$ is the shape parameter , $\beta > 0$ is the scale parameter and $x > 0$ is the amount of precipitation. $\Gamma(\alpha)$ is the value taken by the standard mathematical function known as the Gamma function defined by the integral:

$$\Gamma(\alpha) = \lim_{n\to\infty} \prod_{v=0}^{n-1} \frac{n!n^{y-1}}{y+v} \equiv \int_0^{\infty} y^{\alpha-1} e^{-y} dy \tag{2}$$

In general, the Gamma function is evaluated either numerically or using the values tabulated depending on the value taken by parameter α.

In order to model the data observed with a gamma distributed density function, it is necessary to estimate appropriately parameters α and β.

Different methods have been suggested in literature for the estimate of these parameters, for example, in Edwards and McKee (1997), the Thom (1958) approximation is used for the maximum probability

$$\hat{\alpha} = \frac{1}{4A}\left(1 + \sqrt{1 + \frac{4A}{3}}\right) \tag{3}$$

$$\hat{\beta} = \frac{\bar{x}}{\hat{\alpha}} \tag{4}$$

where for n observations

$$A = \ln\left(\bar{x}\right) - \frac{\sum \ln(x)}{n} \tag{5}$$

The estimate of the parameters can be further improved by using the interactive approach suggested in Wilks (1995).

After estimating coefficients α and β the density of probability function g(x) is integrated with respect to x and we obtain an expression for cumulative probability G(x) that a certain amount of rain has been observed for a given month and for a specific time scale.

$$G(x) = \int_0^x g(x)\,dx = \frac{1}{\hat{\beta}^{\hat{\alpha}}\,\Gamma\left(\hat{\alpha}\right)} \int_0^x x^{\hat{\alpha}-1} e^{-x/\hat{\beta}} dx \tag{6}$$

Taking $t = x/\hat{\beta}$ this equation becomes the incomplete Gamma function

$$G(x) = \frac{1}{\Gamma(\hat{\alpha})} \int_0^x t^{\hat{\alpha}-1} e^{-t} dt \tag{7}$$

The Gamma function is not defined by $x = 0$ and since there may be no precipitation the cumulative probability becomes:

$$H(x) = q + (1-q)\, G(x) \tag{8}$$

where q is the probability of no precipitation. The cumulative probability is then transformed into a normal standardized distribution with null average and unit variance from which we obtain the SPI index. For the details, see Edwards and McKee (1997) or Lloyd-Hughes and Saunders (2002).

The above approach, however, is neither practical nor numerically simple to use if there are many grid points or many stations on which to calculate the SPI index. In this case, an alternative method was described in Edwards and McKee (1997) using the technique of approximate conversion developed in Abramowitz and Stegun (1965) that converts the cumulative probability into a standard variable Z. The SPI index is then defined as:

$$Z = SPI = -\left(t - \frac{c_0 + c_1 t + c_2 t^2}{1 + d_1 t + d_2 t^2 + d_3 t^3}\right) \quad \text{for } 0 < \mathrm{H}(x) < 0.5 \tag{9}$$

$$Z = SPI = +\left(t - \frac{c_0 + c_1 t + c_2 t^2}{1 + d_1 t + d_2 t^2 + d_3 t^3}\right) \quad \text{for } 0.5 < \mathrm{H}(x) < 1 \tag{10}$$

where

$$t = \sqrt{\ln\left[\frac{1}{(H(x))^2}\right]} \quad \text{for } 0 < \mathrm{H(x)} < 0.5 \tag{11}$$

and

$$t = \sqrt{\ln\left[\frac{1}{(1 - H(x))^2}\right]} \quad \text{for } 0.5 < \mathrm{H(x)} < 1 \tag{12}$$

where x is precipitation, H(x) is the cumulative probability of precipitation observed and c^0, c^1, c^2, d^0, d^1, d^2 are constants with the following values:

$$c^0 = 2.515517 \quad c^1 = 0.802853 \quad c^2 = 0.010328$$

$$d^0 = 1.432788 \quad d^1 = 0.189269 \quad d^2 = 0.001308$$

Other types of statistical distribution succeed, in some cases, in providing a better model of precipitation for some seasons or for particular time scales such as, for example, the Poisson-Gamma distribution or the log-normal distribution. If the value of parameter α is very high, the Gamma distribution tends towards a normal distribution and, therefore, it may be more effective, at the computational level, to estimate the SPI index using a Normal distribution with average $\hat{\mu}$ and standard deviation $\hat{\sigma}$, for both the values estimated by the sample used. Then, the SPI index will be defined by:

$$Z = SPI = \frac{(x - \hat{\mu})}{\hat{\sigma}} \tag{13}$$

CHAPTER 3

DEVELOPMENT OF THE PIEDMONT REGION HYDROLOGICAL BULLETIN AS A SUPPORT TO WATER RESOURCES MONITORING AND MANAGEMENT

C. RONCHI, D. RABUFFETTI, A. SALANDIN, A. VARGIU, S. BARBERO AND R. PELOSINI

ARPA Piedmont, Environment Forecast and Monitoring Area, Torino, Italy

Abstract: The Regional Hydrological Bulletin is an operational instrument studied to supply quantitative hints of the Piedmont Region main hydrological river basins state with simple and exhaustive communication modalities. The aim is to evidence the potential established of drought conditions and the insufficient availability of water resource on a regional territory. In order to realize this tool finalized to support the water resource management, contents, methodologies and an appropriate language have been developed, in a framework of climatologic reference. Moreover, the Bulletin supplies, through the statistic analysis of the SPI (Standardized Precipitation Index), an assessment of the short-term drought conditions over river basin spatial scale. This monitor includes all the resources flowed to the environment system in terms of rain volumes, water reservoirs stored in the not-perennial snowy mantle and the streaming capacities of the main watercourses

Keywords: Bulletin, drought, monitoring, SPI, forecast

1. INTRODUCTION

Water Resources Management is the integrating concept for a number of water sub-sectors such as hydropower, water supply and sanitation, irrigation, drainage, and environment. Moreover the evidences of regional climate changes such as the outstanding low water of the Po river during 2003, demonstrated that drought impacts involve areas usually rich of water resources too.

In this framework, ARPA Piedmont – Environment forecast and monitoring Area has deepened its own knowledge about drought sharing the European Community cooperation plans such as SEDEMED I and II – Interreg IIIB – MedOcc focused on this specific theme and identifying the best methodologies to monitor and forecast drought with other European partner.

Thanks to these experiences, specific reports about water resources assessment and concerning meteorology, hydrology and snowfalls state, have been realized and provided to the Direzione Pianificazione delle Risorse Idriche della Regione Piemonte (*Piedmont Region Hydrologicl Resources Planning Department*) during

G. Rossi et al. (eds.), Methods and Tools for Drought Analysis and Management, 49–65.

the years 2003 and 2005 (Direzione Pianificazione delle Risorse Idriche, ARPA Piemonte, 2005).

The succession of water deficit outlined the need of setting an accurate framework about regional hydrological conditions that has to be immediately submitted to the decision makers in order to manage emergencies.

Drought is a common and periodic characteristic of the hydrologic cycle. It is estimated in relationship with the local budget between precipitation and the evapotranspiration (evaporation + transpiration) and according to the period in which it happens: the season, the delay regarding the beginning of the rainfalls time, and the effectiveness of rains, that is their intensity and the number of events. Obviously, other parameters such as the temperature and the soil humidity are important for the definition of the drought conditions.

An operative definition of "drought" has to be able mainly to identify the length and the severity of the phenomena. For this reason, the Regional Hydrological Bulletin considers both meteorological and hydrological drought.

In order to supply a quantitative outlook concerning the state of the main Hydrological river basins of the Po river's tributaries and with the purpose to evidence the beginning of drought conditions, it turns out particularly effective to use climatic and hydrological indices. For ours purposes SPI index (Standardizes Precipitation Index) has been considered due to the fact that it was designed to quantify the precipitation deficit for multiple time scales.

Rainfalls represent the major forcing of system and they are the main indicator of the river basin state in the short and medium period. On the other side, accounting for snow cover accumulation during wintertime is very important for Alpine catchments where snowmelt can play a crucial role up to late spring and summer. To this purpose a numerical physically based hydrological model for snow accumulation and melt is used.

Finally river discharge and water volumes stores in lakes or artificial reservoirs directly represent the real availability of water mainly used for irrigation.

This paper is organized in the following way: rainfall data and treatments are analysed in paragraph 2; in paragraph 3 the development of the bulletin is presented; paragraph 4 presents the application to the case of June 2006 with a detailed description of the pieces of information issued; in the last paragraph conclusion about the main scientific issues and a comment outlining the future developments of this activity are provided.

2. REGIONAL RAINFALL DATA AND TREATMENT METHODOLOGIES

The time series used in this study are registered by the old network managed by the SIMI for the period 1913–2002 joined with the time series coming from the regional monitoring network from the year 1990 up to now. In particular in this work we analysed precipitation from 121 stations recorded by network of the Servizio Idrografico e Mareografico Nazionale (SIMN) with at least 60 years long data set.

2.1 From Point to Area Precipitation

The analysis of the meteoric flows is carried out at the spatial scale of hydrographical basin by means of the mean areal rainfall. Such approach allows considering the effective water volume flowed into the basin, therefore passing problems related to insufficient spatial representativeness connected to the analysis of the point precipitation series. Also the evolution of the network doesn't affect these calculations once assured that the network can be considered homogeneous.

Using the "inverse distance method", the precipitation intensity in a point P (x, y) out of the rain gauge network is inversely proportional to the square distance among that point and the stations according to the formula (Wei and McGuinnes, 1973):

$$P(x, y) = \frac{1}{\sum_{i=1}^{N} \frac{1}{d_i^2}} \sum_{i=1}^{N} \frac{1}{d_i^2} P_i \qquad (1)$$

where N is the number of stations (here we took the 5 nearest rain gauge to each point in the region), d is the distance of point P(x, y) from the i-th station. This method is picked especially for its simplicity, mainly when the network configuration can change in time.

From the point of view of the results reliability, Singh and Chowdhury (1986) studying different methods for the calculation of the mean areal precipitation (inverse distance, Thiessen polygons, isohyets, etc.), found that all the methods give comparable results especially for long time series. Therefore it is possible to estimate the precipitation field and to calculate the mean areal rainfall on any basin of interest by means of all available observations.

2.2 Analysis of the Rain Gauge Network Evolution

Given the method adopted for rainfall interpolation, it is important to understand in which way the evolution in the station number affected the soundness of the mean areal rainfall estimation.

We therefore investigated the density of the gauges calculating for each point of the territory the average distance from the 5 nearest gauges (those used for the interpolation). This average distance is normalized with the average distance similarly calculated for a reference network, constituted by a virtual 5 Km regular grid of gauges, thus defining a relative 'density index'.

Calculation was provided for the whole Po river catchment closed at Becca cross-section (37484 Km^2) and to a number of sub-catchments, based on the river network, whose surfaces range from about 1000 to 3000 Km^2. In this way one can highlight either if the density is homogeneous and how it changed in time.

Table 1 shows the sub-catchment partitioning. About the homogeneity in space of the network density one can affirm that the region is overall well covered; one

Table 1. Sub-catchments definition and properties

	Sub-catchment	Surface [km^2]	Density Index for 2005
1	Ticino svizzero	4747	0,26
2	Toce	1784	0,49
3	Sesia	1132	0,47
4	Cervo	1019	0,47
5	Dora Baltea	3939	0,39
6	Orco	913	0,47
7	Stura Lanzo	886	0,55
8	Dora Riparia	1337	0,62
9	Pellice	975	0,56
10	Alto Po	717	0,48
11	Varaita	601	0,45
12	Maira	1214	0,45
13	Stura Demonte	1472	0,46
14	Tanaro	1812	0,60
15	Bormida	1733	0,54
16	Orba	776	0,46
17	Scrivia Curone	1364	0,51
18	Agogna Terdoppio	1598	0,34
19	Residual Po at Tanaro inlet	2021	0,32
20	Residual Po at Dora Baltea inlet	781	0,50
21	Residual Po at Dora Riparia inlet	1778	0,50
22	Residual Tanaro at Po inlet	2403	0,49
	Po closed at Becca	37484	0,44

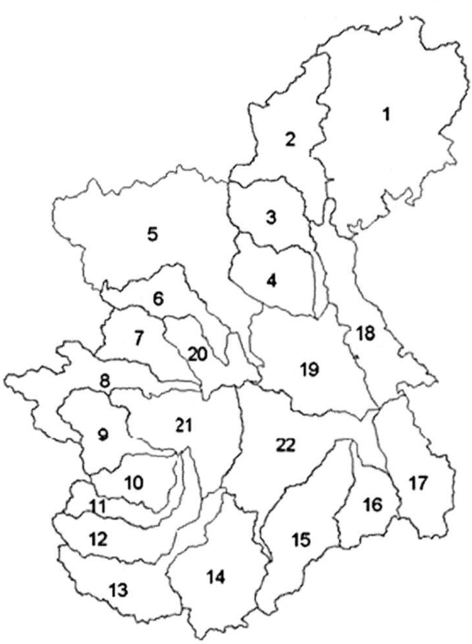

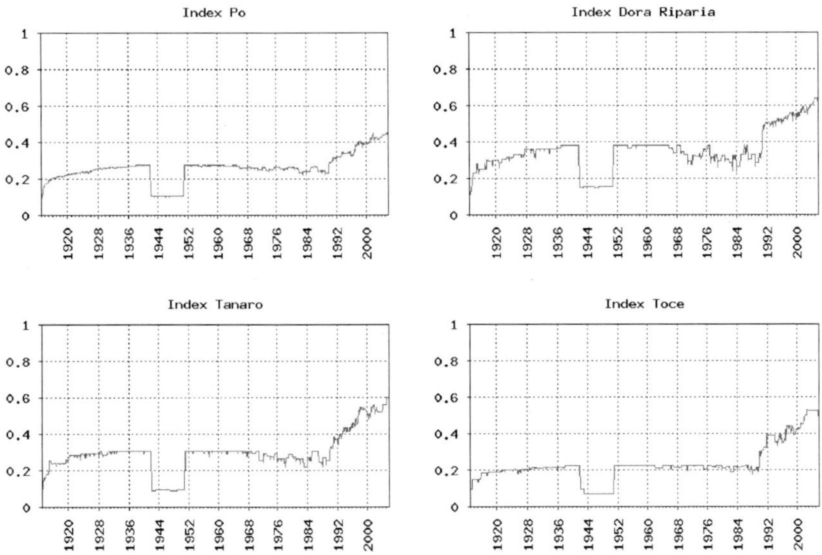

Figure 1. Comparison of the network density index for the whole area and for some sub-catchments

only underline that the north-eastern part of the region and the lowlands are little less covered by the stations.

As shown in Fig. 1, the most important feature to focus on is the important lack of rain gauge density in the years 1944–1950 caused by the II World War. This suggests to consider the period successive to 1950 for any statistical analysis. It is also very clear how in the last 15 year the development of the regional monitoring network significantly improved the monitoring capacity.

2.3 SPI Temporal Analysis

The SPI index, calculated for each time series available of "basin average precipitation", allows evaluating, also for different climatic regimes, the precipitation deficit of each series. Such SPI flexibility allows using this method in various geographic regions (McKee et al., 1995, Komuscu 1999).

SPI is calculated as follow:

$$SPI = -\left(K - \frac{A_0 + A_1 K + A_2 K^2}{1 + b_1 K + b_2 K^2 + b_3 K^3}\right) \quad for\, 0 < H(x) \leq 0.5 \tag{2}$$

$$SPI = +\left(K - \frac{A_0 + A_1 K + A_2 K^2}{1 + b_1 K + b_2 K^2 + b_3 K^3}\right) \quad for\, 0.5 < H(x) < 1 \tag{3}$$

where A_n and b_n are constants, K is calculated from

$$K = \sqrt{\ln\left(\frac{1}{(H(x))^2}\right)} \quad for\, 0 < H(x) \leq 0.5 \tag{4}$$

$$K = \sqrt{\ln\left(\frac{1}{1-(H(x))^2}\right)} \quad for\, 0.5 < H(x) < 1 \tag{5}$$

and H(x) is the excedance probability of a given value of precipitation whose total series follows a Pearson distribution (see McKee et al. 1993).

SPI usually is calculated on 3, 6, 12 and 24 months time scales and characterizes the monthly tendency of the precipitation deficit for those time scales.

In particular, each of those scales considers the impact of the deficit on the availability of several types of water resources. As SPI is a standardized index, compares drought state for various areas, independently from the monitoring point. Positive SPI values indicate greater than median precipitation of the particular station, and negative values indicate less than median precipitation. Each class of SPI is linked to the frequency of drought events during the data time series, so, for each class of SPI, is possible to define a class of drought severity (McKee et al. 1993).

In this article the SPI series were computed from the precipitation time series representing Piedmont Region as described in paragraph 2.1 from December 1950 to October 2006, as following from paragraph 2.2 considerations, at temporal scales of 3 months and 12 months.

3-month SPI is a suitable representation of short and medium term moisture conditions and offers a seasonal estimation of precipitation deficit or surplus particularly useful in agriculture activities. On the other side, the 12 months time scale avoids intra-annual frequency variations and let us identify main hydrological droughts and dry/wet periods (Vicente-Serrano 2005).

Fig. 2a and Fig. 2b show the SPI time series for Piedmont Region computed respectively on 3 months and 12 months time scales for each month.

From these time series, we compute the linear trends by least-square estimates. Linear trend results are summarized in Table 2, both at the yearly and at the seasonal time scale. A standard definition of winter (DJF), spring (MMA), summer (JJA) and autumn (SON) is used.

No significant linear trend is associated with any of these time series, as we verified using both the Pearson correlation index and a Mann-Kendall test, both

Table 2. Linear trends for 3 and 12 months SPI over the region. All units are in standard deviations per year. None of the trends is significant at the 5% level

	Year	DJF	MAM	JJA	SON
3 mo SPI	−0.0004	−0.0032	−0.0022	−0.0016	0.0020
12 mo SPI	−0.0003	−0.0013	−0.0017	−0.0020	−0.0019

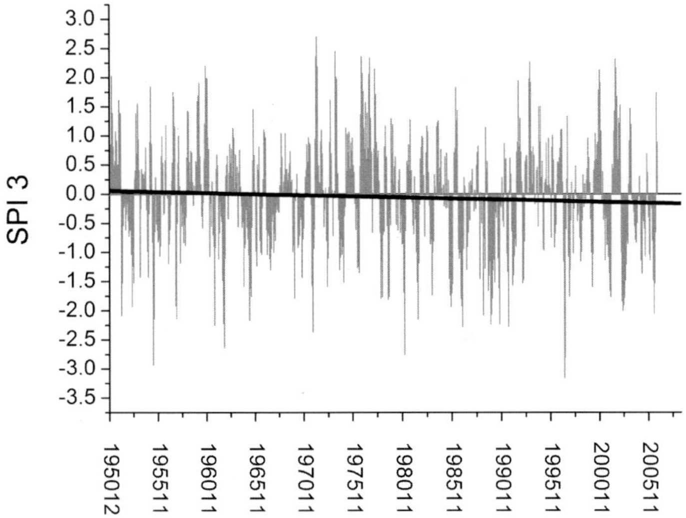

Figure 2a. 3 months SPI evolution (black bars), and linear trend (black solid line) for Piedmont Region from December 1951 to October 2006

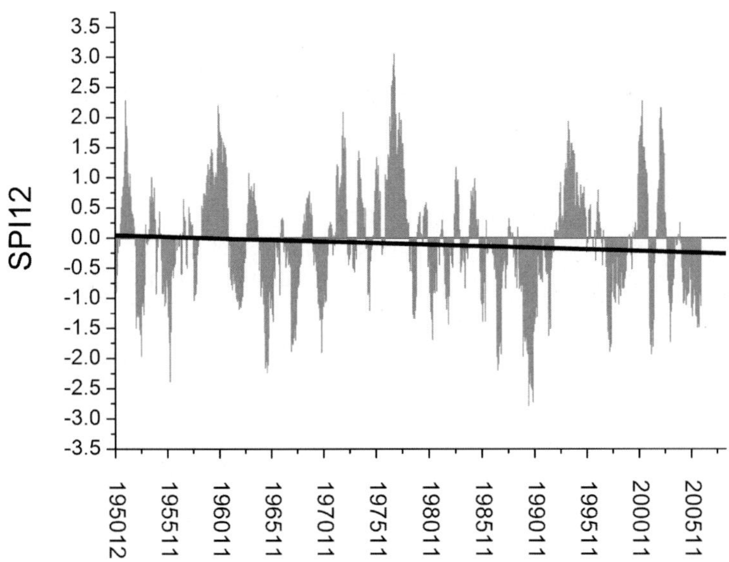

Figure 2b. 12 months SPI evolution (black bars) and linear trend (black solid line) for Piedmont Region from December 1951 to October 2006

at a 5% significance level. This results reflect the great variability in drought trends over Mediterranean areas and agree with what found by Lloyd-Hughes & M. Saunders (2002) that analyzed drought trends in whole Europe during the period 1901–1999.

2.4 Spi Sub-Regional Patterns

Until now, the SPI drought analysis has been performed on a regional scale, considering a unique precipitation time series as representative of about 38000 km^2. In this section we are interested in probing drought features at a more fine scale, separating the Piedmont region in 21 rivers basin as described in paragraph 2.2.

The goal is to understand the answer of sub-regional pattern to drought conditions and to valuate the operative usefulness in taking into account separately smaller portions of the whole area for the purposes of regional hydrological Bulletin illustrated in paragraph 4.

As a result, 3 months and 12 months SPI have been calculated fore each of the 21 river basins over the complete period of analysis and the drought spells are been clustered according to their severity classification as settled by McKee et al. (1993).

This simple method let us to identify major drought spells occurred over Piedmont in the last 50 years and how much portion of the whole area (in percent) has been affected by the phenomenon.

The maximum amount of area suffering drought has been calculated as the sum of single river basin areas in which SPI is less than -1.

As an example, Fig. 3 illustrates the 12 months SPI behaviour for the period 1951–2006. Around 50% of Piedmont area is frequently affected by moderate drought conditions, mainly during the '60ies, from 1985 to 1990 and from 1999 until 2006. Moreover a large part of region suffered high magnitude drought condition during the episode at the end of '80ies.

For the same purposes a Hovmoller diagram of both 3 an 12 months SPI has been plotted in Fig. 4, arranging the river basins in a reasonable order starting from northern basins, coming toward south along Alpine chain and finally arriving on the plain basins. This representation put in evidence the different answer to drought spells related to the peculiar geographic (in sense of latitude and mean altitude) position of each river basin.

This difference is evident for example during the drought spell that affected southern and plain basins stronger and longer than northern areas during 1999.

To sum up the sub-regional variance of drought condition over Piedmont region we computed drought occurrence for five portion of the whole study area, as shown in Table 3. Around the 50% of drought spells affected at least the 40% of total area in the analysis period, considering both 3 months and 12 months SPI values. On the other side, the whole area suffered of drought conditions only for the 15% of total drought spells reordered.

All this results give reason to analyse drought conditions at the sub-regional scale as realized in the regional hydrological Bulletin.

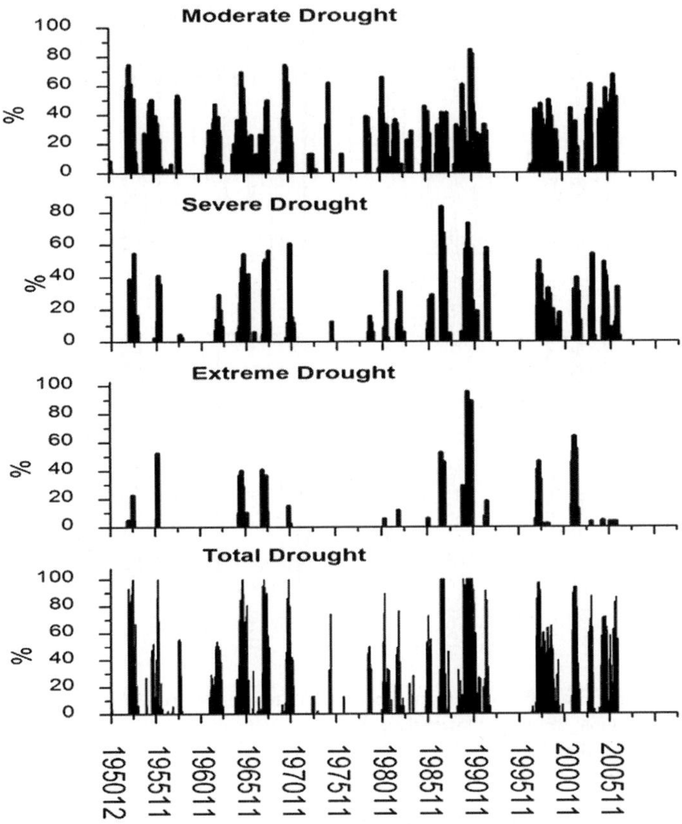

Figure 3. Time series of percentage of Piedmont Region affected by different types of drought severity based on 12 months SPI values. Total Drought is the sum of all drought classes (moderate, severe and extreme)

Table 3. Percentage of total drought spells (SPI 3 and 12 < -1) suffered by different parts of Piedmont region (expressed in percentage of total area) during the period from December 1950 and October 2006

Total Area (%)	Drought Occurrence (%)	
	3 mo SPI	12 mo SPI
20	35	34
40	23	21
60	16	18
80	11	12
100	15	15

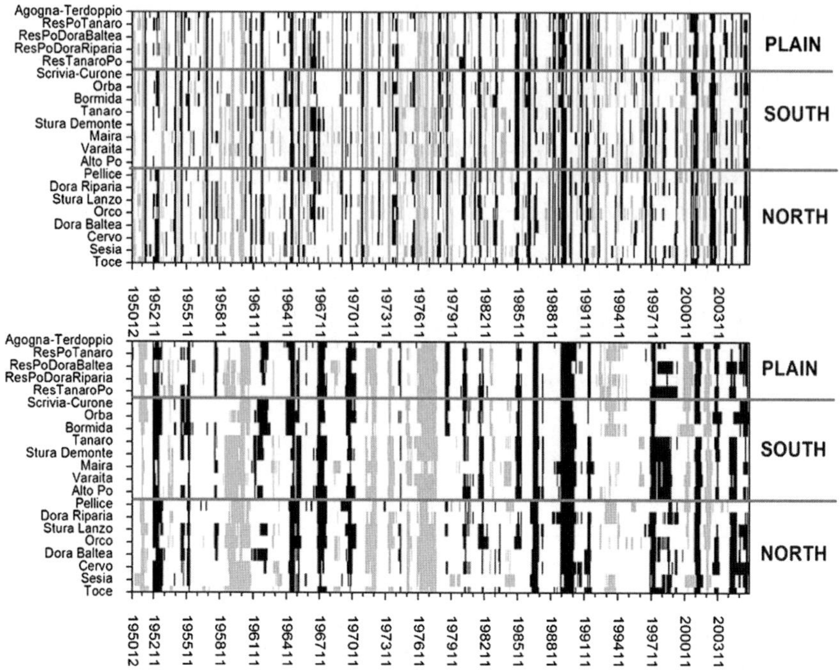

Figure 4. Hovmoller diagram of 3 months SPI (upper panel) and 12 months SPI (lower panel) for the 21 river basins of Piedmont region. SPI values lower than -1 in black and values greater than 1 in light grey. Normal condition are plotted in white

3. DEVELOPMENT OF THE BULLETIN

The 152/2006 legislative decree transfers to Regions the responsibility to collect all useful data in order to describe river basin features, even water availability and to establish the needed measures for quantitative preservation of water availability. In the framework of unitary activities and hydrological balance monitoring aimed to prevent outstanding low water of the Po river, the Po river Basin Authority established the following so-called “macro-components”:

1) Precipitation over basin and discharge into river-bed;
2) Potential and actual water availability into mountain storages;
3) Cumulated water and management legal obligations for lakes;
4) Availability and requirements for the Plain areas;
5) Requirements and discharge limits the Po river.

These “macro-components” are described in terms of measurable variables that could be processed and compared to each other, so the aim of the Regional Hydrological Bulletin is to update the knowledge about the water availability.

From an operational point of view, monitoring and managing a drought cannot be simply based upon lack of precipitation. All the human activities, mainly agriculture and water supply but also industries, that need water have contributed to the development of artificial channel network taking water from streams and rivers. In this context it is important to understand the evolution of the river discharges in order to provide a useful evaluation of water availability in the surface natural network.

Discharge observations are therefore very important data and need to be treated. Due to the great importance of the Alps in the hydrological regimes in Piedmont region, this implies also that the snow cover evolution in winter, spring and in many cases also in summer has to be considered. Snow melt plays a very important role in the formation of river discharges for several months and can help overcoming short spring droughts.

3.1 Precipitation and SPI

In our method, for each basin and for each month, distributions of monthly precipitation and relative percentiles are additionally calculated for the period 1913–2005. In this data classification, the tenth percentile means " severely dry month ", the fiftieth "near normal month", and the ninetieth "very wet month".

Using those climatologic monthly precipitation series it is possible to calculate the 3-months SPI for to the successive (unknown) month with regard to the last recorded, obtaining three different scenarios. Those forecasts, therefore, are useful to estimate drought severity in the short period (a month).

3.2 Snow Cover

For the study of snow cover dynamics a numerical model, based on the hydrological model FEST (Mancini, 1990), is used. It is a physically based, distributed model, already used in a number of applications either for simulation and the forecast of floods and for long-term water budget at catchments scale (Ravazzani et al., 2002).

FEST simulates the following processes:

- infiltration;
- evaporation;
- snow cover accumulation and melt;
- run-off;
- hypodermic fluxes.

In details, the hydrological model of the snow contribution calculating the water volume stored in the snow cover, in terms of snow water equivalent (SWE), the propagation of the water generated by the melting process in the snow mantle and the out-flow from the mantel itself (see Fig. 5).

The meteorological forcing (precipitation, air temperature, solar radiation) comes from the regional survey network; and it is interpolated all over the study area accounting for topographical features such as elevation, slope inclination and orientation

To distinguish the type of precipitation, the air temperature is used. For the melting calculation, two schemes are available. One uses the conceptual degree-day

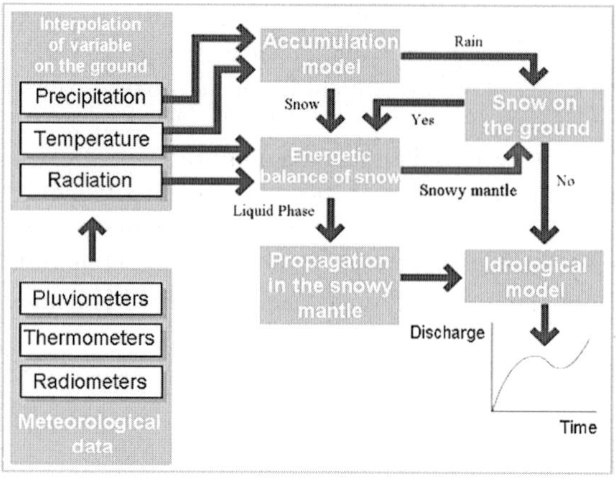

Figure 5. Simplified flow chart of the snow model

method. The second is based on the energetic budget of the snow, the contributions evaluated are: short wave and long-wave radiation, energy from the ground, from water precipitation and convective fluxes.

Once calculated the amount of snow melting for each part of the catchment, the propagation of water in the snow mantle is evaluated. The model uses the conceptual scheme of the linear reservoir. At the end, the outflow from the snow cover is diverted to the runoff propagation module and becomes part of the superficial river network.

The model here described has been successfully applied to the Anza river catchment, in Monte Rosa area, to study the development of an ephemeral lake on the Belvedere glacier in the years 2001, 2002 and 2003, showing that the particular phenomena were mainly due to the snow melting process (Rabuffetti et al., 2004).

Analyses related to the performance of the model to big catchment have also been carried out. The Po river catchment in the winter 2004–2005 has been studied comparing model results and satellite imagery coming from MODIS. Also in this case the model showed good potential mainly on mountain areas (Rabuffetti et al., 2006).

3.3 The Status of the River Network

Last but not least, data linked to the hydrologic monitoring of water bodies are synthesized. Water volumes stored in the biggest artificial reservoirs, the Maggiore Lake water level, the biggest and most important in the region, and the discharge in the main rivers are compared with the climatologic reference value on a monthly basis. From a statistical point of view these reference values are not very significant because are derived from brief time series. Nevertheless, together with all the other data in the bulletin, they help to understand the situation of water resources overall the catchment.

4. THE REGIONAL HYDROLOGICAL MONTHLY BULLETIN

The Regional Hydrological Monthly Bulletin is developed with the aim to provide the hydrological status at a regional scale. It contains the results of the data treatment and of the numerical simulation discussed in the previous paragraph.

Great importance is devoted to communication strategy, trying to use a simple but complete technical language and a number of figures to achieve a direct understanding of the people involved in water management of all the important features presented.

The issue is scheduled monthly at the beginning of the month. If requested, it is also possible to provide information more frequently during very dry periods. The bulletin is divided into three sections hereafter described. The figures are extracted from the bulletin issued on July the 3rd, 2006.

In the first the precipitations are analysed both in terms of mean basin rainfall and of SPI. The recorded precipitation for each catchment is synthesised in a table (Fig. 6 containing: the mean value; the corresponding volume; the absolute and relative deficit with respect to the climatologic value, the normalised deficit expressed as SPI – 1 month.

This information is completed with a map of monthly precipitation with isolines highlighted.

The same precipitation data are use for the calculation of SPI – 3,6,12 months. The results are gathered in a table with an evaluation of the situation. The standard language (McKee et al., 1993) is adopted. For a more direct understanding, the table is joined with a map showing the status of the different catchments.

Moreover the SPI calculations are completed with a scenario forecasting to help decision makers in building up their mind with respect to the expected short-term evolution of the situation. Three other maps are presented (e.g. in Fig. 7) showing the SPI-3 months forecasted for the successive month. This forecast is based on three different scenarios ("severely dry month", "near normal month", and "very wet month") derived from climatologic analysis for each catchment as already described.

The second section of the bulletin shows the water resource available in each catchment in terms of stored water volume (SWE). A map showing the spatial distribution of SWE in the region is then displayed (Fig. 8).

Sub-catchment	Precipitation [mm]	Volume [Mm^3]	Difference from average [mm]	Deficit [%]	Normalized deficit
Alto Po	19,5	14	−82, 7	−81%	−2
Pellice	19,7	19,3	−72, 6	−79%	−2
Varaita	27,8	16,7	−59	−68%	−1, 6
Maira	32,4	39,3	−50, 8	−61%	−1, 3
Residual Po at Dora Riparia inlet	25,6	45,5	−56, 4	−69%	−1, 5

(*Continued*)

(*Continued*)

Sub-catchment	Precipitation [mm]	Volume [Mm^3]	Difference from average [mm]	Deficit [%]	Normalized deficit
Dora Riparia	26,7	35,8	−50, 1	−65%	−1, 6
Stura Lanzo	32,3	28,6	−75, 8	−70%	−1, 7
Orco	30,7	28	−77, 5	−72%	−1, 8
Residual Po at Dora Baltea inlet	17,9	14	−76, 4	−81%	−2, 3
Dora Baltea	40,7	160,3	−35, 1	−46%	−1, 2
Cervo	20,2	20,6	−115, 7	−85%	−2, 8
Sesia	34,9	39,6	−103, 9	−75%	−2, 3
Residual Po at Tanaro inlet	17,7	35,7	−52, 9	−75%	−1, 8
Stura Demonte	42,9	63,2	−43, 7	−50%	−0, 9
Tanaro	23,9	43,3	−70, 3	−75%	−1, 6
Bormida	6,5	11,3	−47, 7	−88%	−2, 3
Orba	4,7	3,6	−41, 7	−90%	−2, 1
Residual Tanaro At Po inlet	22,2	53,4	−33, 6	−60%	−1, 1
Scrivia Curone	9,6	13,1	−45, 7	−83%	−2
Agogna Terdoppio	14,9	23,7	−80, 6	−84%	−2, 6
Toce	71,6	127,7	−53, 1	−43%	−1
Ticino svizzero	42,5	201,6	−82, 1	−66%	–

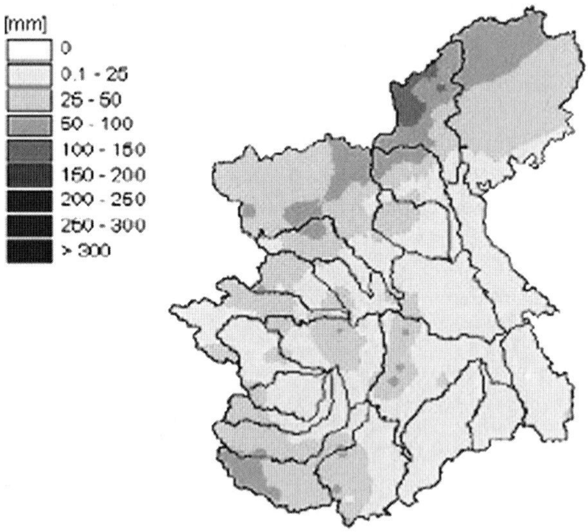

Figure 6. First section of the regional hydrological monthly bulletin: precipitations recorded in June 2006

The last section synthesises the status of the river network and of the main reservoirs. Maggiore lake water level is drawn in a chart compared with climatologic values.

A table illustrates the water stored in artificial reservoirs for each catchment. The percentage of the total amount of potential storing volume is also specified.

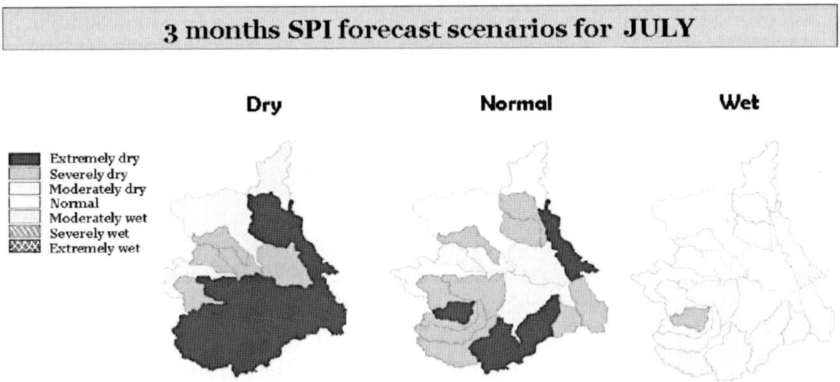

Figure 7. First section of the regional hydrological monthly bulletin: SPI-3 months forecasted for July issued at the beginning of the month. From left to right the different scenarios: "severely dry month ", "near normal month", and "very wet month"

SNOW

Basin	Equivalent Water Volume SWE [mm3]	Equivalent Precipitation Height
Ticino Svizzero	1.9	0.4
Toce	4.9	2.7
Sesia	2.7	2.4
Cervo	0.0	0.0
Dora Baltea	22.1	5.6
Orco	Not significant	-
Stura Lanzo	Not significant	-
Dora Riparia	Not significant	-
Pellice	Not significant	-
Alto Po	Not significant	-
Varaita	Not significant	-
Maira	0.0	0.0
Stura Demonte	0.0	0.0
Tanaro	0.0	0.0

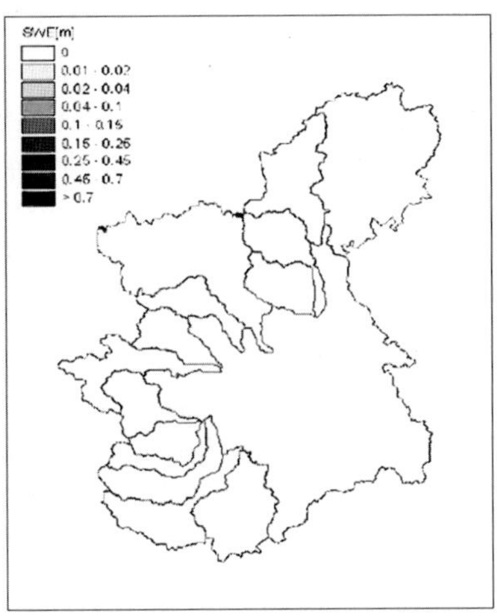

Figure 8. Second section of the regional hydrological monthly bulletin: water stored in the snow mantle at the end of June 2006

A number of charts showing main rivers discharge conclude the Bulletin. These charts allow comparing discharge observation with the minimum recorded water level and the expect minimum.

5. CONCLUSION AND FUTURE DEVELOPMENTS

The analysis of precipitation is carried out at the spatial scale of hydrographical basin by means of the mean areal rainfall. Network density is assessed for the whole Po river catchment closed at Becca cross-section and to a number of sub-catchments for all the period covered by data (1913–2006).

Homogeneity in space is generally well proved while the analysis results, due to significant lacks in the 1940–1950 period, suggest to consider the period successive to 1950 for any statistical analysis.

The 3 months and 12 months SPI temporal analysis computed on singular precipitation time series as representative of the whole Piedmont region shows no significant trend during the period from December 1950 to October 2006, both on yearly and seasonal time scale. Nevertheless it has to be taken in account that the rain gauges network density showed an rapid increase in the last 15 year and the effect of this growth on the rainfall field estimation is still to be better investigated.

An accurate investigation of drought conditions at sub-regional scale confirm a different answer of smaller portion of the total area to a singular drought spell. Particularly, we found that about half of total drought condition recorded in Piedmont affects only 40% of the whole study area. For management purposes, this results suggest to monitor and forecast drought conditions on each river basins of the Po river's tributaries.

The wide and complete data collection provided by the bulletin has been very favourably welcome by all the different regional institutions whose task is related water resources management and planning.

In particular the section presenting the SPI-3months short-term forecasts has a simple and useful operational meaning during droughts allowing a straightforward evaluation of near future possible evolution of the situation.

Precipitation and discharge is representing the theoretical availability of water over a basin referred to hydrological balance. For this reason, the Piedmont Region promoted a experimental operating service based on drought indices such as SPI in order to monitor and forecast drought condition over the river basins.

The Hydrological Bulletin structure has been gone together with Piedmont Region – Water Management Resources Department which represent the end users of such a product. The Bulletin has been used to plan Piedmont water storages, especially mountain ones, for agriculture purposes during the drought spell in summer 2006.

Two are the main development strategies of the Region Hydrological Monthly Bulletin. On the one hand, the simple stochastic model for SPI forecasting will be enriched by a long term precipitation forecast coming from the meteorological models. On the other hand, a more complete hydrological drought index, the Surface Water Supply Index (SWSI), will also be evaluated to account for drought evaluations at catchment scale.

ACKNOWLEDGEMENTS

This work was supported by European Community cooperated plans SEDEMED I and II – Interreg IIIB – MedOcc funds. We wish to thank the following individuals for the suggestions and the substantial contribution to the successful undertaking of the Bulletin: Alfonso Sutera, Isabella Bordi, Carlo Cacciamani, Nicola Loglisci.

REFERENCES

Direzione Pianificazione delle Risorse Idriche, ARPA Piemonte, (2005) Rapporto sulla "Situazione" Idrica Piemontese in Termini di Condizioni Meteoclimatiche, Idrometriche e di Misure Piezometriche.

Komuscu, A.U. (1999) Using the SPI to analyze spatial and temporal patterns of drought in Turkey. *Drought Network News*, 11, pp 7–13.

Lloyd-Hughes, B. and Saunders, M.A. (2002) A drought climatology for Europe. *Int J Climatol*, vol. 22, pp 1571–1592.

McKee, T.B., Doesken, N.J. and Kleist, J. (1993) The relationship of drought frequency and duration to time scales. 8th Conference on Applied Climatology, Anaheim, California, pp 179–184.

McKee, T.B.N, Doesken, J., Kleist, J. (1995) Drought monitoring with multiple time scales. In: (eds) Ninth Conference On Applied Climatology, Dallas, TX. *American Meteorological Society*, pp 233–236.

Mancini, M. (1990) La modellazione della risposta idrologica: effetti della variabilità spaziale e della scala di rappresentazione del fenomeno dell'assorbimento, Phd Thesis, Politecnico di Milano.

Wei, T.C. and McGuinnes, J.L. (1973) Reciprocal distances squared method: a computer technique for estimating areal precipitation. U.S. Department of Agriculture, ARS-NS-8, pp 1–23.

Rabuffetti, D., Salandin, A., Volontè, G., Mancini, M., Barbero, S. and Cordola, M. (2004) Modellazione idrologica del manto nevoso. il caso del lago epiglaciale del ghiacciaio del Belvedere sul Monte Rosa, 29° Convegno di Idraulica e Costruzioni idrauliche, Trento.

Rabuffetti, D., A. Salandin and Cremonini, R. (2006) Hydrological modelling of snow cover in the large upper Po river basin: winter 2004 results and validation with snow cover estimation from satellite. *Geo-Environment and Landscape Evolution II*, ISBN: 1-84564-168-X.

Ravazzani, G., Montaldo, N. and Mancini, M. (2002) Modellistica idrologica distribuita per il caso di studio del bacino del fiume Toce, 28° Convegno di Idraulica e Costruzioni idrauliche, Potenza.

Singh, V.P. and Chowdhury, P.K. (1986) Comparing Some Methods of Estimating Mean Areal Rainfall. *Water Resou Bull*, vol. 22, no. 2, 275–282.

Vicente-Serrano (2005) Differences in Spatial Patterns of Drought on Different Time Scales: An Analysis of the Iberian Peninsula. *Water Resou Manage*, vol. 20, N°1, pp 37–60.

CHAPTER 4

IS DROUGHT OCCURRENCE AND SEVERITY INCREASING DUE TO CLIMATE CHANGE? ANALYSING DROUGHT CLASS TRANSITIONS WITH LOGLINEAR MODELS

E. E. MOREIRA[(1)], A. A. PAULO[(2)] AND L. S. PEREIRA[(2)]

[(1)] *Faculty of Sciences and Technology, New University of Lisbon, Portugal*

[(2)] *Institute of Agronomy, Technical University of Lisbon, Portugal*

Abstract: SPI time series in a 12 month time scale from 6 sites in Alentejo and Algarve, in the south of Portugal, within the period from October 1896 to September 2005, were divided into three or four periods depending on the data set length and a Loglinear modeling approach has been used to investigate differences relative to drought class transitions among these periods. Four drought severity classes were considered. The drought class transitions were computed for the different periods to form a 3-dimensional contingency table. The application of Loglinear modeling to these data allowed the comparison of the periods in terms of probabilities of transition between drought classes in order to detect a possible trend in time evolution of droughts which could be related to climate change

Keywords: Standardized Precipitation Index, 3-dimensional Loglinear models, Drought class transitions, Odds, Impacts of climate change

1. INTRODUCTION

The less predictable characteristics of droughts with respect to their initiation and termination, frequency and severity make drought both a hazard and a disaster: a hazard because it is a natural accident of unpredictable occurrence but of recognizable recurrence; a disaster because it corresponds to the failure of the precipitation regime, causing the disruption of water supply to the natural and agricultural ecosystems as well as to other human activities (Pereira *et al.*, 2002). Forecasting of when a drought is likely to begin or to come to an end is extremely difficult.

Several drought indices have already been used in Portugal, particularly in recent studies applied to the Alentejo region. Comparisons among drought indices show the appropriateness for using the Standardized Precipitation Index (SPI) to characterize droughts in Alentejo (Paulo *et al.*, 2003; Paulo and Pereira, 2006) and the capabilities to use the stochastic properties of the SPI time series for predicting drought class transitions (Paulo *et al.*, 2005).

G. Rossi et al. (eds.), Methods and Tools for Drought Analysis and Management, 67–81.

It is common in our time the idea that water resources have been decreasing in consequence of several causes, mainly due to less precipitation in certain regions of the planet like the Mediterranean basin, as a result of climatic changes. In fact, it is often said that drought events are becoming more frequent and/or more severe due to climate change.

Aiming to investigate drought behavior in the regions of Alentejo and Algarve in southern Portugal, it was decided to analyze the monthly transitions between SPI droughts classes. Loglinear models (Nelder, 1974; Agresti, 1990) were considered to be an adequate tool to perform this analysis because they have shown to be appropriate to perform a monthly prediction of SPI drought class transitions (Paulo *et al.*, 2005). As a result, the SPI at the 12-month time scale was analyzed through fitting of loglinear models to the probabilities of transitions between the SPI drought classes. The results are discussed in this paper.

2. DATA USED, SPI AND DROUGHT CLASSES

Input data to this study consists of SPI monthly values, computed in a 12-month time scale, within the period from Oct. 1896 to Set. 2005, for 6 rainfall sites in Alentejo and Algarve, two water scarce regions in southern Portugal. The 6 sites correspond to the longest time series available at the moment in Alentejo and Algarve. These sites and the size of each correspondent time series are identified in Figure 1: Chouto, Pavia, Évora, Beja, São Brás de Alportel and Faro. As it can be seen in Figure 1, the size of the time series are different for each site and the date of beginning and ending varies from one time series to another.

The goal of this study was to divide the total period with available data, Oct. 1896 to Set. 2005, in shorter periods and compare them in terms of probabilities of monthly drought class transitions using Loglinear models to assess time variability patterns.

The SPI was developed by McKee *et al.* (1993, 1995) for the identification of drought events and to evaluate its severity. Generally multiple time scales, from 3-month to 24-month, are used. The drought classes severity adopted in this study are defined in Table 1, where the severe and extremely severe classes are grouped.

Based on the graphical representation of drought classes time series for the 6 sites (Figure 2), the total period for each time series was divided into 4 periods with different sizes, as it is shown in Table 2. It can be observed in Figure 2, that for almost every sites there is a period from 1947 to 1981 with much less events of moderate and severe/extreme drought compared with the time before and the time after this period. The same can be observed before 1917. So it was decided to consider those dates for ending/beginning of the intermediate periods (2nd and 3rd periods).

In a previous study, a different set of time series, all with 67 years length, were divided into 3 periods of equal size; the results have shown that a similarity exists when comparing the first with the third period (Moreira *et al.*, 2006). However, in the present study for some of the sites the first period happened to be very small

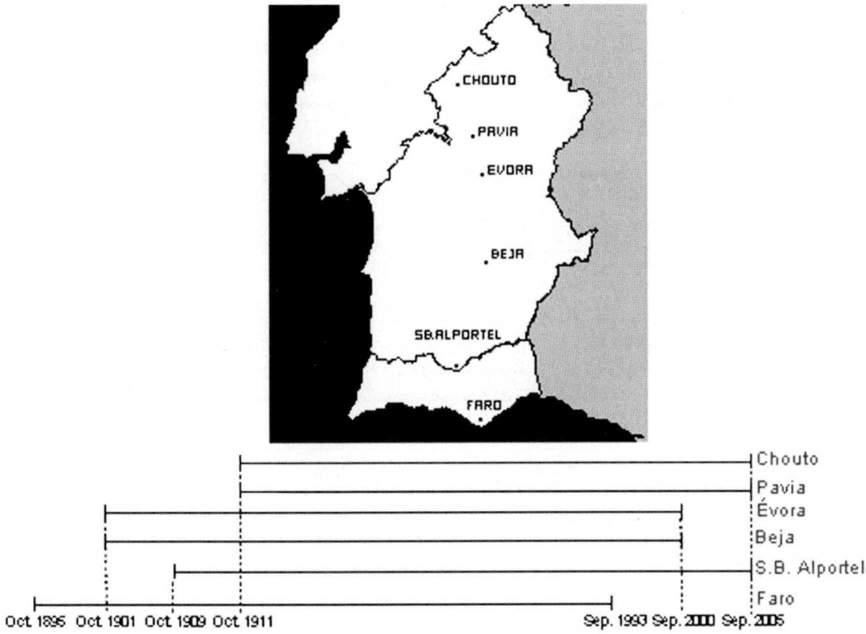

Figure 1. The Alentejo and Algarve regions with identification of the rainfall sites utilized in the study and length of correspondent time series

(6 and 8 years) to be representative of a first period (see Table 2). So it was decided not to consider the first 6 years for Pavia and Chouto and 8 years for S.B. Alportel, thus only 3 periods for these sites.

For each site and time period, the monthly drought classes were calculated based on Table 1. Then, aiming at obtaining the observed frequency counts for building a 3-dimensional contingency table, the monthly transitions between all drought classes were computed.

Table 1. Drought class classification of SPI (modified from McKee *et al.*, 1993)

Code	Drought classes	SPI values
1	Non-drought	$SPI \geq 0$
2	Near normal	$-1 < SPI < 0$
3	Moderate	$-1.5 < SPI \leq -1$
4	Severe/Extreme	$SPI \leq -1.5$

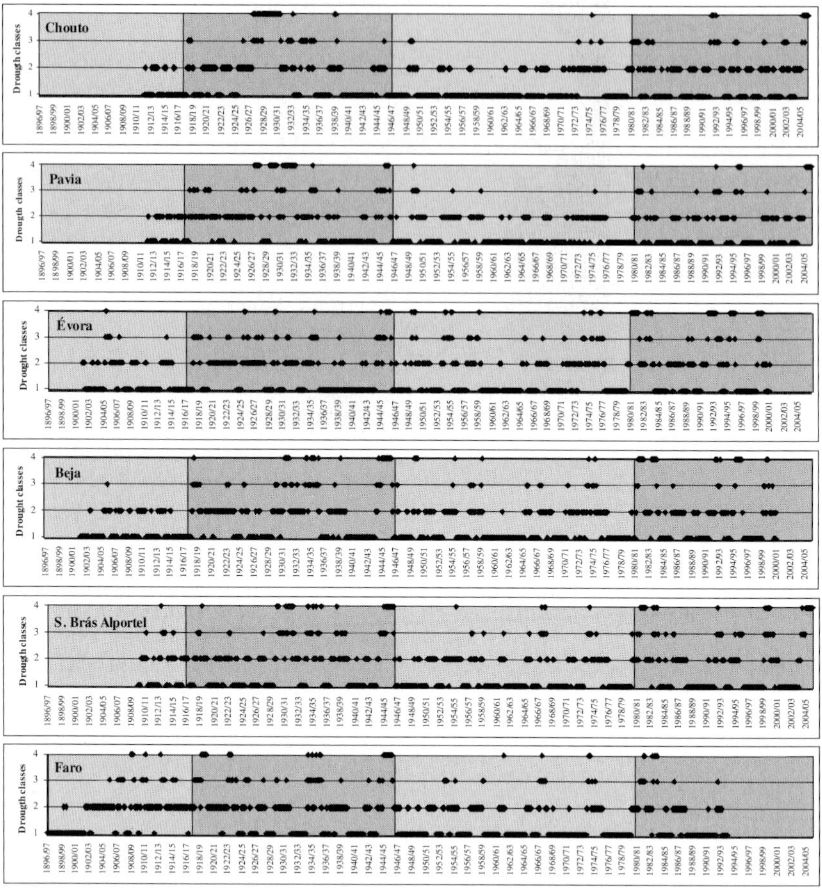

Figure 2. Drought classes in each month for each site. Drought classes: 1 – Non-drought; 2 – Near normal; 3 – Moderate; 4 – Severe and Extreme

Table 2. Time Period division

Period	Pavia and Chouto		Évora and Beja		S.B. Alportel		Faro	
	Year of beginning/ ending	Period size (years)	Year of beginning/ ending	Period size (years)	Year of beginning/ ending	Period size (years)	Year of beginning/ ending	Period size (years)
1st	1911–1917	6	1901–1917	16	1909–1917	8	1896–1917	21
2nd	1917–1946	29	1917–1946	29	1917–1946	29	1917–1946	29
3rd	1946–1980	34	1946–1980	34	1946–1980	34	1946–1980	34
4th	1980–2005	25	1980–2000	20	1980–2005	25	1980–1993	13

3. LOGLINEAR MODELS WITH 3 DIMENSIONS

3.1 Model Presentation

Loglinear models describe association patterns among categorical variables. Modelling is done upon the cell counts in contingency tables. The aim for using the 3-dimensional loglinear models (Agresti, 1990) is to fit the observed frequencies, which are the number of transitions between each drought class, denoted as O_{ijk}, and to model the corresponding expected frequencies, denoted as E_{ijk}, (which are the estimates of the observed frequencies) for each cell of a 3-dimensional contingency table (Table 3).

This 3-dimensional contingency table has three categories (A, B e C) with levels i, j and $k (i = 1, \ldots, 4)$, $(j = 1, \ldots, 4)$ and $(k = 1, \ldots, 4)$, respectively. Category A refers to drought classes at month t, category B refer to drought classes at month t' 1. The levels 1, 2, 3, 4 are associated to drought classes 1, 2, 3 and 4. The category C represents the time period and levels 1, 2, 3 and 4 are associated to the 1st, 2nd, 3rd and 4th time periods defined above.

The observed frequencies (O_{ijk}) refer to number of transitions between the drought class i at month t and the drought class j at month $t+1$ in each period k. For example, the observed O_{111} is the number of times that a given site remains for two consecutive months in drought class 1 (Non-drought) during the 1st period.

In this study, since the periods have different sizes, the observed frequencies had to be weighted in order to better compare the periods. The fit of the model was done after this readjustment of observed frequencies.

Several models for 3-dimensional contingency tables were fitted to the observed frequencies but the quasi-association model (Agresti, 1990) was the one that better fitted the observed frequencies. The **quasi-association** model is defined by:

$$Log\,E_{ijk} = \lambda + \lambda_i^A + \lambda_j^B + \lambda_k^C + \beta u_i v_j + \delta_i I(i = j) + (\lambda_k^C \beta) u_i v_j + \lambda_k^C \delta_i I(i = j) \quad (1)$$

with $i, j, k = 1, \ldots, 4$, where

λ - constant term

λ_i^A - parameter associated with i-th level of the category A

λ_j^B - parameter associated with j-th level of the category B

Table 3. Three-dimensional contingency table of transitions between drought classes relative to the four periods

	1st PERIOD				2nd PERIOD				3rd PERIOD				4th PERIOD			
	Drought Class at month t+1				Drought Class at month t+1				Drought Class at month t+1				Drought Class at month t+1			
Drought Class at month t	1	2	3	4	1	2	3	4	1	2	3	4	1	2	3	4
1	O_{111}	O_{121}	O_{131}	O_{141}	O_{112}	O_{122}	O_{132}	O_{142}	O_{113}	O_{123}	O_{133}	O_{143}	O_{114}	O_{124}	O_{134}	O_{144}
2	O_{211}	O_{221}	O_{231}	O_{241}	O_{212}	O_{222}	O_{232}	O_{242}	O_{213}	O_{223}	O_{233}	O_{243}	O_{214}	O_{224}	O_{234}	O_{244}
3	O_{311}	O_{321}	O_{331}	O_{341}	O_{312}	O_{322}	O_{332}	O_{342}	O_{313}	O_{323}	O_{333}	O_{343}	O_{314}	O_{324}	O_{334}	O_{344}
4	O_{411}	O_{421}	O_{431}	O_{441}	O_{412}	O_{422}	O_{432}	O_{442}	O_{413}	O_{423}	O_{433}	O_{443}	O_{414}	O_{424}	O_{434}	O_{444}

λ_k^C - parameter associated with k-th level of the category C
u_i, v_j - i-th level of category A, j-th level of category B scores (usually we take $u_i = i$, $v_j = j$)
β - linear association parameter
δ_i - parameters associated with the i-th diagonal element
$I(condition)$ - indicator function defined as

$$I(condition) = \begin{cases} 0 \text{ if } condition = true \\ 1 \text{ if } condition = false \end{cases}.$$

3.2 Fitting of the Models

Loglinar models are adjusted using Maximum Likelihood Estimation. Estimation errors of loglinear models are assumed to have the Poisson distribution, and the residual deviances, resulting of the fitting of the models, have asymptotic Chi-Square distribution with the same degrees of freedom of the residual deviance. To evaluate the fitting of a loglinear model, a test proposed by Nelder (1974) and Agresti (1990) was used, which assumes that models are considered well fitted when the residual deviance does not exceed the quantile of the Chi-Square variable for a probability of $1\text{-}\alpha = 0.95$, with the same degrees of freedom of the residual deviance. Degrees of freedom for the residual deviance correspond to the number of observations less the number of the estimated parameters of the fitted model. In other words, all the models presenting a test *p-value* greater than the chosen significance level of $\alpha = 0.05$, are considered well fitted. Only the quasi-association (QA) and the quasi-symmetry models were not rejected, confirming the results by Paulo *et al.* (2005). The QA model (1) proved to be the most adequate in the 6 sites. For each site, the respective degrees of freedom, residual deviance and p-values are presented in Table 4. For the sites Pavia, Chouto and São Brás de Alportel, the model has less parameters, given that the category C (time period) has just 3 levels instead of 4.

A backward elimination method was applied to the fitted QA models for each site in order to reduce the number of parameters without significant loss of information. The backward elimination allows the selection of an alternative sub-model by the elimination of parameters of the initial model. All possible sub-models were rejected

Table 4. Selected Loglinear models for each site

Site	Selected Model	Period division	Degrees of Freedom	Residual Deviance	p-value
Évora			34	27.83	0.7932
Beja		4	34	15.79	0.9967
Faro			34	36.13	0.3693
	Quasi-association				
Pavia			24	26.72	0.3177
Chouto		3	24	12.21	0.9774
S.B. Alportel			24	36.13	0.3693

for 6 sites, thus the initial QA model (1) was kept for these sites. The parameters of the QA model were then estimated and the expected frequencies for each cell were obtained. As an example, in Table 5 are presented the results of the observed and expected frequencies for the site Évora. The results for other sites are similar and thus not shown in this paper.

3.3 Odds

An odds is a ratio of expected frequencies, ranges from 0 to $+\infty$, and represents the number of times that it is more (less) probable the occurrence of a certain event instead of another event different from the first one.

The selected *odds* for the 3-dimensional models were defined by:

$$\Omega_{kl|ij} = E_{ijk} / E_{ijl}, k \neq l \quad (2)$$

The selection of these odds allows the comparison of the probabilities that, one month from now, the site will be in drought class j given that at present it is in class i, when comparing two among the three or four time periods. The odds $\Omega_{12|ij}$ compare the 1st time period with the 2nd time period, the odds $\Omega_{23|ij}$ compare the 2nd time period with 3rd time period and the odds $\Omega_{13|ij}$ compare the 1st time period with the 3rd time period. For example $\Omega_{12|34} = 6.169$ means that it is 6.169 times more probable the transition from 'moderate drought' class (i = 3) to 'severe/extreme drought' class (j = 4) in the 1st period (k = 1) than in the 2nd period (i = 2).

Table 5. Drought class transitions from time t to time t+1: Observed versus Expected frequencies in Évora

	1st PERIOD				**2nd PERIOD**				**3rd PERIOD**				**4th PERIOD**			
	Drought Class at month t+1				Drought Class at month t+1				Drought Class at month t+1				Drought Class at month t+1			
Drought Class at month t	1	2	3	4	1	2	3	4	1	2	3	4	1	2	3	4
	Observed Frequencies															
1	192	16	0	0	136	12	1	0	172	15	0	0	113	16	0	0
2	18	94	5	0	12	106	18	1	14	83	10	3	16	86	9	3
3	0	7	11	2	1	18	24	4	0	9	18	7	0	7	14	10
4	0	0	2	0	0	1	4	10	0	3	6	7	0	4	9	61
	Expected Frequencies															
1	191.9	16.1	0.2	0.0	136.0	15.8	0.3	0.0	171.9	13.6	0.2	0.0	113.0	13.5	0.2	0.0
2	16.3	94.0	6.4	0.9	16.1	106.1	11.7	2.1	13.8	83.0	11.1	2.1	13.7	86.0	12.3	2.5
3	0.2	6.5	11.0	1.3	0.3	11.8	24.1	6.0	0.2	11.1	18.1	6.5	0.2	12.3	14.0	8.6
4	0.0	0.8	1.3	0.0	0.0	2.1	5.8	10.8	0.0	2.0	6.3	7.1	0.0	2.4	8.2	62.6

* Drought classes: 1 – Non-drought; 2 – Near normal; 3 – Moderate drought; 4 – Severe/Extreme drought

The estimates of the corresponding odds are thus obtained by exponentiation of the result obtained by replacing the parameters in $Log\,\Omega_{kl|ij} = Log\,\frac{E_{ijk}}{E_{ijl}} = Log\,E_{ijk} - Log\,E_{ijl}$ by their estimates obtained from the fit of the loglinear model. In these expressions $i,j = 1, 2, 3, 4$; $k, l = 1, 2, 3, 4$ with $k \neq l$.

In the *quasi-association* model the logarithm of the odds may be written as

$$Log\,\Omega_{kl|ij} = Log\,E_{ijk} - Log\,E_{ijl} = \lambda_k^C - \lambda_l^C + \left(\lambda_k^C \beta\right) u_i v_j - \left(\lambda_l^C \beta\right) u_i v_j + \lambda_k^C \delta_i I(i=j) - \lambda_l^C \delta_i I(i=j) \tag{3}$$

Asymptotic confidence intervals associated with a probability of $1\text{-}\alpha = 0.95$ are then obtained for these odds by exponentiation of the corresponding asymptotic confidence intervals for the logarithm of the odds

$$\left[Log\,\Omega_{kl|ij} - z_{1-\alpha/2}\sqrt{Var\left(Log\,\Omega_{kl|ij}\right)},\, Log\,\Omega_{kl|ij} + z_{1-\alpha/2}\sqrt{Var\left(Log\,\Omega_{kl|ij}\right)}\right] \tag{4}$$

where $z_{1-\alpha/2}$ is the $1-\alpha/2$ quantile of a standard normal random variable.

When the confidence intervals for a given odds include the value 1 it means that the drought transitions from class i to class j in the time period k and in the time period l are not significantly different, i.e., the transitions from class i to class j are equally probable, with a probability equal to 0.95, in both time periods k and l. If the value 1 is not included in the confidence interval of a given odds, it means that the transition is more or less probable in one time period than in the other, according to the situation, also with a probability equal to 0.95. For example, the estimate for the odds $\Omega_{12|11} = E_{111}/E_{112} = 0.6542$ means that it is 0.6542 times less probable that the site will stay 2 months in the 'Non-drought' (i=1, j=1) class in the time period 1 (k=1) than in the time period 2 (k=2). All the *odds* may be interpreted in this way.

4. RESULTS AND DISCUSSION

Results in Tables 6 to 11 concern the odds estimates and the respective confidence intervals for each site.

When comparing the 1st and the 2nd period for the sites with 4 periods - Évora, Beja and Faro -, it can be observed (Table 6) in general terms that there are numerous odds values < 1 with confidence interval not including the value 1, specially for the transitions involving classes 3 and 4. So, it can be said that there is a probability p = 0.95 that those odds are really < 1.

These results in Table 6 may be interpreted as that more frequent moderate and severe/extreme droughts occur for the 2nd period than for the 1st period at those sites. However, this conclusion is limited by the extent of the time series, given that the 1st period has a smaller size (21 years for Faro and 16 years for Évora and Beja) than the 2nd period (29 years), thus making it less representative of a full size period when compared with the 2nd period.

Table 6. Comparing the 1st and 2nd periods: Estimates of $\Omega_{12|ij} = \mathrm{E}_{ij1}/\mathrm{E}_{ij2}, i, j = 1, 2, 3, 4$ and its confidence intervals for the 6 sites. (In grey the cases when the confidence interval includes the value 1)

		Odds estimates				Confidance interval at 95%							
		Drought Class at month t+1				Left bound				Right bound			
	Drought Class at month t	1	2	3	4	1	2	3	4	1	2	3	4
Évora													
	1	1.41	1.01	0.87	0.75	1.13	0.59	0.51	0.43	1.76	1.74	1.49	1.28
	2	1.01	0.89	0.55	0.40	0.59	0.56	0.32	0.24	1.74	1.40	0.94	0.69
	3	0.87	0.55	0.46	0.22	0.51	0.32	0.14	0.13	1.49	0.94	1.53	0.37
	4	0.75	0.40	0.22	0.00	0.43	0.24	0.13	0.00	1.28	0.69	0.37	7E+19
Beja													
	1	1.99	1.51	0.81	0.43	1.58	0.84	0.45	0.24	2.50	2.71	1.45	0.77
	2	1.51	0.76	0.12	0.03	0.84	0.33	0.07	0.02	2.71	1.76	0.22	0.06
	3	0.81	0.12	0.00	0.00	0.45	0.07	0.00	0.00	1.45	0.22	5E+34	0.01
	4	0.43	0.03	0.00	0.00	0.24	0.02	0.00	0.00	0.77	0.06	0.01	5E+34
Faro													
	1	0.99	0.96	0.87	0.80	0.75	0.55	0.50	0.46	1.30	1.66	1.51	1.38
	2	0.96	1.37	0.66	0.55	0.55	0.92	0.38	0.32	1.66	2.03	1.14	0.95
	3	0.87	0.66	1.08	0.38	0.50	0.38	0.40	0.22	1.51	1.14	2.89	0.65
	4	0.80	0.55	0.38	0.15	0.46	0.32	0.22	0.02	1.38	0.95	0.65	0.97
S.B. Alportel													
	1	0.66	1.09	1.23	1.39	0.53	0.64	0.72	0.81	0.84	1.86	2.09	2.36
	2	1.09	0.84	1.76	2.25	0.64	0.55	1.04	1.32	1.86	1.26	3.01	3.82
	3	1.23	1.76	4.25	3.64	0.72	1.04	1.40	2.14	2.09	3.01	12.85	6.19
	4	1.39	2.25	3.64	20.97	0.81	1.32	2.14	1.72	2.36	3.82	6.19	254.96
Pavia													
	1	0.50	0.96	1.20	1.50	0.40	0.56	0.70	0.88	0.64	1.63	2.04	2.56
	2	0.96	1.02	2.37	3.73	0.56	0.61	1.39	2.19	1.63	1.72	4.03	6.35
	3	1.20	2.37	10.33	9.23	0.70	1.39	1.96	5.42	2.04	4.03	54.33	15.73
	4	1.50	3.73	9.23	9E+04	0.88	2.19	5.42	0.00	2.56	6.35	15.73	1E+29
Chouto													
	1	0.54	0.84	1.13	1.53	0.44	0.45	0.60	0.82	0.67	1.57	2.13	2.87
	2	0.84	1.36	2.80	5.11	0.45	0.74	1.49	2.72	1.57	2.49	5.25	9.59
	3	1.13	2.80	5.00	17.06	0.60	1.49	0.91	9.09	2.13	5.25	27.35	32.02
	4	1.53	5.11	17.06	2E+05	0.82	2.72	9.09	0.00	2.87	9.59	32.02	2E+45

Doing the same comparison for the other 3 sites with 3 periods (Table 6), it can be observed in general terms, that there are numerous odds greater than 1 with confidence interval not including the value 1, specially for the transitions involving the highest drought classes (class 3 and 4). So, there is a probability p = 0.95 of those odds to be really greater than 1. It means that there are more frequent moderate and severe/extreme droughts during the 1st period than during the next period. These results are compatible with those obtained for 6 other sites in Alentejo reported in Moreira *et al.* (2006).

When comparing the 1st and the 3rd period (Table 7), for Faro there are numerous odds equal to 1 with p = 0.95. For Évora, the number of odds with value 1 is lesser, and much lesser for Beja. Thus, there is some similarity between the 1st and the 3rd period for Faro, while for Évora and Beja this similarity is not evidenced. However, the small size of the 1st period limits these conclusions.

Table 7. Comparing the 1st and 3rd periods: Estimates of $\Omega_{13|ij} = E_{ij1}/E_{ij3}, i, j = 1, 2, 3, 4$ and its confidence intervals for the 6 sites studied. (In grey the cases when the confidence interval includes the value 1)

		Odds estimates				Confidance interval at 95%							
		Drought Class at month t+1				Left bound				Right bound			
	Drought Class at month t	1	2	3	4	1	2	3	4	1	2	3	4
Évora													
	1	1.12	1.18	0.99	0.83	0.91	0.68	0.57	0.48	1.37	2.05	1.72	1.44
	2	1.18	1.13	0.58	0.41	0.68	0.71	0.33	0.23	2.05	1.81	1.01	0.71
	3	0.99	0.58	0.61	0.20	0.57	0.33	0.18	0.12	1.72	1.01	2.10	0.35
	4	0.83	0.41	0.20	0.00	0.48	0.23	0.12	0.00	1.44	0.71	0.35	1E+20
Beja													
	1	1.18	1.24	0.76	0.46	0.97	0.70	0.43	0.26	1.44	2.18	1.33	0.81
	2	1.24	0.91	0.17	0.06	0.70	0.39	0.10	0.04	2.18	2.12	0.30	0.11
	3	0.76	0.17	0.00	0.01	0.43	0.10	0.00	0.01	1.33	0.30	2E+35	0.02
	4	0.46	0.06	0.01	0.00	0.26	0.04	0.01	0.00	0.81	0.11	0.02	5E+35
Faro													
	1	0.49	1.01	1.09	1.17	0.39	0.56	0.60	0.65	0.62	1.81	1.96	2.11
	2	1.01	1.97	1.36	1.59	0.56	1.22	0.76	0.88	1.81	3.20	2.45	2.86
	3	1.09	1.36	2.55	2.15	0.60	0.76	0.71	1.19	1.96	2.45	9.10	3.87
	4	1.17	1.59	2.15	1.33	0.65	0.88	1.19	0.11	2.11	2.86	3.87	16.07
S.B.Alportel													
	1	0.80	1.19	1.19	1.19	0.63	0.71	0.71	0.71	1.01	2.01	2.01	2.01
	2	1.19	1.29	1.19	1.19	0.71	0.87	0.71	0.70	2.01	1.93	2.00	2.00
	3	1.19	1.19	1.21	1.18	0.71	0.71	0.54	0.70	2.01	2.00	2.74	1.99
	4	1.19	1.19	1.18	0.75	0.71	0.70	0.70	0.20	2.01	2.00	1.99	2.84
Pavia													
	1	0.63	1.10	1.15	1.21	0.49	0.66	0.69	0.72	0.80	1.84	1.92	2.01
	2	1.10	1.11	1.32	1.44	0.66	0.73	0.79	0.86	1.84	1.69	2.20	2.41
	3	1.15	1.32	1.41	1.72	0.69	0.79	0.53	1.03	1.92	2.20	3.76	2.87
	4	1.21	1.44	1.72	2.69	0.72	0.86	1.03	0.52	2.01	2.41	2.87	13.77
Chouto													
	1	0.98	0.56	0.62	0.68	0.76	0.33	0.36	0.40	1.25	0.95	1.05	1.16
	2	0.56	1.05	0.83	1.02	0.33	0.70	0.49	0.60	0.95	1.58	1.41	1.73
	3	0.62	0.83	0.71	1.52	0.36	0.49	0.28	0.89	1.05	1.41	1.83	2.57
	4	0.68	1.02	1.52	3.70	0.40	0.60	0.89	0.72	1.16	1.73	2.57	19.08

Table 8. Comparing the 2nd and 3rd periods: Estimates of $\Omega_{23|ij} = E_{ij2}/E_{ij3}, i, j = 1, 2, 3, 4$ and its confidence intervals for the 6 sites studied (In grey the cases when the confidence interval includes the value 1)

		Odds estimates				Confidance interval at 95%							
		Drought Class at month t+1				Left bound				Right bound			
	Drought Class at month t	1	2	3	4	1	2	3	4	1	2	3	4
Évora													
	1	0.79	1.16	1.14	1.11	0.63	0.69	0.67	0.66	0.99	1.97	1.93	1.88
	2	1.16	1.28	1.06	1.01	0.69	0.86	0.63	0.60	1.97	1.90	1.80	1.72
	3	1.14	1.06	1.33	0.92	0.67	0.63	0.51	0.55	1.93	1.80	3.46	1.56
	4	1.11	1.01	0.92	1.53	0.66	0.60	0.55	0.63	1.88	1.72	1.56	3.75
Beja													
	1	0.60	0.82	0.94	1.07	0.47	0.49	0.56	0.64	0.75	1.37	1.57	1.79
	2	0.82	1.19	1.40	1.84	0.49	0.79	0.84	1.10	1.37	1.82	2.35	3.08
	3	0.94	1.40	3.14	3.15	0.56	0.84	1.79	1.88	1.57	2.35	5.48	5.28
	4	1.07	1.84	3.15	9.86	0.64	1.10	1.88	3.84	1.79	3.08	5.28	25.30
Faro													
	1	0.50	1.05	1.24	1.47	0.39	0.59	0.70	0.83	0.63	1.86	2.20	2.61
	2	1.05	1.44	2.07	2.90	0.59	0.90	1.17	1.64	1.86	2.30	3.66	5.13
	3	1.24	2.07	2.36	5.72	0.70	1.17	0.71	3.24	2.20	3.66	7.92	10.12
	4	1.47	2.90	5.72	8.99	0.83	1.64	3.24	1.01	2.61	5.13	10.12	80.21
S.B. Alportel													
	1	1.20	1.10	0.97	0.86	0.97	0.63	0.56	0.49	1.49	1.91	1.69	1.50
	2	1.10	1.55	0.67	0.53	0.63	1.00	0.39	0.30	1.91	2.38	1.17	0.92
	3	0.97	0.67	0.29	0.32	0.56	0.39	0.09	0.19	1.69	1.17	0.89	0.56
	4	0.86	0.53	0.32	0.04	0.49	0.30	0.19	0.00	1.50	0.92	0.56	0.44
Pavia													
	1	1.24	1.15	0.96	0.80	1.01	0.67	0.56	0.46	1.53	2.00	1.66	1.39
	2	1.15	1.09	0.56	0.39	0.67	0.64	0.32	0.22	2.00	1.86	0.96	0.67
	3	0.96	0.56	0.14	0.19	0.56	0.32	0.02	0.11	1.66	0.96	0.75	0.32
	4	0.80	0.39	0.19	0.00	0.46	0.22	0.11	0.00	1.39	0.67	0.32	4E+19
Chouto													
	1	1.80	0.67	0.54	0.45	1.45	0.38	0.31	0.25	2.24	1.18	0.96	0.79
	2	0.67	0.77	0.30	0.20	0.38	0.43	0.17	0.11	1.18	1.41	0.53	0.35
	3	0.54	0.30	0.14	0.09	0.31	0.17	0.03	0.05	0.96	0.53	0.76	0.16
	4	0.45	0.20	0.09	0.00	0.25	0.11	0.05	0.00	0.79	0.35	0.16	1E+35

In general terms, for the other remaining 3 sites having only 3 periods for the comparative analysis (Table 7), the odds values are equal to 1 with p = 0.95 for almost all the transitions between drought classes. This may be interpreted as not existing significant differences for those sites among frequencies of drought events, including moderate and severe/extreme droughts, between the 1st and the

3rd period. These results are comparable with those obtained in the previous study by Moreira *et al.* (2006).

When comparing the 2nd and the 3rd period (4 periods, Table 8), excepting for Évora, there are numerous odds greater than 1 with p = 0.95, especially for the transitions involving the highest drought classes (class 3 and 4). Thus, for Beja and Faro, there are more frequent moderate and severe/extreme droughts during the 2nd period than during the 3rd period, while for Évora those periods are quite similar.

For the sites Pavia, Chouto and São Brás de Alportel (3 periods, Table 8), in general terms there are numerous odds less than 1 with p = 0.95, mainly for the transitions involving the highest drought classes (class 3 and 4). Thus, there are less transitions to moderate and severe/extreme drought classes during the 2nd period, than during the 3rd period. Once again, these results are similar to those obtained by Moreira *et al.* (2006).

When comparing the 1st and the 4th period (Table 9), for Faro there are numerous odds equal to 1 with p = 0.95. For Évora and Beja the number of odds with value 1 is much lesser. Thus, there are similarity between the 1st and the 4th period for Faro, while for Évora and Beja this similarity is not evidenced. However, the small size of the 1st period and of the 4th period for Faro (13 years), limits conclusions and may also explain the different behavior for Faro.

When comparing the 2nd and the 4th period (4 periods, Table 10), for Faro and Beja, there are numerous odds equal to 1 with p = 0.95. For Évora, the odds values

Table 9. Comparing the 1st and 4th periods: Estimates of $\Omega_{14|ij} = E_{ij1}/E_{ij4}, i, j = 1, 2, 3, 4$ and its confidence intervals for Evora, Beja and Faro (In grey the cases when the confidence interval includes the value 1)

		Odds estimates				Confidance interval at 95%							
		Drought Class at month t+1				Left bound				Right bound			
	Drought Class at month t	1	2	3	4	1	2	3	4	1	2	3	4
Évora													
	1	1.70	1.19	0.97	0.79	1.35	0.68	0.56	0.45	2.14	2.06	1.68	1.37
	2	1.19	1.09	0.52	0.35	0.68	0.69	0.30	0.20	2.06	1.74	0.91	0.60
	3	0.97	0.52	0.79	0.15	0.56	0.30	0.23	0.09	1.68	0.91	2.73	0.27
	4	0.79	0.35	0.15	0.00	0.45	0.20	0.09	0.00	1.37	0.60	0.27	1E+19
Beja													
	1	1.78	1.39	0.79	0.45	1.43	0.78	0.45	0.25	2.22	2.48	1.41	0.80
	2	1.39	0.85	0.15	0.05	0.78	0.37	0.08	0.03	2.48	1.97	0.26	0.08
	3	0.79	0.15	0.00	0.01	0.45	0.08	0.00	0.00	1.41	0.26	2E+35	0.01
	4	0.45	0.05	0.01	0.00	0.25	0.03	0.00	0.00	0.80	0.08	0.01	3E+34
Faro													
	1	0.60	1.15	1.09	1.03	0.47	0.64	0.61	0.58	0.76	2.06	1.95	1.85
	2	1.15	1.75	0.93	0.83	0.64	1.14	0.52	0.47	2.06	2.68	1.66	1.49
	3	1.09	0.93	2.15	0.67	0.61	0.52	0.71	0.38	1.95	1.66	6.56	1.20
	4	1.03	0.83	0.67	0.25	0.58	0.47	0.38	0.03	1.85	1.49	1.20	1.87

Table 10. Comparing the 2nd and 4th periods: Estimates of $\Omega_{24|ij} = E_{ij2}/E_{ij4}, i, j = 1, 2, 3, 4$ and its confidence intervals for Évora, Beja and Faro (In grey the cases when the confidence interval includes the value 1)

		Odds estimates				Confidance interval at 95%							
		Drought Class at month t+1				Left bound				Right bound			
	Drought Class at month t	1	2	3	4	1	2	3	4	1	2	3	4
Évora													
	1	1.20	1.17	1.11	1.06	0.94	0.69	0.66	0.63	1.55	1.98	1.88	1.78
	2	1.17	1.23	0.96	0.86	0.69	0.84	0.57	0.51	1.98	1.82	1.61	1.46
	3	1.11	0.96	1.71	0.70	0.66	0.57	0.65	0.42	1.88	1.61	4.50	1.19
	4	1.06	0.86	0.70	0.17	0.63	0.51	0.42	0.07	1.78	1.46	1.19	0.41
Beja													
	1	0.89	0.92	0.98	1.05	0.69	0.55	0.59	0.63	1.16	1.54	1.65	1.76
	2	0.92	1.12	1.20	1.37	0.55	0.75	0.71	0.82	1.54	1.66	2.01	2.29
	3	0.98	1.20	3.25	1.78	0.59	0.71	1.94	1.06	1.65	2.01	5.45	2.98
	4	1.05	1.37	1.78	0.57	0.63	0.82	1.06	0.23	1.76	2.29	2.98	1.40
Faro													
	1	0.60	1.20	1.25	1.30	0.47	0.69	0.71	0.74	0.77	2.10	2.18	2.27
	2	1.20	1.28	1.41	1.53	0.69	0.86	0.81	0.87	2.10	1.92	2.46	2.67
	3	1.25	1.41	2.00	1.79	0.71	0.81	0.72	1.03	2.18	2.46	5.57	3.13
	4	1.30	1.53	1.79	1.69	0.74	0.87	1.03	0.34	2.27	2.67	3.13	8.27

concerning the transitions expressing the maintenance of class 4 is < 1 with p = 0.95. Thus, for Faro and Beja there are similarity between the 2nd and the 4th period, but for Évora there are less frequent severe/extreme drought events during the 2nd period than during the 4th.

When comparing the 3rd and the 4th period (4 periods, Table 11) for Évora and Beja relative to the transitions involving the highest drought classes (class 3 and 4), the odds values are less than 1, with p = 0.95. However, for Faro the odds value relative to the transition that implies the maintenance of class 4 is equal to 1. Thus, excluding Faro, there are less frequent severe/extreme drought events in the 3rd than in 4th period. For Faro, the small size of the 4th period (13 years) limits the formulation of similar conclusions.

In general terms, these results indicate a similarity between the period from Oct. 1917 to Sep. 1946 and the period from Oct. 1980 to Sep. 2005 in terms of transitions between drought classes that involve the highest drought classes (3 and 4). However, for the site Évora the behavior is somewhat different.

For the site Faro (the site with the longest 1st period, 21 years) there is also some similarity between the period from Oct. 1896 to Sep. 1917 and the period from Oct. 1946 to Sep. 1980. For the other sites, the 1st period is not long enough

Table 11. Comparing the 3rd and 4th periods: Estimates of $\Omega_{34|ij} = E_{ij3}/E_{ij4}, i, j = 1, 2, 3, 4$ and its confidence intervals for Evora, Beja and Faro. (In grey the cases when the confidence interval includes the value 1)

		Odds estimates				Confidance interval at 95%							
		Drought Class at month t+1				Left bound				Right bound			
	Drought Class at month t	1	2	3	4	1	2	3	4	1	2	3	4
Évora													
	1	1.52	1.01	0.98	0.95	1.20	0.59	0.57	0.56	1.93	1.72	1.67	1.63
	2	1.01	0.96	0.90	0.85	0.59	0.65	0.53	0.50	1.72	1.44	1.54	1.46
	3	0.98	0.90	1.28	0.76	0.57	0.53	0.48	0.45	1.67	1.54	3.47	1.30
	4	0.95	0.85	0.76	0.11	0.56	0.50	0.45	0.05	1.63	1.46	1.30	0.27
Beja													
	1	1.50	1.13	1.05	0.98	1.20	0.67	0.63	0.59	1.89	1.88	1.76	1.64
	2	1.13	0.94	0.85	0.74	0.67	0.61	0.51	0.45	1.88	1.45	1.43	1.24
	3	1.05	0.85	1.04	0.56	0.63	0.51	0.62	0.34	1.76	1.43	1.73	0.94
	4	0.98	0.74	0.56	0.06	0.59	0.45	0.34	0.02	1.64	1.24	0.94	0.17
Faro													
	1	1.21	1.14	1.00	0.88	0.99	0.63	0.55	0.48	1.49	2.08	1.83	1.61
	2	1.14	0.89	0.68	0.53	0.63	0.54	0.37	0.29	2.08	1.46	1.24	0.96
	3	1.00	0.68	0.85	0.31	0.55	0.37	0.23	0.17	1.83	1.24	3.14	0.57
	4	0.88	0.53	0.31	0.19	0.48	0.29	0.17	0.02	1.61	0.96	0.57	1.87

to conclude for such similarity between those periods. Excepting for Évora, other comparisons between periods did not show evidence of similarities.

It can be concluded that, with the available data, there is no evidence of a growing tendency for an increase of droughts occurrence and severity in Alentejo and Algarve, which could be attributed to a climate change. This conclusion confirms those suggested by Moreira *et al.* (2006) but still requires better confirmation using other and desirably longer time series if those could be available.

These results are not in disagreement with the works by Bordi *et al.* (2004) and Bordi and Sutera (2001), which reveal a linear trend characterizing the climatic signals during the last fifty years in regions like the Mediterranean basin and central Europe. But, as it was underscored by some authors, this linear trend can be just a part of a long-term periodicity because of the limited length of the data; thus, analysing longer time series should clarify the origin of the revealed trend.

5. CONCLUSIONS

The Loglinear models show to be a powerful tool to compare drought class transitions among different time periods through 3-dimensional contingency tables.

As an overall conclusion it can be said that, with the available data and in general terms, there is no evidence of a growing trend for occurrence of more and more severe droughts in Alentejo and Algarve. This contradicts a common tendency to attribute recent drought events to climate change. However, more complete data continues to be needed to reinforce this conclusion, which was drawn from the previous study by Moreira *et al.* (2006).

ACKNOWLEDGEMENTS

Data used in this study were made available by the Institute for Water (INAG), Portugal. This study was first funded through the national project PEDIZA 1999.64.006326.1, and is now part of the research contract INTERREG III B MEDOC 2002-02-4.4-1-084, project SEDEMED II.

REFERENCES

Agresti, A. (1990) *Categorical Data Analysis*. J. Wiley & Sons, New York.

Bordi, I., Sutera, A. (2001) Fifty Years of Precipitation: Some Spatially Remote Teleconnections. *Water Resour. Manage.* 15: 247–280.

Bordi, I., Fraedrich, K., Gerstengarbe, F.W., Werner, P.C., Sutera, A. (2004) Potential predictability of dry and wet periods: Sicily and Elbe-Basin (Germany). *Theor. Appl. Climatol.* 77: 125–138.

McKee, T.B., Doesken, N.J., Kleist, J. (1993) The relationship of drought frequency and duration to time scales. In: *8th Conference on Applied Climatology*. Am. Meteor. Soc., Boston, pp. 179–184.

McKee, T.B., Doesken, N.J., Kleist, J. (1995) Drought monitoring with multiple time scales. In: *9th Conference on Applied Climatology*, Am. Meteor. Soc., Boston, pp. 233–236.

Moreira, E.E., Paulo, A.A., Pereira, L.S., Mexia, J.T. (2006) Analysis of SPI drought class transitions using loglinear models. *J. Hydrology* 331: 349–359.

Nelder, J.A. (1974) Loglinear models for contingency tables: a generalization of classical least squares. *Appl. Statistics*, 23: 323–329.

Paulo, A.A., Pereira, L.S. (2006) Drought Concepts and Characterization. Comparing Drought Indices Applied at Local and Regional Scales, *Water Int.* 31: 37–49.

Paulo, A.A., Pereira, L.S., Matias, P.G. (2003) Analysis of local and regional droughts in southern Portugal using the theory of runs and the Standardized Precipitation Index. In: Rossi, G., Cancelliere, A., Pereira, L.S., Oweis, T., Shatanawi, M., Zairi, A. (Eds.) *Tools for Drought Mitigation in Mediterranean Regions*, Kluwer, Dordrecht, pp. 55–78.

Paulo, A.A., Ferreira, E., Coelho, C., Pereira, L.S. (2005) Drought class transition analysis through Markov and Loglinear models, an approach to early warning. *Agric. Water Manag.* 77: 59–81.

Pereira, L.S., Cordery, I., Iacovides, I. (2002) Coping with Water Scarcity. UNESCO IHP VI, Technical Documents in Hydrology No. 58, UNESCO, Paris, 267 p. (available on line: http://unesdoc.unesco.org/images/0012/001278/127846e.pdf).

CHAPTER 5

STOCHASTIC FORECASTING OF DROUGHT INDICES

A. CANCELLIERE, G. DI MAURO, B. BONACCORSO AND G. ROSSI

Department of Civil and Environmental Engineering, University of Catania, Italy

Abstract: Unlike other natural disasters, drought events evolve slowly in time and their impacts generally span a long period of time. Such features do make possible a more effective mitigation of the most adverse effects, provided a timely monitoring of an incoming drought is available

Among the several proposed drought monitoring indices, the Standardized Precipitation Index (SPI) has found widespread application for describing and comparing droughts among different time periods and regions with different climatic conditions. However, limited efforts have been made to analyze the role of the SPI for drought forecasting

The aim of the chapter is to provide two methodologies for the seasonal forecasting of SPI. In the first methodology, the transition probabilities from a current drought condition to another in the future, and from a single value of current SPI to a drought class are derived as functions of the statistics of the underlying monthly precipitation process. The proposed analytical approach appears particularly valuable from a practical stand point in light of the difficulties of applying a frequency approach or a Markov chain approach, due to the limited number of transitions generally observed even on relatively long SPI records. In the second methodology, SPI forecasts at a generic time horizon M are analytically determined, in terms of conditional expectation, as a function of a finite number of past values of monthly precipitation. Forecasting accuracy is estimated through an expression of the Mean Square Error, which enables to derive confidence intervals of prediction. Validation of the derived expressions is carried out by comparing theoretical forecasts and observed SPI values. The methodologies have been applied to the series of SPI, based on monthly precipitation observed in Sicily over 40 rain gauges in the period 1921-2003. Results seem to confirm the reliability of the proposed methodologies, which therefore can find useful application within a drought monitoring system

Keywords: drought, SPI, stochastic techniques, transition probabilities, forecast

1. INTRODUCTION

Due to a slow evolution in time, drought is a phenomenon whose consequences take a significant amount of time with respect to its inception in order to be perceived by the socio-economic systems. Taking advantage of this feature, an effective mitigation of the most adverse drought impacts is possible, more than in the case of other extreme hydrological events such as floods, earthquakes, hurricanes, etc., provided a drought monitoring system, able to promptly warn of the onset of a

G. Rossi et al. (eds.), Methods and Tools for Drought Analysis and Management, 83–100.

drought and to follow its evolution in space and time, is in operation (Rossi, 2003). To this end, an accurate selection of indices for drought identification, providing a synthetic and objective description of drought conditions, represents a key point for the implementation of an efficient drought watch system.

Among the several proposed indices for drought monitoring, the Standardized Precipitation Index (SPI) has found widespread application (McKee et al., 1993; Heim, 2000; Wilhite et al., 2000; Rossi and Cancelliere, 2002). Guttman (1998) and Hayes et al. (1999) compared SPI with Palmer Drought Severity Index (PDSI) and concluded that the SPI has advantages of statistical consistency, and the ability to describe both short-term and long-term drought impacts through the different time scales of precipitation anomalies. Also, due to its intrinsic probabilistic nature, the SPI is the ideal candidate for carrying out drought risk analysis (Guttman, 1999). An evaluation of common indicators, according to six weighted evaluation criteria of performance (robustness, tractability, transparency, sophistication, extendability, and dimensionality), indicates strengths of the SPI and Deciles over the PDSI (Keyantash and Dracup, 2002).

Although most of the proposed indices have been developed with the intent to monitor current drought conditions, nevertheless some of them can be used to forecast the possible evolution of an ongoing drought, in order to adopt appropriate mitigation measures and drought policies for water resources management. Within this framework, Karl et al. (1986) assessed the amount of precipitation needed to restore normal conditions after a drought event, with reference to the Palmer Hydrologic Drought Index (PHDI). Cancelliere et al. (1996) proposed a procedure for short-middle term forecasting of the Palmer Index and tested its applicability to Mediterranean regions, by computing the probability that an ongoing drought will end in the following months. Other authors (Lohani et al., 1998) proposed a forecasting procedure of the Palmer index based on the first-order Markov chains, which enables one to forecast drought conditions for future months, based on the current drought class described by the PHDI values. Recently, Bordi et al. (2005) compared two stochastic techniques, namely an autoregressive model and a novel method called Gamma Highest Probability (GAHP), for forecasting SPI series at lag 1. The latter method forecasts precipitation of the next month as the mode of a Gamma distribution fitted to the observed precipitation series. They concluded that the GAHP performs better, especially in spring and summer months.

In the present chapter, a seasonal forecast of the SPI is addressed by means of stochastic techniques. In particular, transition probabilities from a drought class to another and from a current value of SPI to a drought class at different time horizons are analytically derived as a function of the statistical properties of the underlying monthly precipitation. The usefulness of such analytical derivation for estimating transition probabilities is evident in light of the fact that a Markov chain approach appears not adequate to model transitions of SPI values (Cancelliere et al., 2007). Moreover the analytical approach enables one to overcome the difficulties related to a frequency approach, whose reliability may be hindered by the generally limited sample size of the available precipitation series.

Also a model to evaluate SPI forecast on the basis of past values of precipitation has been developed. More specifically, analytical expressions of short-middle term forecasts of the SPI are derived as the expectation of future SPI values conditioned on a finite number of past observations. The accuracy of the model is evaluated in terms of the Mean Square Error (MSE) of prediction (Brockwell and Davis, 1996), which allows confidence intervals for forecasted values to be computed. Forecasting future values in terms of conditional expectation ensures that the corresponding forecasts will have minimum MSE. Validation of the model is carried out based on the historical series observed at 40 precipitation stations in Sicily (Italy), making use of a moving window scheme for parameters estimation.

2. THE STANDARDIZED PRECIPITATION INDEX

The SPI is able to take into account the different time scales at which the drought phenomenon occurs and, because of its standardization, is particularly suited to compare drought conditions among different time periods and regions with different climatic conditions (Bonaccorso et al., 2003).

The index is based on an equi-probability transformation of aggregated monthly precipitation into a standard normal variable. In practice, computation of the index requires fitting a probability distribution to aggregated monthly precipitation series (e.g. $k = 3$, 6, 12, 24 months, etc.), computing the non-exceedence probability related to such aggregated values and defining the corresponding standard normal quantile as the SPI. McKee et al. (1993) assumed an aggregated precipitation gamma distributed and used a maximum likelihood method to estimate the parameters of the distribution.

Although McKee et al. (1993) originally proposed a classification restricted only to drought periods, it has become customary to use the index to classify wet periods as well. Table 1 reports the climatic classification according to the SPI, provided by the National Drought Mitigation Center (NDMC, http://drought.unl. edu). Also, the probabilities ΔP, that the index lies within each class are listed. Since our present work focuses on forecasting drought conditions, the near normal and wet classes have been grouped into one class termed "Non-drought".

Table 1. Wet and drought period classification according to the SPI index

Index value	Class	Probability	ΔP	
$SPI \geq 2.00$	Extremely wet	0.977-1.000	0.023	Non Drought
$1.50 \leq SPI < 2.00$	Very wet	0.933-0.977	0.044	Non Drought
$1.00 \leq SPI < 1.50$	Moderately wet	0.841-0.933	0.092	Non Drought
$-1.00 \leq SPI < 1.00$	Near normal	0.159-0.841	0.682	Non Drought
$-1.50 \leq SPI < -1.00$	Moderate drought	0.067-0.159	0.092	
$-2.00 \leq SPI < -1.50$	Severe drought	0.023-0.067	0.044	
$SPI < -2.00$	Extreme drought	0.000-0.023	0.023	

3. ANALYTICAL DERIVATION OF TRANSITION PROBABILITIES OF DROUGHT CLASSES

3.1 Transition Probabilities From A Drought Class to Another

Let $Z^{(k)}_{\nu,\tau}$ indicate the SPI value at year ν and month $\tau = 1, 2, \ldots 12$, for an aggregation time scale k of monthly precipitation. Also, let's indicate by C_i the generic drought class, for instance C_1 = Extreme, C_2 = Severe, C_3 = Moderate, C_4 = Non-drought. The probability that the SPI value after M months lies within a class C_j given that the SPI value at the current month lies within a class C_i, can be expressed as (Mood et al., 1974):

$$P\left[Z^{(k)}_{\nu,\tau+M} \in C_j \,|\, Z^{(k)}_{\nu,\tau} \in C_i\right] = \frac{\int\int_{C_i,C_j} f_{Z^{(k)}_{\nu,\tau},Z^{(k)}_{\nu,\tau+M}}(t,s)\cdot dt \cdot ds}{\int_{C_i} f_{Z^{(k)}_{\nu,\tau}}(t)\cdot dt} \tag{1}$$

where $f_{Z^{(k)}_{\nu,\tau},Z^{(k)}_{\nu,\tau+M}}(\cdot)$ is the joint density function of $Z^{(k)}_{\nu,\tau}$ and $Z^{(k)}_{\nu,\tau+M}$, $f_{Z^{(k)}_{\nu,\tau}}(\cdot)$ is the marginal density function of $Z^{(k)}_{\nu,\tau}$, t and s are integration dummy variables, and the integrals are extended to the range of each drought class.

Since, by definition, SPI is marginally distributed as a standard normal variable, it is fair to assume the joint density function in eq. (1) to be bivariate normal, namely:

$$f_{Z^{(k)}_{\nu,\tau},Z^{(k)}_{\nu,\tau+M}}(t,s) = \frac{1}{2\pi\,|\Sigma|}\cdot \exp(-\frac{1}{2}\mathbf{X}^{T}\Sigma^{-1}\mathbf{X}) \tag{2}$$

where $X = [t, s]^T$, and Σ represents the variance-covariance matrix:

$$\Sigma = \begin{bmatrix} 1 & \mathrm{cov}\left[Z^{(k)}_{\nu,\tau}, Z^{(k)}_{\nu,\tau+M}\right] \\ \mathrm{cov}\left[Z^{(k)}_{\nu,\tau}, Z^{(k)}_{\nu,\tau+M}\right] & 1 \end{bmatrix} \tag{3}$$

Thus, the computation of transition probabilities in eq. (1) requires the determination of the autocovariance at lag M of $Z^{(k)}_{\nu,\tau+M}$ namely $\mathrm{cov}\left[Z^{(k)}_{\nu,\tau}, Z^{(k)}_{\nu,\tau+M}\right]$. Such autocovariance can be efficiently estimated from an available sample of SPI series. Alternatively, under some hypothesis, it is possible to derive analytical expression of the autocovariance SPI, as a function of the statistics of the underlying precipitation. In general terms, such derivation is not straightforward, because of the equi-probability transformation underlying the SPI computation. However, under the hypothesis of monthly precipitation aggregated at time scale k normally distributed, the corresponding value of SPI can be computed through a simple standardization procedure:

$$Z^{(k)}_{\nu,\tau} = \frac{Y^{(k)}_{\nu,\tau} - \mu^{(k)}_{\tau}}{\sigma^{(k)}_{\tau}} \tag{4}$$

with $Y^{(k)}_{\nu,\tau} = \sum_{i=0}^{k-1} X_{\nu,\tau-i}$ aggregated precipitation at k months.

By assuming precipitation at month τ with mean μ_τ, the mean of the corresponding aggregated precipitation $Y_{\nu,\tau}^{(k)}$ will be:

$$\mu_\tau^{(k)} = \sum_{i=0}^{k-1} \mu_{\tau-i} \tag{5a}$$

Also, if σ_τ^2 is the variance of precipitation at month τ, under the hypothesis of precipitation values uncorrelated in time, the standard deviation of the corresponding aggregated precipitation $Y_{\nu,\tau}^{(k)}$ will be:

$$\sigma_\tau^{(k)} = \sqrt{\sum_{i=0}^{k-1} \sigma_{\tau-i}^2(x)} \tag{5b}$$

Substituting, eq. (4) becomes:

$$Z_{\nu,\tau}^{(k)} = \frac{\sum\limits_{i=0}^{k-1} X_{\nu,\tau-i} - \sum\limits_{i=0}^{k-1} \mu_{\tau-i}}{\sqrt{\sum\limits_{i=0}^{k-1} \sigma_{\tau-i}^2}} \tag{6}$$

Therefore, the autocovariance can be expressed as:

$$\begin{aligned} cov\left[Z_{\nu,\tau+M}^{(k)}, Z_{\nu,\tau}^{(k)}\right] &= \frac{1}{\sqrt{\sum\limits_{i=0}^{k-1} \sigma_{\tau+M-i}^2 \sum\limits_{j=0}^{k-1} \sigma_{\tau-j}^2}} \cdot \sum_{i=0}^{k-1}\sum_{j=0}^{k-1} cov\left[X_{\nu,\tau+M-j}, X_{\nu,\tau-i}\right] = \\ &= \frac{1}{\sqrt{\sum\limits_{i=0}^{k-1} \sigma_{\tau+M-i}^2 \sum\limits_{j=0}^{k-1} \sigma_{\tau-j}^2}} \cdot \sum_{i=0}^{k-M-1} \sigma_{\tau-i}^2 \end{aligned} \tag{7}$$

By substituting eq. (7) in the variance-covariance matrix Σ, it follows:

$$\Sigma = \begin{bmatrix} 1 & \dfrac{\sum\limits_{i=0}^{k-M-1} \sigma_{\tau-i}^2}{\sqrt{\sum\limits_{i=0}^{k-1} \sigma_{\tau+M-i}^2 \sum\limits_{j=0}^{k-1} \sigma_{\tau-j}^2}} \\ \dfrac{\sum\limits_{i=0}^{k-M-1} \sigma_{\tau-i}^2}{\sqrt{\sum\limits_{i=0}^{k-1} \sigma_{\tau+M-i}^2 \sum\limits_{j=o}^{k-1} \sigma_{\tau-j}^2}} & 1 \end{bmatrix} \tag{8}$$

Finally, by combining eq. (8) with eqs. (1) and (2), it is possible to express the SPI transition probabilities, in terms of the variances of monthly precipitation. Although the hypothesis of normality for aggregated monthly precipitation may appear restrictive, it is worth observing that it can be justified, especially for higher values of the aggregation time scale k, as a consequence of the central limit theorem.

3.2 Transition Probabilities from a Spi Value to a Drought Class

Eq. (1) enables to derive transition probabilities from one drought class to another in the future. One may expect however, that such probability should somewhat be affected by the value taken by SPI, and not just by the class it is laying in. For instance, transition toward Non drought class (SPI ≥ -1) from a moderate class ($-1.5 <$ SPI < -1) should be more or less probable depending whether the observed SPI is closer to the class upper bound or to the class lower bound. Thus, it is of interest to derive an analytical expression in order to calculate the transition probabilities from a single value of SPI to a drought class in the future. The latter approach enables, with respect to the previous one, to take into account the influence of the specific past values on the forecasts.

Let's consider the random variable defined as:

$$Z = Z_{\nu,\tau+M} \left| Z_{\nu,\tau} \right.$$

According to the definition of SPI, the variables $Z_{\nu,\tau}$ e $Z_{\nu,\tau+M}$ are distributed as a normal distribution with zero mean and variance one; therefore, according to a well known statistical property of multivariate normal distributions, it is fair to assume Z as a conditional bivariate normal distribution, with mean and variance (Mood et al., 1974):

$$\mu_Z = \mu_{Z_{\nu,\tau+M}} + \left(\rho \cdot \sigma_{Z_{\nu,\tau+M}} \Big/ \sigma_{Z_{\nu,\tau}}\right)\left(z_{\nu,\tau} - \mu_{Z_{\nu,\tau}}\right) = \rho \cdot z_{\nu,\tau} \tag{9}$$

$$\sigma_Z^2 = \sigma_{Z_{\nu,\tau+M}}^2 \left(1-\rho^2\right) = 1-\rho^2 \tag{10}$$

where ρ can be calculated either directly by the observed series (correlation coefficient between SPI series at the present month τ and at month τ' M) or exploiting the previous equation (7) found for the autocovariance.

It follows that the transition probability to pass from a value of SPI Z_0 at the month τ, to a class of drought C_0, M months ahead, will be given by:

$$P[Z_{\nu,\tau+M} \in C_0 \left| Z_{\nu,\tau} = z_{\nu,\tau} \right.] = \int_{C_{0i}}^{C_{0s}} \frac{1}{\sqrt{2\pi}\sigma_Z} \cdot e^{-\frac{1}{2}\left(\frac{x-\rho z_{\nu,\tau}}{1-\rho^2}\right)^2} dx \tag{11}$$

where C_{oi} e C_{os} are respectively the lower and the upper bounds of the particular drought class C_0 considered.

Despite the apparent complexity of the proposed analytical approach for estimation of transition probabilities, it should be pointed out that the methodology yields results that can be considered in general more reliable than those obtained by alternative approaches, such as frequency analysis of observed transitions in an historical sample, or application of a Markov chain scheme (Cancelliere et al., 2007).

4. SPI FORECASTING

From a stochastic point of view, the problem of forecasting future values of a random variable is equivalent to the determination of the probability density function of future values conditioned by past observations. Once the conditional distribution is known, the forecast is usually defined as the expected value or a quantile of such distribution, and confidence intervals of the forecast values can be computed.

Let's consider a sequence of random variables $Y_1, Y_2, \ldots, Y_t$. The interest here lies in determining a function $f(Y_1, Y_2, \ldots, Y_t)$ that forecasts a future value Y_{t+M} with minimum error. The latter is usually expressed as the Mean Square Error (MSE) of prediction, defined as (Brockwell and Davis, 1996):

$$\text{MSE} = \text{E}[(Y_{t+M} - \text{f}(Y_1, Y_2, \ldots, Y_t))^2] \tag{12}$$

It can be shown, that the function $\text{f}(\cdot)$ that minimizes the MSE is the expected value of Y_{t+M} conditioned on $Y_1, Y_2, \cdots, Y_t$, i.e.:

$$\text{f}(Y_1, Y_2, \ldots, Y_t) = \text{E}[Y_{t+M} | Y_1, Y_2, \ldots, Y_t] \tag{13}$$

The above property enables to derive the "best" forecast (in MSE sense), provided the conditional expectation can be computed. Also, it may be worthwhile to note that if Y_{t+M} is independent of $Y_1, Y_2, \cdots, Y_t$, the best predictor of Y_{t+M} is its expected value, and furthermore, the MSE of prediction is just the variance of Y_{t+M}.

Besides MSE, a practical way of quantifying the accuracy of the forecast is by estimating the confidence interval of prediction, i.e. an interval that contains the future observed value with a fixed probability 1–α (e.g. 95 %). Obviously, the wider the interval, the less is the accuracy of the forecast and vice-versa. Confidence intervals of prediction for SPI can be estimated by capitalizing on the intrinsic normality of the index and by observing that, since the predictor is unbiased, its variance coincides with the MSE. Thus, the upper and lower confidence limits $Y_{1,2}$ of fixed probability 1– α can be computed as:

$$Y_{1,2} = \tilde{Y} \pm \sqrt{MSE} \cdot u_{1-\alpha/2} \tag{14}$$

where, for the sake of brevity, $\tilde{Y}$ represents the generic forecast and $\text{u}(\cdot)$ is the quantile of a standard normal (random) variable of probability 1–α/2.

Let's define $Z_{\nu,\tau}^{(k)}$ as the SPI value at year ν and month $\tau = 1, 2, \ldots, 12$, for an aggregation time scale k of monthly precipitation. The interest here lies in determining the conditional distribution (and its expectation) of a future value M months ahead $Z_1 = Z_{\nu,\tau+M}^{(k)}$, conditioned on the vector of θ past observations $\mathbf{Z}_2 = \left[\mathbf{Z}_{\nu,\tau}^{(k)}, Z_{\nu,\tau-1}^{(k)}, \ldots Z_{\nu,\tau-\theta+1}^{(k)}\right]^T$, i.e. the distribution of the following conditioned variable:

$$Z_{1|2} = Z_{\nu,\tau+M}^{(k)} \left| Z_{\nu,\tau}^{(k)}, Z_{\nu,\tau-1}^{(k)}, \ldots Z_{\nu,\tau-\theta+1}^{(k)} \right.$$

The conditional density function of $Z_{1|2}$ is defined as:

$$f_{Z_{1|2}}\left(z_{1|2}\right)=\frac{f_Z(z)}{f_{Z_2}(z_2)} \tag{15}$$

where $f_Z(\mathbf{z})$ is the multivariate pdf of the random vector $\mathbf{Z}=\left[\mathbf{Z}^{(k)}_{\nu,\tau+M}, Z^{(k)}_{\nu,\tau}, Z^{(k)}_{\nu,\tau-1}, \ldots Z^{(k)}_{\nu,\tau-\theta+1}\right]^T$ and $f_{Z2}(\mathbf{z}_2)$ is the joint density function of $\mathbf{Z}_2$.

As already mentioned, since, by definition, the SPI is marginally distributed as a standard normal variable, it is fair to assume that the vector $\mathbf{Z}$ is multivariate normal with zero mean and variance-covariance matrix Σ. The latter can be partitioned as follows:

$$\Sigma=\begin{bmatrix}\Sigma_{11} & \Sigma_{12}\\ \Sigma_{21} & \Sigma_{22}\end{bmatrix} \tag{16}$$

where:

$$\Sigma_{11}=Var\left[Z_1\right]=1 \tag{17}$$

$$\Sigma_{12}=\Sigma_{21}^{\mathrm{T}}=Cov\left[\mathbf{Z_1},\mathbf{Z_2}\right] \tag{18}$$

According to the property shown before, the conditional density function of $Z_{1|2}$ is normal itself. Besides, reminding that the SPI has zero mean, it can be shown that the expected value of $Z_{1|2}$ is (Kotz et al., 2000):

$$\mathrm{E}\left[Z_{1|2}\right]=\Sigma_{12}\cdot\Sigma_{22}^{-1}\cdot\mathbf{z} \tag{19}$$

where $\mathbf{z}=\left[z^{(k)}_{\nu,\tau}, z^{(k)}_{\nu,\tau-1}, \ldots z^{(k)}_{\nu,\tau-\theta+1}\right]^{\mathrm{T}}$ is the vector of past observations. Equation (19) yields the best forecast (in MSE sense) of a future value $\tilde{Z}^{(k)}_{\nu,\tau+M}$ given θ past observations $\mathbf{z}=\left[z^{(k)}_{\nu,\tau}, z^{(k)}_{\nu,\tau-1}, \ldots z^{(k)}_{\nu,\tau-\theta+1}\right]^T$. Furthermore, it is easy to see that the above forecast is unbiased and therefore the Mean Square Error of forecast coincides with the variance of $Z_{1|2}$, which is given by (Kotz et al. 2000):

$$MSE=Var\left[Z_{1|2}\right]=1-\Sigma_{12}\cdot\Sigma_{22}^{-1}\cdot\Sigma_{21} \tag{20}$$

Obviously, if $Z^{(k)}_{\nu,\tau+M}$ is uncorrelated with $Z^{(k)}_{\nu,\tau}, Z^{(k)}_{\nu,\tau-1}, \ldots, Z^{(k)}_{\nu,\tau-\theta}$, the best forecast of $Z^{(k)}_{\nu,\tau+M}$ is its expected value, and furthermore, the MSE of prediction is just the variance of $Z^{(k)}_{\nu,\tau+M}$.

If $\theta=1$, i.e. the forecast is based only on the present value, eq. (19) simplifies as:

$$\tilde{Z}^{(k)}_{\nu,\tau+M}=\mathrm{E}\left[Z^{(k)}_{\nu,\tau+M}|Z^{(k)}_{\nu,\tau}=z^{(k)}_{\nu,\tau}\right]=\rho\cdot z^{(k)}_{\nu,\tau} \tag{21}$$

while the Mean Square Error becomes:

$$MSE = \mathrm{Var}\left[Z^{(k)}_{\nu,\tau+M}\left|Z^{(k)}_{\nu,\tau} = z^{(k)}_{\nu,\tau}\right.\right] = 1 - \rho^2 \tag{22}$$

where ρ is the correlation coefficient between $Z^{(k)}_{\nu,\tau}$ and $Z^{(k)}_{\nu,\tau+M}$.

5. APPLICATIONS TO PRECIPITATION SERIES OBSERVED IN SICILY

The proposed methodologies have been applied to monthly precipitation observed from 1921 until 2003, for 40 precipitation stations in Sicily. The selected stations are included in the drought monitoring bulletin published on the website of the Sicilian Regional Hydrographic Office (Rossi and Cancelliere, 2002, http://www.uirsicilia.it).

5.1 Evaluation of Transition Probabilities

By fixing several combinations of forecasting time horizon M (months) and aggregation time scales k, for each station the transition probabilities from a class of SPI at month τ to another one at month $\tau + M$ have been computed by eq. (1), for every month. In order to compute the double integral in eq. (1), which expresses the normal joint cdf (cumulative distribution function), the algorithm MULNOR has been adopted (Schervish, 1984).

The mean values of transition probabilities corresponding to the 40 stations, for $M = 6$ and $k = 24$, are presented in Figure 1, as a function of the current month τ. In particular, transition probabilities related to different combinations of initial and final drought conditions (extreme Ex, severe Se, moderate Mo, non-drought N) have been considered. In order to show also the variability of transition probability among different stations, in the same plots, the limits related to ± 1 standard deviation are indicated by dashed lines. It can be observed that transition probabilities vary from one month to another, and for some transitions, also from one station to another (as indicated by the width of the limits). In particular, the mean probability value, indicated by the continuous line, of remaining in the extreme class (Ex/Ex) ranges from 60% (for February-March) to less than 25% (for August-September). The mean probability value of remaining in the non-drought class (N/N) presents a limited variability across the months, since it ranges from 95% (for February-March) to about 90% (for August-September). The transition probabilities from one class to another, especially in the lower triangular part of the matrix constituted by the graphics, are very low, at least for the considered time horizon, namely $M = 6$ months. In general, it can be concluded that starting from a wet month there is an higher probability to remain in the same drought class M months ahead (plots along the diagonal), than when starting from dry months, and conversely a lower probability to return to normal conditions.

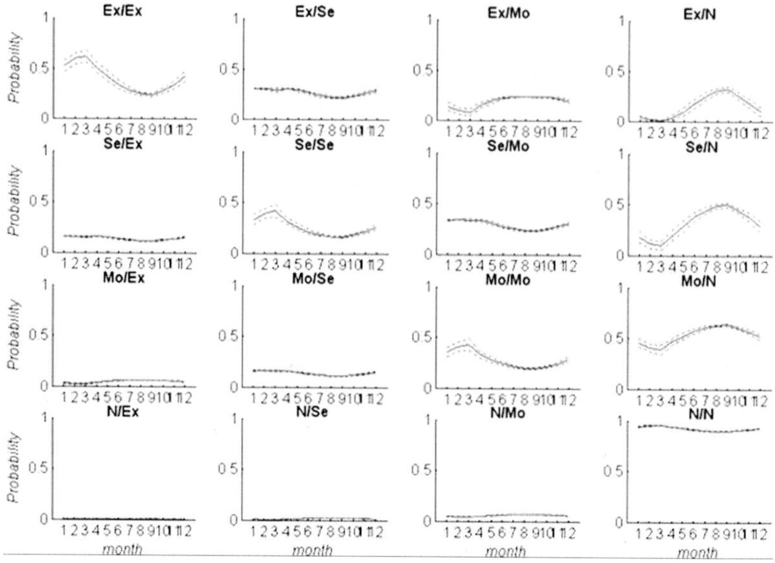

Figure 1. Mean of transition probabilities (continuous line) for $M = 6$ and $k = 24$ computed on 40 stations, and limits corresponding to ± 1 standard deviation (dashed line)

Such different behavior can be justified by considering that starting from a wet month and considering a 6 months time horizon, the occurrence probability of precipitation events, able to restore normal conditions, is very low. On the contrary, starting from a dry month, there is a high chance to observe values of precipitation such to modify drought conditions during the next 6 months.

In order to analyze the effects of the aggregation time scale k and of the forecasting time horizon M, transition probabilities computed for the 40 stations have been averaged and represented on a 3D-plot for a specific starting month and class, as a function of the final class and of the M values. In Figure 2 the case corresponding to extreme drought as starting class, August as starting month and

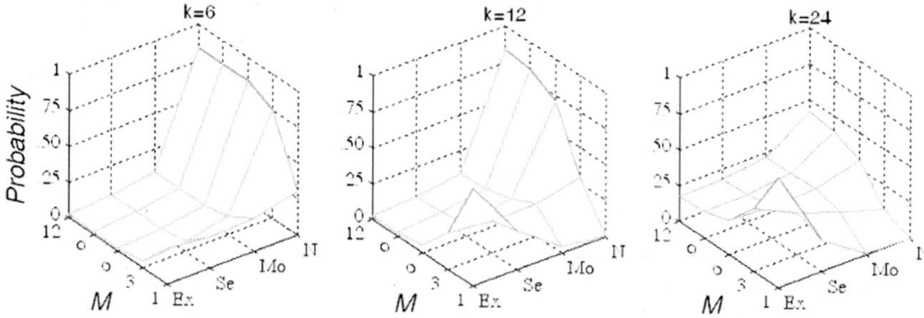

Figure 2. Mean transition probabilities (computed on the 40 stations) as a function of forecasting time horizon M (starting class: extreme drought (Ex), starting month: August)

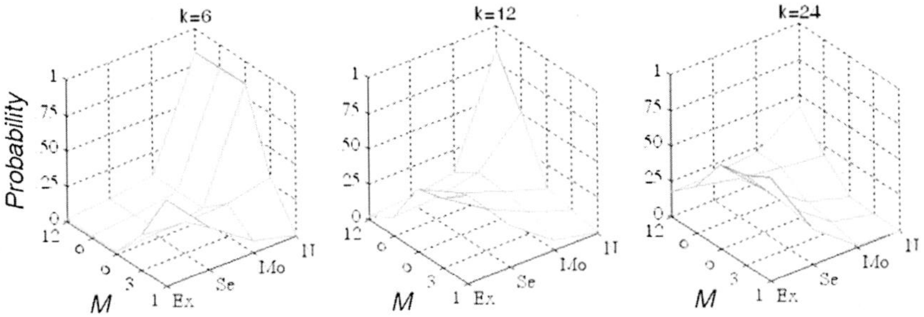

Figure 3. Mean transition probabilities (computed on the 40 stations) as a function of forecasting time horizon M (starting class: extreme drought (Ex), starting month: February)

$k = 6$, 12 and 24 are illustrated. In Figure 3 similar 3D-plots for February as starting month are shown.

It can be seen that probabilities to remain in the same class generally decrease as the forecasting time horizon M increases, while transition probabilities to non-drought condition show an opposite behaviour. Further, as the aggregation time scale k increases, the probabilities of remaining in the same class increase, whereas transition probabilities to return to non-drought condition generally decrease. Finally, by comparing transition probabilities related to August to those corresponding to February, it can be inferred that it is generally easier to recover from drought conditions starting from August than starting from February.

When the interest turns to the computation of transition probabilities from one value of current SPI to a drought class in the future, equation (11) is applied. In particular, by fixing the starting month τ, the aggregation time scale k (3, 6, 9, 12 and 24 months), the time horizon M (1, 3, 6, 9, 12 and 24 months) and the drought class C_0 in the future, it is possible to derive the corresponding transition probabilities for different current SPI values z_τ.

Figure 4 and Figure 5 illustrate the mean transition probabilities to different drought classes (Ex, Se, Mo, N) as a function of current SPI value, computed based on the precipitation series of the considered 40 stations. For the sake of clarity, only the portion of the plots corresponding to $-3 \leq Z_\tau \leq 0$ have been plotted. Besides, each application is performed with reference to two starting months, namely: February and August.

More specifically, in Figure 4 transition probabilities parametrized with respect to different time horizons M (1, 3, 6 months) are reported for an aggregation time scale k=9 months. Figure 5 shows the transition probabilities corresponding to the aggregation time scale k (9, 12, 24 months), for a fixed time horizon M=6 months.

From both figures it is possible to infer similar results to those previously obtained for the transition probability from one class to another. In fact, the probability to remain in the same class decreases as the forecasting time horizon M increases, while transition probabilities to non-drought condition show an opposite behaviour. For instance, from the first plots at the left hand side in Figure 4, representing the transitions to a drought class Extreme, it can be clearly observed that for a current extremely

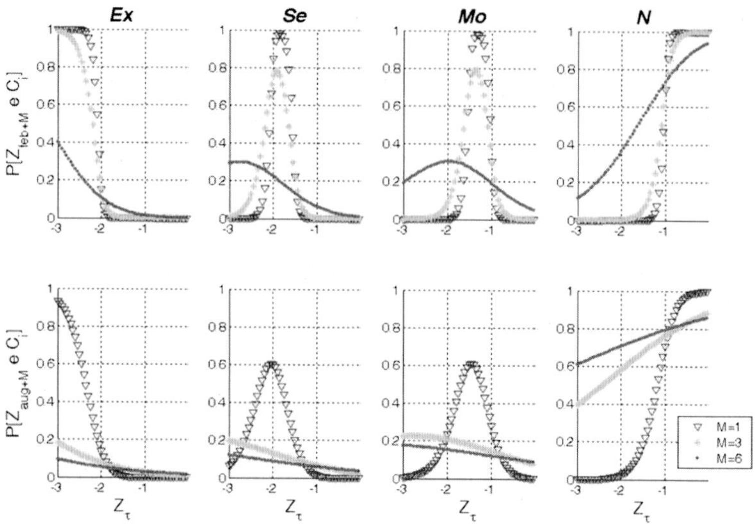

Figure 4. Mean transition probabilities (computed on the 40 stations) from the starting SPI Z_τ to a drought class C_i 1 month ahead (column), as a function of forecasting horizon time *M* (starting months: February, August – rows; aggregation scale *k*=9 months)

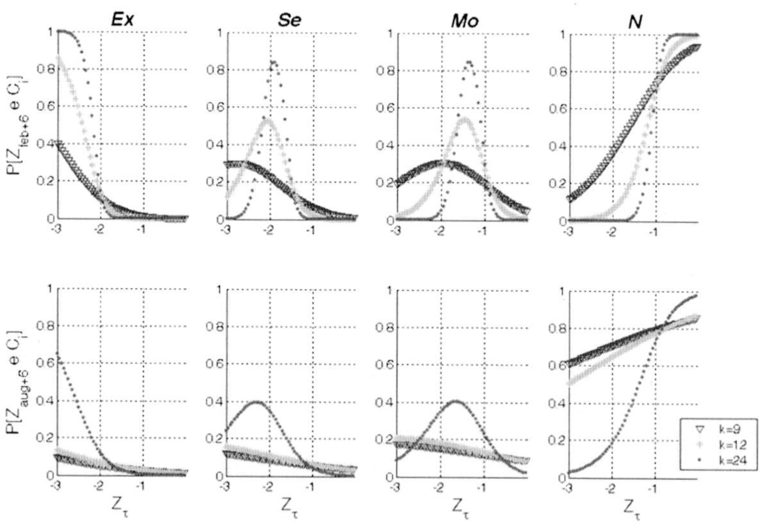

Figure 5. Mean transition probabilities (computed on the 40 stations) from the starting SPI Z_τ to a drought class C_i (column), as a function of aggregation scale *k* (starting months: February, August–rows)

dry condition, namely SPI< -2, the probabilities decrease as M increases, while for all the other hydrometeorological conditions, namely SPI> -2, the probabilities increase as M increases, even if in this case such values are very low (less than ~ 0.20).

On the contrary, from the last plots at the right hand side, representing the transitions to a Normal condition, it can observed that, as M increases, the probabilities increase for SPI< -1 and decrease for SPI> -1.

Figure 5 shows the influence of the aggregation time scale k on the transition probabilities for M $= 6$ month. In particular, as it was expected, it can be noticed a general increase of the probabilities to remain in the same class and a decrease of transition probabilities to different classes, as k increases.

From both figures, it is worth pointing out that transition probabilities are almost independent of M and k, for current values of SPI close to the lower limits of the future classes of transition.

Another important result consists in the high variability of the transition probabilities within the same starting class. In Figure 4, for example, the probability to pass to a Non drought condition 1 month ahead with aggregation scale 9 months starting from August (2nd row, 4th column, line marked by triangles) increases from the value 20% for $Z_\tau = -1.5$ to the value 70% for $Z_\tau = -1$, a considerable difference within the same starting Moderate class. Else by considering Figure 5, the probability to remain in Severe conditions 6 months ahead with aggregation scale 24 months starting from February (1st row, 2nd column, line marked by circles) decreases from the value 80% for $Z_\tau = -2$ to the value 20% for $Z_\tau = -1.5$.

In order to emphasize this feature, in Figure 6 transition probabilities for two different values of SPI inside the same drought class, namely $Z_\tau = -2.95$ and -2.05,

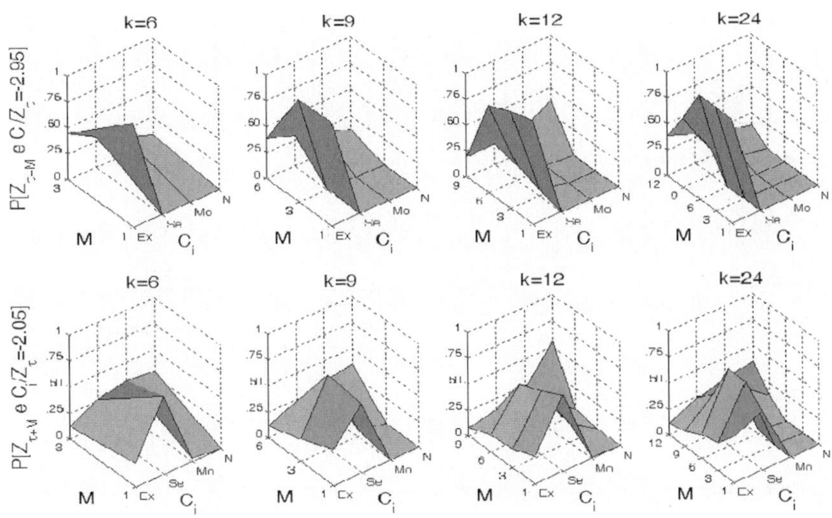

Figure 6. Mean transition probabilities (computed on the 40 stations) as a function of forecasting time horizon M and drought class C_i (starting SPI: -2.95 on the first row, -2.05 on the second row; starting month: February)

are shown. In particular, the three-dimensional graphics represent the variability of transition probabilities (z-axis) versus the time horizon M (x-axis) and the drought class C_i (y-axis) for different aggregation time scale k and by fixing February as starting month. For instance, for an aggregation time scale $k = 12$ months and a horizon time $M = 9$ months, it is possible to note how the probability to pass to non-drought condition increases from the value 35% for $Z_\tau = -2.95$ to the value 60% for $Z_\tau = -2.05$, as it was easily expected, since the second condition is less extreme than the other. Same features can be obtained for other time scales and time horizons.

5.2 SPI Forecast

The forecasting models proposed has been applied by considering several aggregation time scales, $k = 6$, 9, 12 and 24 months, as well as different forecasting time horizons $M = 3$, 6, 9 and 12 months.

First, theoretical MSE values (see eq. (20)) have been computed for all the 40 stations. In particular, for each available series, SPI values have been calculated and the variance- covariance matrix Σ (eq. (16)) has been estimated by means of its sample counterpart for fixed $\tau, k\ \theta$ and M.

Figure 7 illustrates the boxplots of monthly MSE values for different initial month as a function of the number of past observations θ adopted to compute the

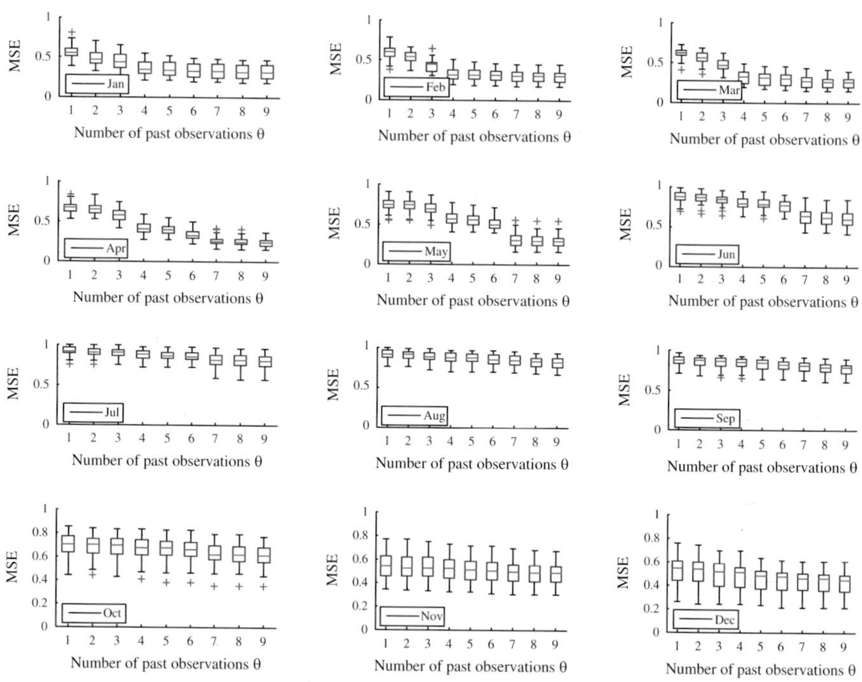

Figure 7. Boxplots of MSE of forecast as a function of the number of past observations θ for different initial month (by considering k=6 and M=3)

forecast. MSE have been computed considering an aggregation time scale $k = 6$ and a forecasting time horizon $M = 3$ months. The overall height of each boxplot indicates the variability of MSE among the different stations. As expected, the performance of the forecasting model generally improves (lower MSE's) as a larger number of past observations is adopted, although the decrease of MSE exhibits a different behaviour according to the considered initial month.

This reflects the intrinsic seasonality in the autocovariance function of SPI, as pointed out by Cancelliere et al., (2005). From the plots, it can be inferred that no significant improvement in the performance of the forecast are obtained for $\theta > 7$ in all cases, and therefore such a value seems to be preferable. Results related to different combinations of k and M (here not shown for brevity), lead to similar conclusions and therefore $\theta = 7$ has been adopted in what follows.

Then, the forecasting model has been validated by comparing observed and forecasted SPI computed by eq. (19) during the period 1921–2003. As an example, Figure 8 shows the comparison between observed and forecasted values SPI for one of the 40 stations, namely Caltanissetta, for different combinations of the time

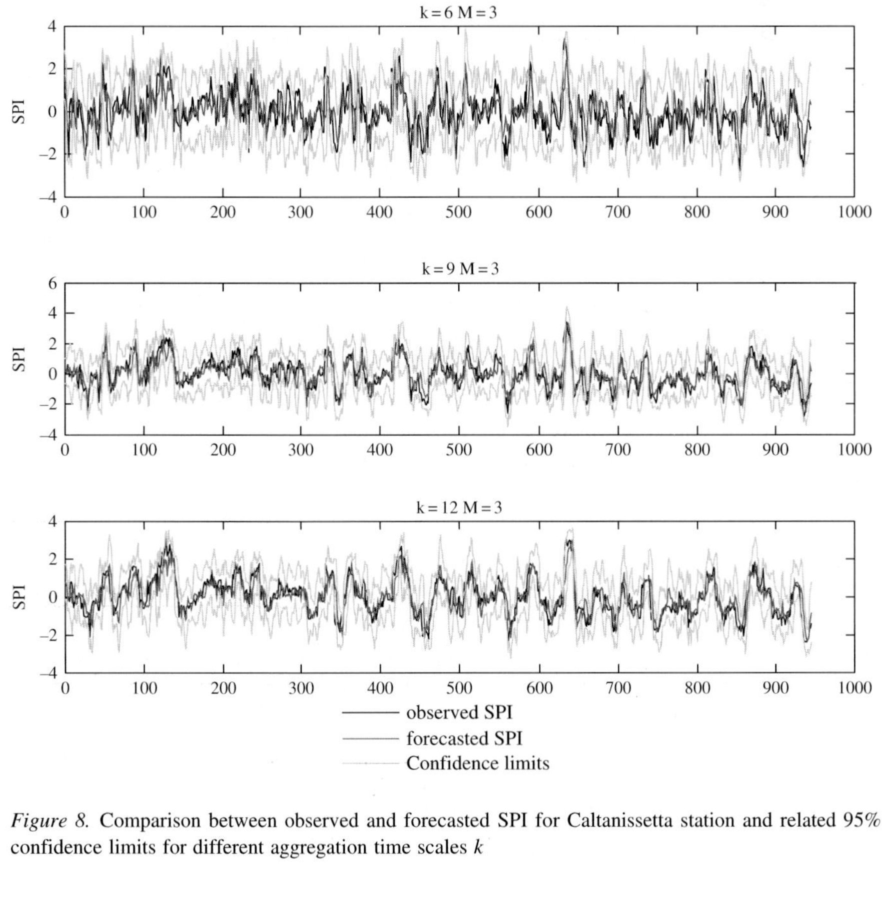

Figure 8. Comparison between observed and forecasted SPI for Caltanissetta station and related 95% confidence limits for different aggregation time scales k

scales k and $M = 3$ months. On the same plots, 95% confidence intervals, estimated by means of eq. (14) are also shown.

From the figure it can be inferred a fairly good agreement between observed and forecasted SPI values as is also evident from the fact that almost all of the observed values lie within the confidence limits. As expected, the accuracy of the forecast decreases as larger k are adopted with respect to the fixed M.

6. CONCLUSIONS

Drought monitoring and forecasting are essential tools for implementing appropriate mitigation measures in order to reduce negative drought impacts. Knowledge of transition probabilities from a drought class to another, for a given site or region, as well as the availability of forecasts of drought indices, and of the related confidence intervals, can help to improve the decision making process for drought mitigation, since appropriate measures can be selected based on the risk associated with the possible evolution of a current drought condition.

In the chapter, stochastic methodologies (i) to compute drought transition probabilities (from one class to another and from a current SPI value to a drought class), and (ii) to forecast future SPI values on the basis of past precipitation, are presented. Analytical expressions of the Mean Square Error of prediction are also presented, which enables one to derive confidence intervals for the forecasted values.

The effects of the aggregation time scale k and of the forecasting time horizon M on transition probabilities have been analyzed for 40 stations, by fixing the starting month and class, as a function of the final class and of the M values. Also, an analysis done by fixing the starting value of SPI, shows the importance of considering a current value of SPI rather than a class in the calculation of transition probabilities. In fact the transition probabilities are strongly affected by the value taken by the SPI within the same starting class. Thus taking into account the starting value rather than the starting class leads to a significant improvement in the identification of most probable drought classes in the future.

The proposed methods to compute transition probabilities is particularly valuable from a practical standpoint, in light of the difficulties of applying a frequency approach due to the limited number of transitions generally observed even on relatively long SPI records. In fact, the number of observed transitions is generally not sufficient to compute reliable frequency estimates. Furthermore, the lack of observed transitions in some cases would lead to the misleading conclusion that such transitions have zero occurrence probability, which is obviously not correct. Application of the analytical approach on the other hand, allows one to always estimate transition probabilities, even from relatively short records, since the whole available precipitation series, and not just the few observed transitions, are utilized. Regarding the applicability of Markov chain hypothesis to model transitions of SPI values from one drought class to another, Cancelliere et al. (2007) have also demonstrated that such an assumption may not be valid in general.

Validation of the forecasting model has been carried out by comparing SPI values computed on precipitation observed in the same 40 stations in Sicily and the corresponding forecasts. The results show a fairly good agreement between observations and forecasts, as it has also been confirmed by the values of some performance indices, which indicates the suitability of the model for short- medium term forecast of drought conditions.

Ongoing research is being carried out to improve the forecasting capabilities of the model, by taking into account exogenous covariates such as large scale climatic indices.

ACKNOWLEDGEMENTS

The present research has been developed with the partial financial support of the EU Projects "Mediterranean drought preparedness and mitigation planning" (MEDROPLAN, EuropeAid MEDA, contract no. ME8/AIDCO/2001/0515/59770-P027), of the Italian project MIUR PRIN 2005 "Previsione e mitigazione della siccità" and of the research contract INTERREG III B MEDOCC project SEDEMED II.

REFERENCES

Bonaccorso, B., Bordi, I., Cancelliere, A., Rossi, G. and Sutera, A. (2003) Spatial variability of drought: an analysis of SPI in Sicily. *Water Resource Management* 17, pp. 273–296.

Bordi, I., Fraedrich, K., Petitta, M. and Sutera, A. (2005) Methods for predicting drought occurrences. *Proc. of the 6th International Conference of the European Water Resources Association*, Menton, France, 7–10 September 2005.

Brockwell, P.J. and Davis, R.A. (1996) *Introduction to Time Series and Forecasting*. Springer-Verlag ed., New York, pp. 1–422.

Cancelliere, A., Di Mauro, G., Bonaccorso, B. and Rossi, G. (2007). Drought forecasting using the Standardized Precipitation Index. *Water Resource Management* 21(5), pp. 801–819.

Cancelliere, A., Di Mauro, G., Bonaccorso, B. and Rossi, G. (2005) Stochastic forecasting of Standardized Precipitation Index. *Proc. of XXXI IAHR Congress "Water Engineering for the future: Choice and Challenges"*, Seoul, Korea, 11–16 September 2005, pp. 3252–3260.

Cancelliere, A., Rossi, G. and Ancarani, A. (1996) Use of Palmer Index as drought indicator in Mediterranean regions. *Proc. IAHR Congress "From Flood to Drought"*, Sun City, South Africa, August 5–7, 1996, pp. S4.12. 1–25.

Guttman, N.B. (1998) Comparing the Palmer Drought Severity Index and the Standardized Precipitation Index. *Journal of the American Water Resources Association* 34(1), pp. 113–121.

Guttman, N.B. (1999) Accepting the Standardized Precipitation Index: a calculating algorithm. *Journal of the American Water Resources Association* 35(2), pp. 311-323.

Hayes, M.J., Svoboda, M., Wilhite, D.A. and Vanyarkho, O. (1999) Monitoring the 1996 drought using the SPI. *Bulletin of the American Meteorological Society* 80, pp. 429–438.

Heim, R.R. (2000) Drought indices: a review. In: D.A. Wilhite (Ed.), *Drought: A Global Assessment*, Routledge, London and New York.

Karl, T., Quinlan, F. and Ezell, D.S. (1986) Drought termination and amelioration: its climatological probability. *Journal of Climate and Applied Meteorology* 26, pp. 1198–1209.

Keyantash, J. and Dracup, J.A. (2002) The quantification of drought: an evaluation of drought indices. *Bulletin of the American Meteorological Society* 83(8), pp. 1167–1180.

Kotz, S., Balakrishnan, N. and Johnson, N.L. (2000) *Continuous multivariate distributions*, Volume 1. John Wiley and Sons, Canada.

Lohani, V.K., Loganathan, G.V. and Mostaghimi, S. (1998) Long-term analysis and short-term forecasting of dry spells by Palmer Drought Severity Index. *Nord. Hydrol.* 29(1), pp. 21–40.

McKee, T.B., Doesken, N. J. and Kleist, J. (1993) The relationship of drought frequency and duration to time scales. *Proc. 8th Conference on Applied Climatology,* Anaheim, California, pp. 179–184.

Mood, A. M., Graybill, F. A. and Boes (1974) *Introduction to the Theory of Statistics.* McGraw-Hill, New York.

Rossi, G., Cancelliere, A. (2002) Early warning of drought: development of a drought bullettin for Sicily. *Proc. 2nd International Conference "New trends in Water and Environmental Engineering for Safety and Life: Eco-compatible Solutions for Aquatic Environments"*, Capri, Italy, June 24–28, 2002, pp. 1–12.

Rossi, G. (2003) Requisites for a drought watch system. In: G. Rossi et al. (Eds.), *Tools for Drought Mitigation in Mediterranean Regions*, Kluwer Academic Publishing, Dordrecht, pp. 147–157.

Schervish, M.J. (1984) Algorithm AS 195. Multivariate Normal Probabilities with Error Bound. *Applied Statistics* 33(1), pp. 81–94.

Wilhite, D.A., Hayes, M. J. and Svoboda, M.D. (2000) Drought monitoring and assessment: status and trends in the United States. In: J.V.Vogt, F. Somma (Eds.), *Drought and Drought Mitigation in Europe*, Kluwer Academic Publishers, Dordrecht, Boston, London, pp. 149–160.

PART II

DROUGHT DESCRIPTION THROUGH AGROMETEOROLOGICAL INDICES

CHAPTER 6

USE OF A NEW AGRICULTURAL DROUGHT INDEX WITHIN A REGIONAL DROUGHT OBSERVATORY

A. MATERA, G. FONTANA,V. MARLETTO, F. ZINONI, L. BOTARELLI AND F. TOMEI

Hydrometeorological Service of ARPA Emilia Romagna-Bologna, Italy

Abstract: Drought management at a regional level necessitates the use of suitable instruments for monitoring, analysis and decisions support. A new index of agricultural drought, DTx, based on the daily transpiration deficit calculated by a water balance model has been defined and tested in three Mediterranean areas giving good results. A new module for DTx calculation has been implemented in the CRITeRIA water balance model as well as new algorithms to simulate cereals yield with the WOFOST methodology. A comparison between DTx values and the local climatology has been introduced in order to quantify the anomaly revealed by the index statistically. Therefore DTx can be associate with pluviometric and hydrological indices to better describe drought effects in different fields. Information on drought are collected in a single database and diffused by a dedicated website to support the regional management program defined with local authorities and water supply consortia. The program goals are to define the reference areas, identify an adequate system of monitors and alerts to help decisions, indicate proactive measures and spread knowledge and awareness on drought

Keywords: drought index, DTx, drought observatory, mitigation, Emilia-Romagna

1. INTRODUCTION

The intensity and frequent occurrence of drought events in the last few years has prompted the institutions in Emilia-Romagna to consider these phenomena very carefully. With the adoption of the Water Protection Plan (Regione Emilia Romagna, 2006a), in conformity with the national L.D. 152/99 and the European Directive 2000/60, now the Region has an instrument to achieve its goals of environmental quality of the inland and coastal waters and to guarantee a sustainable water supply over the long term. In this sphere, the plan calls for the drafting of a Program for drought management, initially preceded by the preparation of the first guidelines. The necessary instruments in support of the Program have been identified in order to monitor, assess and mitigate drought phenomena. Means of analysis, indices and thresholds of attention and intervention are being or have been identified in addition to the networks for monitoring meteorological, hydrological and piezometric variables. ARPA-SIM participates in the drafting of the Program in various

G. Rossi et al. (eds.), Methods and Tools for Drought Analysis and Management, 103–124.

fields of action and study, also devoting energies to the definition of a new index for agricultural drought (DTx) applicable to different lands and situations. The new index has been elaborated from bibliographic studies and experience acquired in the field of modeling, during the INTERREG project SEDEMED II. The DTx index and other elaborations from monitoring networks data are available on the website "Drought Observatory", now used by the regional Program as the first source of information on drought conditions. The website was also developed in the sphere of the Interregional projects. The Program for drought management includes actions of mitigation divided by territorial areas and sectors of intervention.

2. A NEW INDEX FOR AGRICULTURAL DROUGHT

2.1 Description of the New Agricultural Index

Drought is not a simple phenomenon with a univocal description; indeed, for some authors (Boken et al., 2005) the number of definitions found in literature exceeds 150.

The concept of agricultural drought (NDMC, 2003) implies, of course, a continuing water shortage due to insufficient rainfall and/or irrigation that, combined with a high level of atmospheric evaporative demand, causes insufficient moisture in the soil. Agricultural drought links the characteristics of the meteorological (or hydrological drought) to its effects on crops, concentrating on the lack of rainfall, on the differences between real and potential transpiration, on the moisture deficit of the soil, on the reduced availability of subsoil water, and so on (Byun and Wilhite, 1999). A plant's requirement for water depends on the prevailing atmospheric conditions, the biological characteristics of the plant species, its stage of development, the physical and biological properties of the soil and its moisture content (De Wit, 1986).

Insufficient moisture in the soil reduces the effective transpiration (Te) of the plant compared to the maximum transpiration (Tm), by causing the closure of its leaf pores or stomata and thereby causing a noticeable reduction in the growth rate of the crop and final yield, due to the reduction of photosynthetic assimilation (Zinoni and Marletto, 2003). The full production potential can only be achieved effectively when the moisture content of the soil does not limit transpiration, as long as other factors such as a shortage of nutrients, parasites, diseases, weeds or other adverse climatic events do not reduce plant development (De Wit, 1958). Maximum transpiration depends notoriously on both potential evapotranspiration (PET), which depends in turn on atmospheric conditions, and development of the leaf surface, measured through the leaf area index (LAI). Te depends, however, on the intensity of Tm, the type of crop and the moisture available in the soil for the roots suction (Dorenboos and Kassam, 1979).

A good definition of agricultural drought should therefore be able to represent the different crop sensitivity during different stages of development, from seedling to mature plant. There have been many studies on the link between transpiration and the production of dry matter, and they have a long history; it has always been found that this ratio tends significantly to remain linear under conditions of moisture deficits as well as under optimum moisture conditions (Steduto and Albrizio, 2005).

The use of the transpiration deficit (DT) is here proposed to assess the agricultural droughts. The DT is defined as the difference between the maximum and the effective transpiration, daily calculated by a water balance model such as CRITeRIA (Zinoni and Marletto, 2003).

Compared with other techniques for the assessment of the agricultural drought, this method appears sensitive to the drought just when it expresses its negative effect, that is, when the soil is close to exhaust its moisture reserves available for the plants (wilting point). In effect, as the moisture in the soil decreases, the variations in moisture due to the evapotranspiration are hindered by the suction applied by the matrix and are always numerically inferior to those caused by the same PET effect under conditions of greater moisture. The indices based on the variations of moisture in the soil, however, show strong oscillations when the moisture is high and small ones when it is low; while the indices based on residual moisture in the soil do not take into account the curve of moisture retention and thus the different moisture availability in different soil types at the same moisture level. The transpiration deficit calculated by a water balance model takes into account these factors and places more emphasis on the effective risk for crop production caused by the drought (Zinoni and Marletto, 2003). The transpiration deficit (DT) is significant if it remains high for a prolonged period.

The DTx, or the transpiration deficit integrated on previous periods of greater or lesser length (x = 30, 60, …, 180 days), has been therefore proposed as an index of agricultural drought. Such an index is expressed as:

$$DTx = \sum_{\text{today}-\text{X}}^{\text{today}} (T_m - T_e)$$

where: T_m is the maximum transpiration and T_e is the effective transpiration. Both are calculated daily.

The index was used to characterize the drought of the summer 2003 in Emilia-Romagna (Marletto et al., 2005) and was also compared with the SPI, the standardized precipitation index (McKee et al., 1993), showing noticeable differences, because of different aspects and factors of the drought considered by the two indices.

The DTx is calculated by a water balance model that takes account of the soil, crop and weather condition; any temperature anomalies and their effects on evapotranspiration must be considered by the system (Marletto et al., 2005). The CRITeRIA model determines the moisture balance of a bare field or a cultivated one, calculating the variation of the moisture content in the soil profile including all the moisture additions and losses. The model uses the Driessen (1986) and the Driessen and Konijn approaches (1992) to estimate the phenomena.

2.2 Case Studies

A collaboration with the Joint Research Center of the European Union (CCR, Ispra) has been established to verify the applicability of the DTx index to the other regions of the Medocc area. The CCR uses the WOFOST model (WOrld FOod Studies, Supit et al., 1994) to estimate crop yields in the European countries and in a vast section of the Mediterranean through a spatial large grid. The WOFOST model estimates effective production related to the potential production by calculating the difference between the maximum and the effective transpiration, necessary also to assess the DTx index. In this way transpiration and output data processed over the period 1975–2005 were acquired for wheat and barley, referring to different points of the grid on representative regions for the Medocc area: Aragona (Spain), Emilia-Romagna and Sicily (Italy). After acknowledging the direct relationship between the reduction of transpiration capacity and reduced production, the DTx index estimation represents a valid instrument to measure the fallout of agricultural drought on production results. The same WOFOST model was then used to compare the difference between potential and water limited production, assuming that all other factors are not limiting, and the DTx index, first on the area of study and later on other Mediterranean areas. The results show that an increase in the transpiration deficit corresponds to a decrease in production. The graphs in Figures 1, 2 and 3 show a close correlation between the reduction in transpiration and the yield loss. The chosen DTx indices cover the entire growth period for the crops: 180 days for wheat and 120 days for barley.

The WOFOST model was also introduced in the CRITeRIA modeling system to allow the quantitative estimation of the crop yields. Wheat growth simulation has been provided in Cadriano (BO) with the CRITeRIA-WOFOST balance model under potential and water limited condition, analyzing the correlation between the

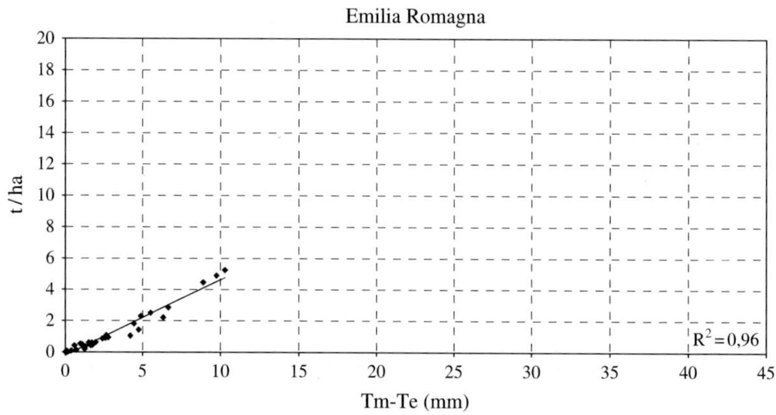

Figure 1. Correlation between difference in yield [y-coordinate] and DTx [x-coordinate] for wheat (Emilia Romagna; 1975–2005 period)

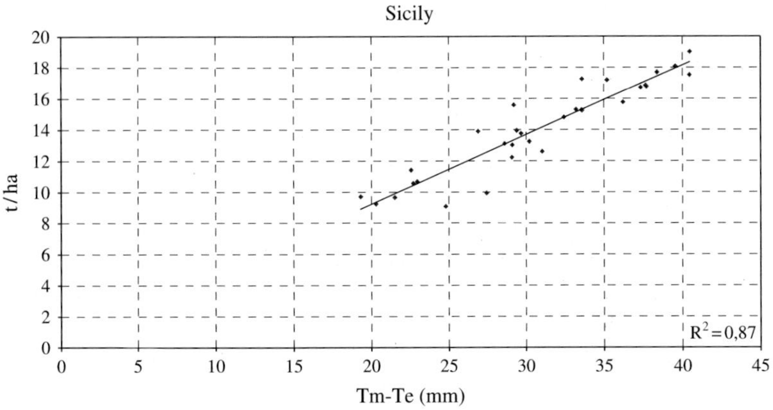

Figure 2. Correlation between difference in yield [y-coordinate] and DTx [x-coordinate] for wheat (Sicily; 1975–2005 period)

transpiration and yield loss: the results were similar to those obtained with the WOFOST of the JRC (Figure 4). Real yield data of wheat were collected in Cadriano (Emilia Romagna) for the years from 1975 to 1998 and then compared with the simulation results to improve the performance of the CRITeRIA – WOFOST model (Figure 5). In this period some wheat yields were affected by pest attacks or extreme weather events besides the water stress.

The ratio between DT180 and the cumulated maximum transpiration with the ratio of observed yield loss and potential yield was compared every year. The lost yield due to pest attacks or extreme weather events from recorded data and literature was estimated to assess the overall yield loss. Unfavorable events occurred

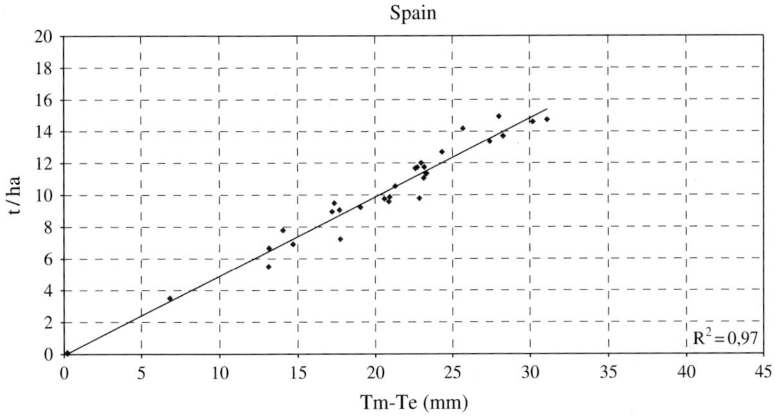

Figure 3. Correlation between difference in yield [y-coordinate] and DTx [x-coordinate] for barley (Aragona; 1975–2005 period)

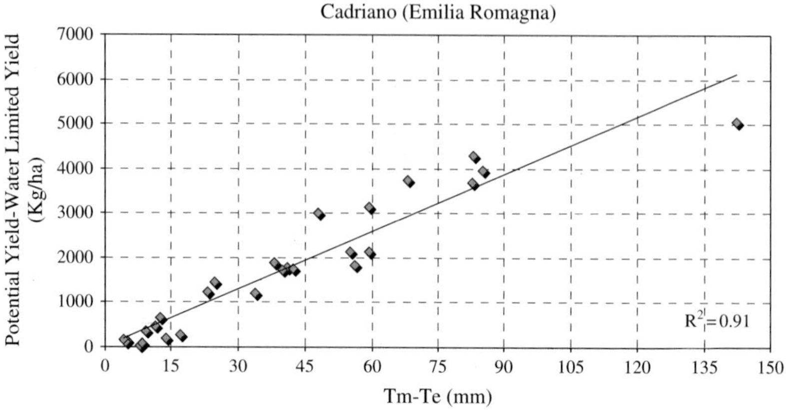

Figure 4. Correlation between difference in yield [y-coordinate] and DTx [x-coordinate] for wheat at Cadriano (Emilia Romagna, Italy); results of the simulation made with the CRITeRIA – WOFOST model

in following years: 1975, 1977, 1978, 1986, 1988, 1989, 1990, 1991, 1995 and 1998. The graph in figure 5 shows how the cumulated percentage of all stress factors (DT180/Tm, pest percentage stress and extreme events percentage stress) and the ratio of observed yield loss to potential yield are similar, particularly during the years when other negative factors affecting yield are not present besides water deficit.

The DT180/Tm and the percentage ratio of observed yield loss to potential yield was correlated just for those years. The graph in Figure 6 shows how high the correlation is ($R^2 = 0.66$) and how accurately CRITeRIA can describe the phenomenon.

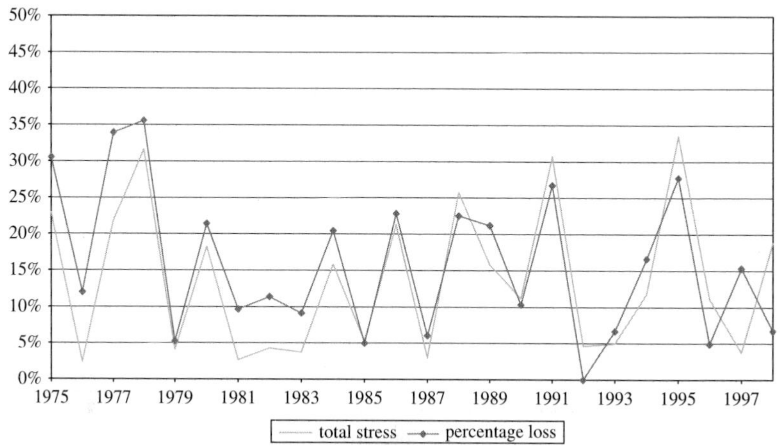

Figure 5. Comparison between the cumulated percentage of all stress factors (DT180/Tm, pest percentage stress and extreme events percentage stress) [in green] and the percentage ratio of observed yield loss to potential yield [in red]in all years

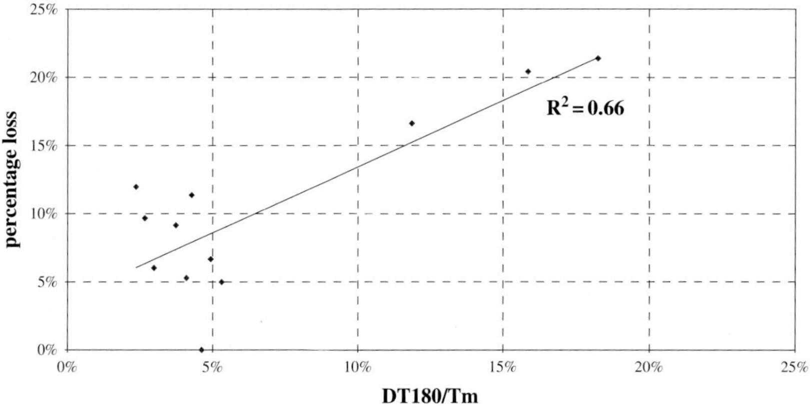

Figure 6. Correlation between the ratio of DT180 to maximum cumulative transpiration ratio [x-coordinate] with the percentage ratio of observed yield loss to potential yield [y-coordinate] in years without pest attacks or extreme meteorological events

For all the years of the case study the correlation was repeated including other stress factors as pest attacks and extreme meteorological events (Figure 7).

The DT180 performance was then calculated in three Mediterranean areas, applying a specific calculation module for the agricultural drought index to the CRITeRIA water balance model. Aragona for Spain, Emilia Romagna and Sicily for Italy were the selected areas where the water balance was simulated for meadows and alfalfa on regional typical soils. The graphs in Figures 8, 9 and 10 illustrate the index oscillation, with a minimum in springtime and a maximum at the summer end. The index shows a clear rising trend during the 1975–2005 period and for all the areas.

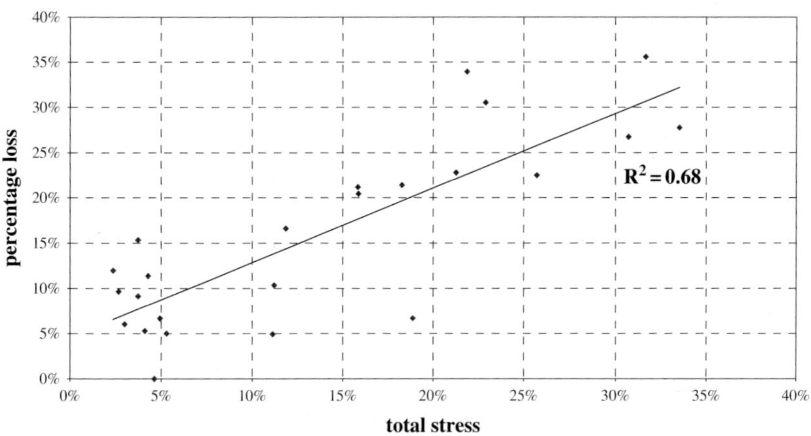

Figure 7. Correlation between the cumulated percentage of all stress factors (DT180/Tm, pest percentage stress and extreme events percentage stress) [x-coordinate] with the percentage ratio of observed loss to potential yield [y-coordinate] in all years

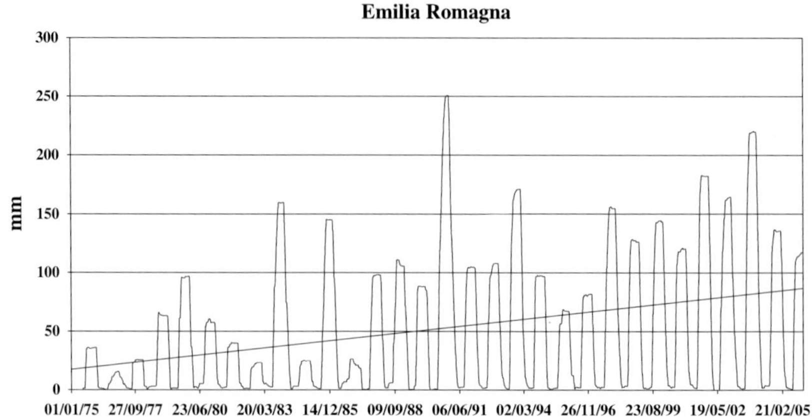

Figure 8. Emilia-Romagna: performance of DT180 on meadow (USDA soil classification: FAL; JRC meteorological data:1975–2005)

The graphs referring to Sicily and Aragona (Figure 9 and Figure 10), show that for several years the index does not return to zero in spring due to the scarcity of precipitation in fall and winter.

A comparison between DTx values and the local climatology has been introduced in order to quantify the anomaly revealed by the index statistically.

Calculation of the percentiles enables to determine the average temporal recurrence of the event. For example, values of DTx in the 75th percentile represent drought phenomena that occur in four year cycles, while values in the 90th percentile occur every ten years and values in the 95th percentile every twenty. The 100th percentile symbolizes a drought event with inensity never occurred in the reference period.

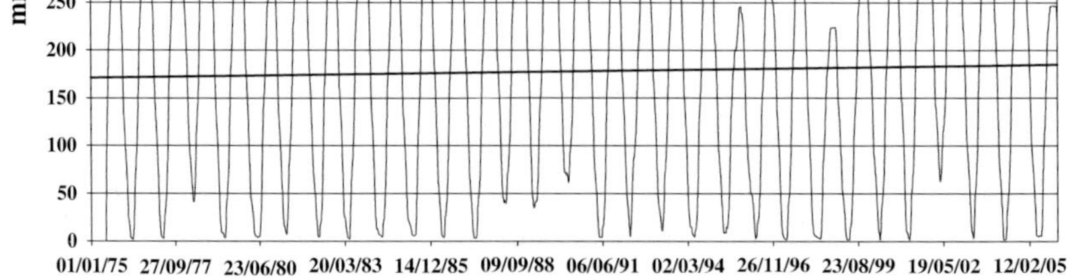

Figure 9. Sicily: performance of DT180 for alfalfa (USDA soil classification: AL-FAL; JRC meteorological data:1975–2005)

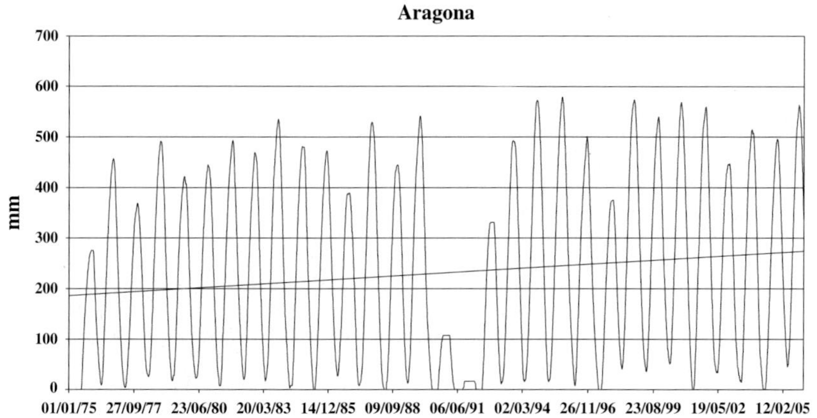

Figure 10. Aragona: performance of DT180 for meadow (USDA soil classification: FL; JRC meteorological data:1975–2005)

The percentiles method allows the comparisons among locations characterized by diverse climatic conditions.

A module for calculating DTx percentiles was implemented in the CRITERIA system to produce thematic maps with percentile values related to a reference climatology on the territory of the Emilia-Romagna region.

Figures 11, 12 and 13 show the regional maps of DT180 obtained by simulation made with CRITeRIA on the same day, July 20, in three different years, 2003, 2005 and 2006; the reference crop is grassland.

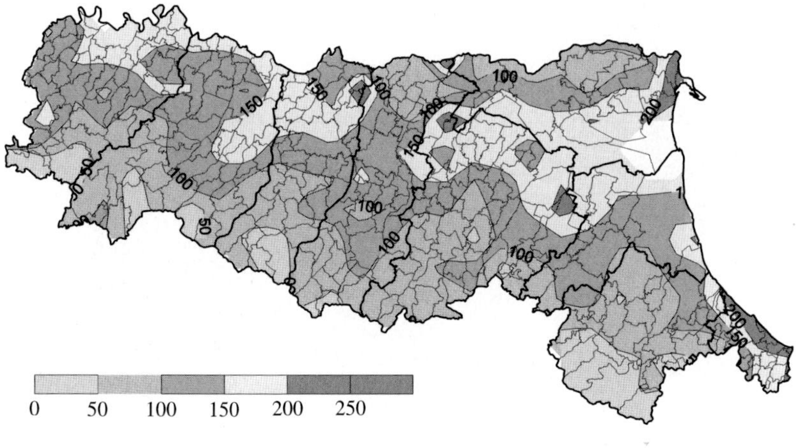

Figure 11. DT180 values (mm) for grassland (Emilia-Romagna; July 20, 2003)

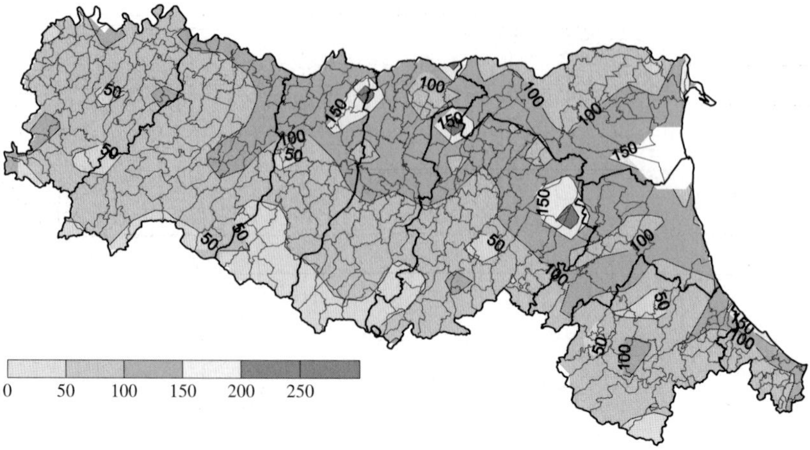

Figure 12. DT180 values (mm) for grassland (Emilia-Romagna; July 20, 2005)

The maps of DT180 percentiles, calculated with respect to a climatological reference period (1951–2001), are presented in Figures 14, 15 and 16 for the same day and the same years. Only the percentiles representative of a normal situation and those characterizing drought anomalies (from the 50th to the 100th percentile) are shown.

The comparison of the maps, which have the same chromatic scale, enables the pinpoint of the drought that struck the region during the spring and summer of 2003; the map of the DT180 index for that year represents the vast regional area

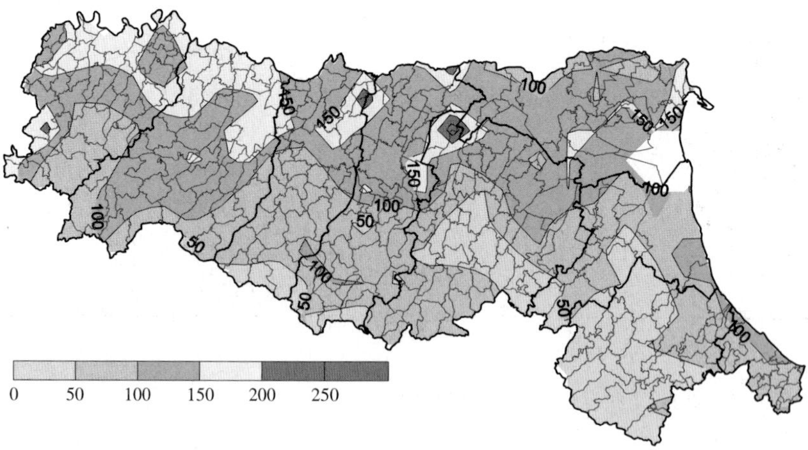

Figure 13. DT180 values (mm) for grassland (Emilia-Romagna; July 20, 2006)

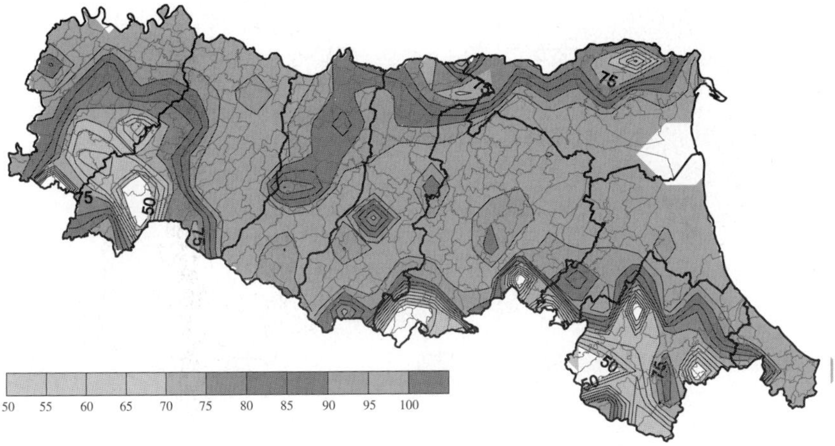

Figure 14. Percentile of DT180 (%) for grassland (Emilia-Romagna; July 20, 2003)

with very high transpiration deficit and percentile value over the 95th, indicative of drought events tending to recur every twenty years.

According with the study cases results, the DT method appears capable of pinpointing and describing drought conditions effectively for agricultural purposes, considering the interaction between atmosphere, plants and soil. It can be also used to evaluate the intensity of the drought event through comparison with an adequate reference period, in which the historical values of the index are calculated.

The DTx index is applicable even in situations other than those of the experiment; it can therefore be considered a suitable instrument for providing sufficiently

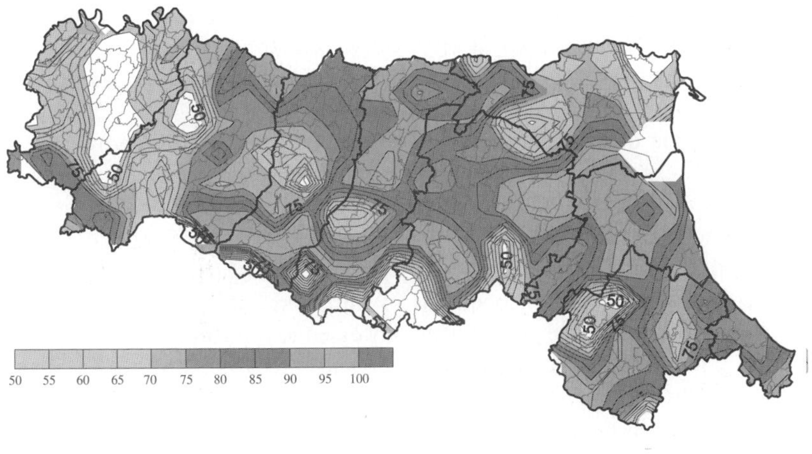

Figure 15. Percentile of DT180 (%) for grassland (Emilia-Romagna; July 20, 2005)

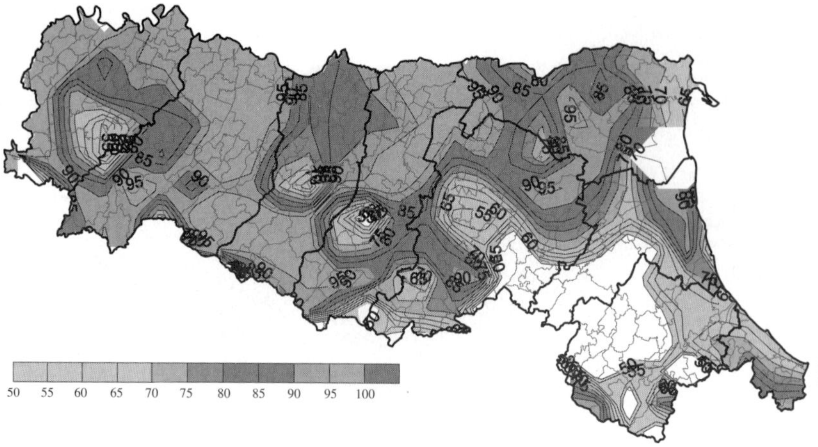

Figure 16. Percentile of DT180 (%) for grassland (Emilia-Romagna; July 20, 2006)

complete and comparable information about agricultural drought in the Mediterranean basin areas.

3. DROUGHT OBSERVATORY

At this time of heightened awareness of problems connected with meteorological phenomena, including drought, it is important to improve our techniques of monitoring and assessment, and take measures to prevent and mitigate such events.

ARPA's Hydrometeorological Service has developed and integrated a drought information system based on a website implemented in the sphere of the Interregional projects, that has become the regional site of choice for information about the subject (www.arpa.emr.it/ia_siccita/osservatorio.htm). This system allows the forecast of water shortages and the exchange of information on the effects of phenomena through the selection of appropriate indices.

The focal point of the website is the Drought Observatory, that provides instruments and data at the regional level for studying the problem, and that gathers updated documentation useful for evaluating situations and providing the institutions, operators and citizens with reliable information. The Observatory is able to supply all the information from the hydro-meteorological monitoring networks of the region as well as the conditions of the water table (regional piezometric network) in a single website. All data are processed to produce the reference indices with brief comments on the current situation and forecasts. Products developed in the sphere of the Interreg projects and connections to the national and international websites engaged in monitoring, preventing and countering drought conditions are available in the other sections of the website. In addition to the main themes, drought and desertification, information on related topics, such as safeguarding the water supply, remote measurement techniques, climate changes can be found in the website. The first product of the Observatory is the drought monitoring bulletin (see Figure 17).

The bulletin is posted weekly during the periods of greater intensity of drought events directly on the website and sent to a specific list of institutional users and to the press. It contains information about the current weather conditions, the amount of rainfall, the status of the main rivers, the situation in agriculture, the weather forecast for the coming week and the effects expected on agriculture and hydrology.

The indices used to describe the drought events are those defined by the regional group in charge of managing drought phenomena, appointed in the sphere of the Water Protection Plan, with other specific hydrologic elaborations. The rest of the bulletin is divided into two sections: "data and data processing" and "documentation". The former reports the results of data processing and the values of the indices selected to describe the phenomenon; the second contains documentation and in-depth explanation of the methods and indices selected for the description of the events and the assessment of the drought; case studies and applications of particular interest at the regional level are also available.

In particular, the first part of the section "data and data processing" has information on the hydro-meteorological and piezometric monitoring networks with a description of the methods, the location of the stations, the types of data available and the methods used for acquisition; it also offers links with the sites where data originate. The second part of the section contains maps and tables of reference indices, data on monitoring of the main regional rivers and simulations on the moisture content of agricultural soils. The indices are: SPI, deciles, DTx and water flow status on the main regional rivers.

The second section, "documentation", contains several reports presented in the sphere of the Program for management of drought phenomena in Emilia-Romagna, the definitions of the drought indices and the relative background and detailed documents. This section also contains data on specific and intense events affecting the regional territory.

When the website goes into full activity, the state of surface water resources will be documented on the basis of the following points:

1) examination of average cumulated rainfall in the last 1, 3 and 6 months on the individual areas/basins, mountain – hill and plain, and the entire Po basin, and comparison with the relative deciles highlighting the drought/moisture levels per area/basin;
2) calculation of the regional SPI indices for different time scales;
3) hypothetical future scenarios based on the SPI index simulation;
4) analyses of temperature data and derived measurements, such as the ETP, for plain and hills and calculation of the agricultural drought index, DTx, or comparison with the subsoil water table levels;
5) analysis of the measured runoff values on the Po, the main Apennine rivers and comparison with the averages or deciles if available;

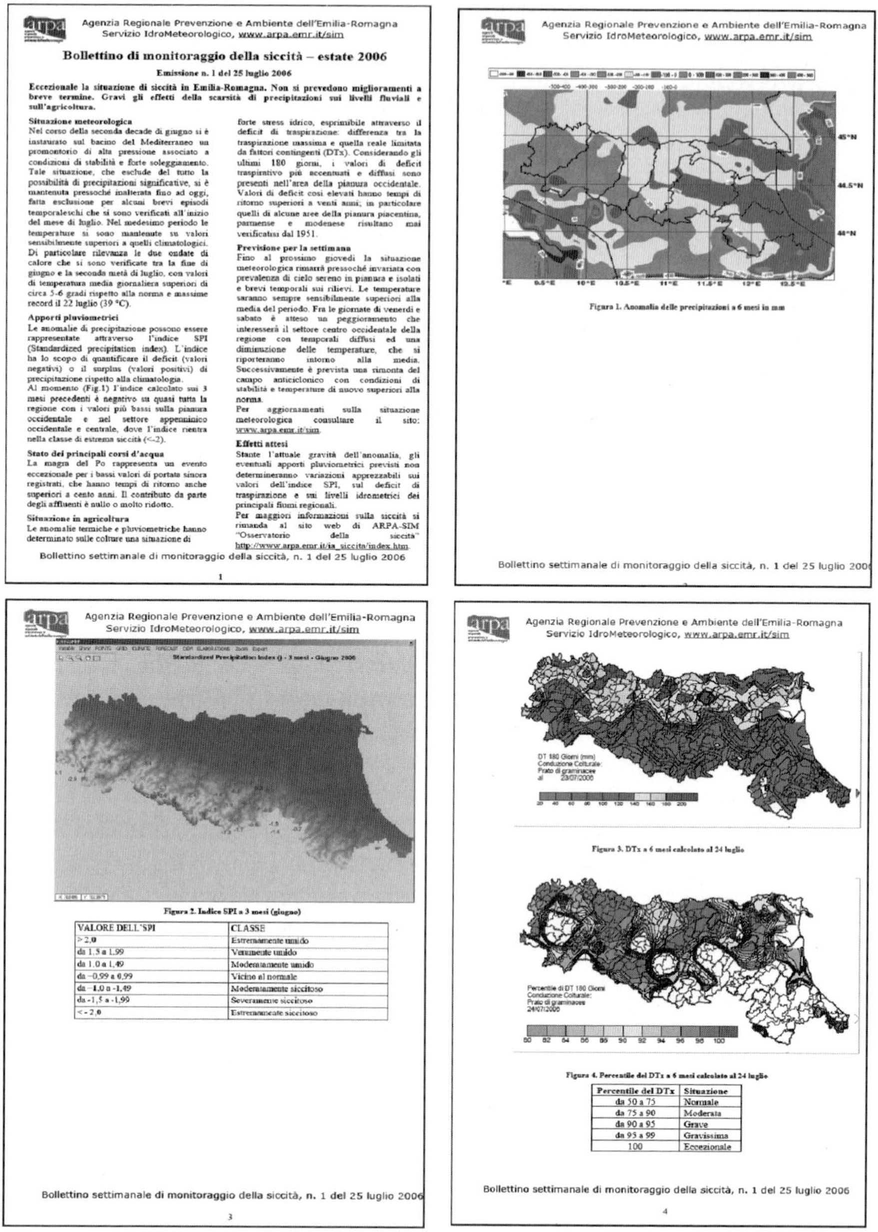
Agenzia Regionale Prevenzione e Ambiente dell'Emilia-Romagna
Servizio IdroMeteorologico, www.arpa.emr.it/sim

Bollettino di monitoraggio della siccità – estate 2006

Emissione n. 1 del 25 luglio 2006

Eccezionale la situazione di siccità in Emilia-Romagna. Non si prevedono miglioramenti a breve termine. Gravi gli effetti della scarsità di precipitazioni sui livelli fluviali e sull'agricoltura.

Situazione meteorologica
Nel corso della seconda decade di giugno si è instaurato sul bacino del Mediterraneo un promontorio di alta pressione associato a condizioni di stabilità e forte soleggiamento. Tale situazione, che esclude del tutto la possibilità di precipitazioni significative, si è mantenuta pressoché inalterata fino ad oggi, fatta esclusione per alcuni brevi episodi temporaleschi che si sono verificati all'inizio del mese di luglio. Nel medesimo periodo le temperature si sono mantenute su valori sensibilmente superiori a quelli climatologici. Di particolare rilevanza le due ondate di calore che si sono verificate tra la fine di giugno e la seconda metà di luglio, con valori di temperatura media giornaliera superiori di circa 5-6 gradi rispetto alla norma e massime record il 22 luglio (39 °C).

Apporti pluviometrici
Le anomalie di precipitazione possono essere rappresentate attraverso l'indice SPI (Standardized precipitation index). L'indice ha lo scopo di quantificare il deficit (valori negativi) o il surplus (valori positivi) di precipitazione rispetto alla climatologia.
Al momento (Fig.1) l'indice calcolato sui 3 mesi precedenti è negativo su quasi tutta la regione con i valori più bassi sulla pianura occidentale e nel settore appenninico occidentale e centrale, dove l'indice rientra nella classe di estrema siccità (<-2).

Stato dei principali corsi d'acqua
La magra del Po rappresenta un evento eccezionale per i bassi valori di portata sinora registrati, che hanno tempi di ritorno anche superiori a cento anni. Il contributo da parte degli affluenti è nullo o molto ridotto.

Situazione in agricoltura
Le anomalie termiche e pluviometriche hanno determinato sulle colture una situazione di forte stress idrico, esprimibile attraverso il deficit di traspirazione: differenza tra la traspirazione massima e quella reale limitata da fattori contingenti (DTx). Considerando gli ultimi 180 giorni, i valori di deficit traspirativo più accentuati e diffusi sono presenti nell'area della pianura occidentale. Valori di deficit così elevati hanno tempi di ritorno superiori a venti anni; in particolare quelli di alcune aree della pianura piacentina, parmense e modenese risultano mai verificatisi dal 1951.

Previsione per la settimana
Fino al prossimo giovedì la situazione meteorologica rimarrà pressoché invariata con prevalenza di cielo sereno in pianura e isolati e brevi temporali sui rilievi. Le temperature saranno sempre sensibilmente superiori alla media del periodo. Fra le giornate di venerdì e sabato è atteso un peggioramento che interesserà il settore centro occidentale della regione con temporali diffusi ed una diminuzione delle temperature, che si riporteranno intorno alla media. Successivamente è prevista una rimonta del campo anticiclonico con condizioni di stabilità e temperature di nuovo superiori alla norma.
Per aggiornamenti sulla situazione meteorologica consultare il sito: www.arpa.emr.it/sim.

Effetti attesi
Stante l'attuale gravità dell'anomalia, gli eventuali apporti pluviometrici previsti non determineranno variazioni apprezzabili sui valori dell'indice SPI, sul deficit di traspirazione e sui livelli idrometrici dei principali fiumi regionali.
Per maggiori informazioni sulla siccità si rimanda al sito web di ARPA-SIM "Osservatorio della siccità" http://www.arpa.emr.it/ia_siccita/index.htm.

Bollettino settimanale di monitoraggio della siccità, n. 1 del 25 luglio 2006

1

Agenzia Regionale Prevenzione e Ambiente dell'Emilia-Romagna
Servizio IdroMeteorologico, www.arpa.emr.it/sim

Figura 1. Anomalia delle precipitazioni a 6 mesi in mm

Bollettino settimanale di monitoraggio della siccità, n. 1 del 25 luglio 2006

Agenzia Regionale Prevenzione e Ambiente dell'Emilia-Romagna
Servizio IdroMeteorologico, www.arpa.emr.it/sim

Figura 2. Indice SPI a 3 mesi (giugno)

VALORE DELL'SPI	CLASSE
> 2,0	Estremamente umido
da 1,5 a 1,99	Veramente umido
da 1,0 a 1,49	Moderatamente umido
da –0,99 a 0,99	Vicino al normale
da –1,0 a -1,49	Moderatamente siccitoso
da -1,5 a -1,99	Severamente siccitoso
< - 2,0	Estremamente siccitoso

Bollettino settimanale di monitoraggio della siccità, n. 1 del 25 luglio 2006

3

Agenzia Regionale Prevenzione e Ambiente dell'Emilia-Romagna
Servizio IdroMeteorologico, www.arpa.emr.it/sim

Figura 3. DTx a 6 mesi calcolato al 24 luglio

Figura 4. Percentile del DTx a 6 mesi calcolato al 24 luglio

Percentile del DTx	Situazione
da 50 a 75	Normale
da 75 a 90	Moderata
da 90 a 95	Grave
da 95 a 99	Gravissima
100	Eccezionale

Bollettino settimanale di monitoraggio della siccità, n. 1 del 25 luglio 2006

4

Figure 17. Example of the drought monitoring bulletin (July 25, 2006)

6) indication of proactive and counter measures applicable at the central and the local level;
7) piezometric status of the plain water table;
8) forecasts of runoff deficit probability in the succeeding month or months.

4. IMPLEMENTATION OF A REGIONAL PROGRAM FOR DROUGHT MANAGEMENT

ARPA-SIM participates in the implementation of the "Program for management of drought phenomena" included in the regional plan for water protection and to which the Provinces and other territorial institutions of the Emilia-Romagna region contribute (Regione Emilia Romagna, 2006a). A first synthesis of the Regional Program of drought management (Regione Emilia Romagna, 2006b) has been drafted to start the discussion about the entire regional program.

The goals of the Program are to:

- define the reference areas/basins;
- identify an adequate system of monitors and alerts, with specific levels of attention and alarm on the basis of chosen thresholds, and capable of providing an evaluation of the extension and intensity of the event, so as to take the appropriate decisions regarding the measures to be applied;
- indicate the proactive measures to apply in order to reduce the vulnerability to drought of the area, to limit the droughts effects and to maintain prepared the institutions involved in the emergency situations;
- assess all possible measures of attenuation of the droughts effects by spreading knowledge of the problem and heightening the awareness of the people in areas and sectors with water supplies more at risk, keeping them informed of developments in the situation.

The Program defines the activities to be carried out at the regional level on drought, with a view to address local actions and indicates the contents of the documents on drought management measures at the local level, to be drafted by the ATOs (Optimum Territorial Areas) and the Water Supply Consortia. The drought management program sets the territorial spheres of action, the methods for monitoring the situation, the indices of critical factors for estimating current and future situations, the levels of regional and local risk and the lines of action connected to preventive and counter measures for the event control.

The first draft of the Program addresses the item of alerts preparation and experimentation, establishes rules for the evaluation of the risk level and for the preventive and counter measures implementation, and emphasizes the importance of public information on actions and results, through the publication of documents by the ATOs and Water Supply Consortia.

For drought management at the regional level, 8 macro-areas were identified, from east to west, including the basins of the main rivers and their tributaries. Each of these macro-areas was divided into basic hydrological areas, including the hydrographic basins flowing into the Po or the Adriatic Sea, or their aggregations, and the part of the Po floodplain currently irrigated with its water. This solution allows the aggregation of the data from the meteorological monitoring network for each of the mountain areas related to the runoff of the main basins and for the different zones of the plain that require irrigation, without however limiting excessively the reference territories and maintaining an acceptable number of monitoring stations of the fiduciary network for each area. So the smaller surrounding catchment basins can be included in the larger ones, usually limited by mountain zones. The basic hydrological areas also included the parts outside the region of the main regional catchment basins.

The main variables to be monitored for drought event control are precipitation, temperatures, water levels with the relative conversions in discharge, piezometric levels of both the deep and phreatic water tables. The relative information has to be found in real time at least for the first three variables. Equally main irrigation and aqueduct reservoirs have to be monitored.

The monitoring networks involved are listed in Table 1.

Besides monitoring at the regional level, the Provinces, ATOs and Water Supply Consortia have to define their own local networks. Specific synthetic indices for the current available water resources have to be defined and maintained updated; specific thresholds for those indices have also to be identified, first for the alerts and then for the counter measures application.

Subsequently to the monitoring operations, the data collected will be processed and compared with the historical ones; indices will be prepared for different alarm levels and the emergency development; a meteorological and hydrological data bank will be completed to contain the historical data sets on precipitation and temperature for the entire Po basin and the remaining part of the region, as well as runoff data sets on the Emilia-Romagna and Lombardy sections of the Po and on the regional rivers and piezometric data.

Reports for any area/basin will be produced on a regular basis with the total amount of precipitation in the last month and in the last 3, 6 and 12 months,

Table 1. Monitoring networks

Measurement network	Size	Quality subsystem	Distribution of stations	Acquisition of data
Meteorological network	200 stations active	60 stations of drought fiduciary network	2/4 stations per area	continuous
Hydrometric network	240 stations	40 primaries with discharge scales	at least 5 stations on PO floodplain	continuous
Piezometric network	470 stations	not initially provided	mainly in the plain	2 measurements per year (min)
Subsoil water table network	113 stations	not initially provided	in the plain	variable

compared with the climatological data and the most suitable pluviometric indices (SPI, deciles) values.

The daily temperatures will be compared with those recorded in the previous decades and with the climatological data so as to monitor the agricultural meteorological data derived therefrom, such as potential evapotranspiration; comparison with the relative historical data is made to reveal any positive thermal anomaly contributing to the intensification of drought phenomena, especially in the agricultural field.

Water levels of the main rivers will be compared with historical data in order to lead to the identification of the proper thresholds of attention and the definition of indices for specific critical factors. The assessment of the possible shortage of precipitation and/or runoff in the floodplain will be determined from the joint examination of rainfall data and river discharges distinguished by area/basin, respectively cumulated and averaged and compared with the historical data sets. The correlations between monthly rainfall and data for the Emilia-Romagna section of the Po floodplain and other regional basins will be assessed to define forecast models based on rainfall measurements.

The data regarding the phreatic water table, if gathered at least every two weeks in summer, will permit a significant assessment/calibration by area of any moisture deficit in the soil, helping calibrate the agricultural drought indices (e.g. DTx).

The following indices are deemed interesting for:

- pluviometric drought (deciles and SPI) and river discharges by regional areas, basins and groups of basins;
- agricultural drought (crop moisture balance, DTx) for the regional areas of the plain and foothills;
- overall availability of water resources at the regional or sub-regional level.

For the main waterways a discharge (q_0) is determined below which the river is considered to be "in drought conditions". According to the analysis carried out over 11 years of reconstructed runoff data (1991–2001), the threshold of 4.7% of the average historical discharge has been indicated for the larger rivers. For them, with this threshold, the average number of drought events lasting 7 or more days per year, was 1.5 or fewer. For Romagna, where streams have more evident torrential characteristics, the limit was raised to 2. Therefore the threshold 2% of the average runoff was assumed as correct for smaller basins, so as to obtain an average number of drought events of 7 days or more contained generally in the range of 1.5 to 3.

These percentages indicated, respectively of 4.7 and 2%, will be increased by the percentages connected to the MVR (minimum vital runoff), that represents not usable runoff, except in case of special permits. Until 2008 this additional percentage of minimum hydrologic release, at the closure sections of the mountain basins is fixed at 1/3 of the average discharge; after 2008 it will be variable around 6–7% of the average discharge.

As regards the piezometry of the main plain aquifers, it will be necessary to highlight local situations with excessive deficits on annual basis, and their development year by year, particularly as regards the average recovery suitability between a drought year and the following one.

Finally, let us look at the two additional measurements deemed interesting: the temperatures and the levels of the phreatic water table. Positive anomalies of temperature cause a higher evapotranspiration (ET) rate and thus the rapid reduction of the moisture content in agricultural soils, and consequently of the phreatic water table.

On the basis of the values reached by the reference indices (SPI and deciles) the thresholds for alert on rainfall in different spheres are established. The DTx values and thresholds will be assumed to better describe the drought situation in the agricultural sector. Reaching these thresholds will trigger an alert, while persistence beyond a fixed period of the rainfall deficits will signal the beginning of a crisis situation. The thresholds can be adjusted later in relation to the response obtained during the stage of implementation of mitigation measures.

The ways for assessing the deficit risk in the water supply during future months will be developed through the simulation of the reference indices progress.

Drought events can be classified on a scale of regional or local gravity, if we consider the extension of the involved area, the intensity of the phenomenon, i.e. its duration mainly, and the vulnerability of the water supply and distribution system, that is the adequacy of the existing infrastructures for average or moderately dry years conditions. A classification example for drought events areas of interest of drought events is shown in Table 2.

In these areas, the first lines of the Program of drought management identify levels of action for short term preventive measures, counter measures to apply upon phenomenon appearance and mitigation measures of the risk level for the medium-long term. The main actions for the different sectors (A=Municipal supply, U=Urban area and I=Irrigation supply) and for the different decision-makers and implementers are shown in Tables 3a, 3b and 3c.

The mitigation measures applicable upon phenomenon appearance will be coordinated by the Regional Civil Protection authorities for events of regional importance.

Table 2. Risk level

Intensity – on inflow – on runoff (event duration)	Low < 3rd decile at 1 month Less than 15–20 g		Medium < 3rd decile at 3 months Less than 30–40 g		High < 3rd decile at 6 months months Over 30–40 g	
Extension Vulnerability	**Low**	**High**	**Low**	**High**	**Low**	**High**
Mountain or hill-plain portion of an area	Local	Local	Local	Regional	Local	Regional
Extended area or an entire macro area	Local	Local	Regional	Regional	Regional	Regional
"Po area" (irrigated land)	Local		Regional		Regional	

Table 3a. Action lines for short term measures

Sector	Description of measures	Decision-makers/implementers
	SHORT TERM MEASURES	
A	Identification of potential water resources and the infrastructural characteristics of the network, to pinpoint any critical technical aspects	ATO, Management agencies (M.Ag.)
	Adjustment of the volume of mountain reservoirs	M.Ag.
	Maintenance of reservoir sources	M.Ag.
	Preparation of agreements for importing water in case of emergency	ATO, M.Ag.
	Prior identification of local structures for management of the emergency and the procedures for implementing measures	ATO, M.Ag., municipalities
	Informing the users of drought events, rationing measures and behavior designed to save the resource	ATO, M.Ag., Municipalities
I	Recognition of potential water resources and infrastructures to highlight the main critical technical elements and meet the minimum irrigation needs	WS Consortia
	Analysis of the possibility of exploiting the refillable aquifers from the Po, to supply the zones of the medium-low plain	Region
I	Definition of the areas: with moisture balance deficit; at risk of repeated water shortage; other contexts of water supply at risk of shortage, also connected to application of the MVR	Region, WS Consortia
	For the least irrigating organisms, redefinition of the maximum thresholds for uptake in case of drought, on the basis of the irrigated crops grown in the area served	Region, WS Consortia
	Preparation of agreements for importing water in case of emergencies	WS Consortium
	Prior identification of local structures for management of the emergency and the procedures for implementing measures	WS Consortia
	Informing the agricultural users of drought events, rationing measures and behavior designed to save the resource	WS Consortia
	Publication of use of the website CER-IRRINET for assessment of the irrigation needs of crops	Region, WS Consortia
	Informing the farmers of the availability of insurance coverage against drought damage	WS Consortia

Table 3b. Counter measures to apply upon appearance of the phenomenon

Sector	Description of measures	Decision-makers/implementers
	COUNTER MEASURES TO APPLY UPON APPEARANCE OF THE PHENOMENON	
A	Using water tank trucks to integrate scarce local resources	Provinces, munic., M.Ag.
	Preparation of "mobile" accumulation and compensation tanks	Provinces, munic., M.Ag.
	Preparation and use of additional sources and preparation of treatment facilities for drinking water	M.Ag.
	Transfer of water from outside systems	M.Ag.
	Temporary acquisition of sources granted to private users	Provinces, Municipalities
	Ordinances to prohibit non-essential private uses	Municipalities
	Use of treated waste water for "public" uses that do not require drinking water (street and sewer cleaning, public parks, etc.)	Municipalities, M.Ag.
	Reduction of mains pressure to minimum acceptable levels	M.Ag.
	Using different networks and suspending service at night	M.Ag.
A	Temporary updating of limit values at fixed parameters relative to additional sources, after informing the population	Region
	Reduction or cut of non-domestic supplies if available from other sources	Municipalities, M.Ag.
I	Importing/drawing water from outside sources	WS Consortia
	Over exploitation of aquifers refillable from the Po	WS Consortia
	Over-exploitation of the medium-high plain aquifers	WS Consortia
	Privilege in supply to users with irrigations systems at medium-high output	WS Consortia
	Optimum management of resources through annual knowledge of areas with irrigation crops and the measure/estimate of quantities effectively supplied	WS Consortia
	Control and limitation of water drawn/provided to autonomous irrigation organism in relation to the effective irrigated crops in the area	Provinces, WS Consortia
	Privilege to more valuable crops	WS Consortia
	Information to the users about availability and irrigation needs	Region, WS Consortia
	Relaxation of water quality parameters	Region
U	More frequent street cleaning to reduce dust	Municipalities

Table 3c. Mitigation measures for the medium-long term risk level

Sector	Description of measures	Decision-makers/implementers
	MEASURES FOR ATTENUATION OF THE MEDIUM-LONG TERM RISK LEVEL	
A	Increase of accumulation and compensation tanks	ATO, M.Ag.
	Reduction of leaks in the water mains	ATO, M.Ag.
	Pursuit of policies to encourage the users to save water, through widespread publication of ways to save water, production cycles requiring less water, adequate utility rates	Region, Prov., Munic., ATO, M.Ag.
	Division of the network into districts and adjusting the pressure	ATO, M.Ag.
	Providing alternative/emergency sources and connecting them to the existing mains	ATO, M.Ag.
	Increasing interconnections to use "in case of emergency"	ATO, M.Ag.
	Creation of additional/summer supply systems	ATO
I	Creation of accumulation tanks with low environmental impact to compensate the MVR and drought events of lesser gravity	Provinces, WS Consortia
	Increase refilling operations of the aquifers from former quarry ponds	Provinces, WS Consortia
	Reduction of leaks in the pipelines	WS Consortia
	Creation of consortium systems for distribution under pressure and use of irrigation techniques of less water-consuming adaptation	WS Consortia
	Programming of crops in areas with frequent limitation of water resources and preparation of supports for emergency irrigation under conditions of controlled moisture stress	Region, Provinces
U	Promote the extension of large "wooded" areas adjacent to cities	Municipalities
	Accentuate the growth of native trees other than evergreens	Municipalities

ACKNOWLEDGMENTS

We wish to thank the Scientific Committee of the SEDEMED2 project for their suggestions during the project development and Prof. Giuseppe Rossi and his staff particularly for their precious comments during the writing of this paper.

REFERENCES

Boken, V., Cracknell, A., Heathcote, R. (eds.), 2005. *Monitoring and Predicting Agricultural Drought: A Global Study*, Oxford University Press, New York, 472 pp.

Byun, H., Wilhite, D.A., 1999. Objective quantification of drought severity and duration. *J. Clim.* 12: 2747–2756.

De Wit, C.T., 1958. *Transpiration and Crop Yelds.* Agricultural Reseach Reports 64.6, Wageningen, Pudoc, 88pp.

De Wit, C.T., 1986. Introduction. In: Van Keulen, H., Wolf, J., (eds). *Modelling of agricultural production: weather, soils and crops.* Pudoc, Wageningen, 479 pp.

Dorenboos, J., Kassam, A.H., 1979. *Yield response to water.* FAO Irrigation and drainage paper n. 33, Roma, 193 pp.

Driessen, P.M., 1986. The water balance of the soil. In: Van Keulen, H., Wolf, J., (eds). *Modelling of agricultural production: weather, soils and crops.* Pudoc, Wageningen, 479 pp.

Driessen, P.M., Konijn, N.T., 1992. *Land – use systems analysis.* Wageningen Agricultural University, 230 pp.

Marletto, V., Zinoni, F., Botarelli, L., Alessandrini, C., 2005. Studio dei fenomeni siccitosi in Emilia-Romagna con il modello di bilancio idrico CRITERIA (extended abstract). *Quaderno dei riassunti Convegno AIAM* Vasto/Caramanico 3–5 maggio 2005. Riv. Ital. Agromet. 9: 32–33.

McKee, T.B., Doesken, N.J., Kleist, J., 1993. The relationship of drought frequency and duration of time scales. Proc. *Eighth Conference on Applied Climatology, Amer. Meteor. Soc.*, Jan 17–23, 1993, Anaheim CA, 179–186.

NDMC (National Drought Mitigation Center), 2003. What is drought? http://www.drought.unl.edu/whatis/what.htm.

Regione Emilia Romagna, (2006a) *Piano regionale di Tutela delle Acque,* BUR n. 20 del 13/2/ 2006, Bologna.

Regione Emilia Romagna, (2006b) *Prime linee del "Programma per la gestione del fenomeno della siccità".* Assessorato all'Ambiente ed allo Sviluppo Sostenibile, Bologna.

Steduto, P., Albrizio R., 2005. Resource use efficiency of field grown sunflower, sorghum, wheat and chickpea. II. Water use efficiency and comparison with radiation use efficiency. *Agricultural and Forest Meteorology.* 130: 269–281.

Supit, L., Hooijer, A.A., Van Diepen C.A., (eds.), 1994. *System description of the WOFOST 6.0 crop simulation model implemented in CGMS*, vol.1. JRC European Commission.

Zinoni, F., Marletto, V., 2003. Prime valutazioni di un nuovo indice di siccità agricola. *Atti convegno Aiam 2003*, Bologna 24–25 maggio, 232–238.

CHAPTER 7

DISTRIBUTED ESTIMATION OF ACTUAL EVAPOTRANSPIRATION THROUGH REMOTE SENSING TECHNIQUES

G. CALCAGNO [(1)], G. MENDICINO [(1)], G. MONACELLI [(2)], A. SENATORE [(1)] AND P. VERSACE [(1)]

[(1)] *Department of Soil Conservation, University of Calabria, Cosenza, Italy*
[(2)] *APAT – Agency for Environmental Protection and Technical Services, Rome, Italy*

Abstract: Evapotranspiration (*ET*) is one of the main water balance components, and its actual value appears to be the most difficult to measure directly. Therefore, the choice of reliable models capable of predicting spatially distributed actual *ET* represents a drought monitoring fundamental aspect.

This chapter presents a brief introduction to the main remote sensing methods for *ET* estimate and, by means of ground *ET* measurements carried out through eddy covariance systems at three different sites in southern Italy, analyzes the performance given by the Surface Energy Balance Algorithm for Land (SEBAL) model using images of the Moderate Resolution Imaging Spectroradiometer (MODIS) on areas characterized by different physiographic and vegetative conditions (sparse vegetation, crop canopy and high mountain vegetation). The distributed results obtained for different days from summer 2004 to summer 2006 on a wide southern Italian area pointed out generally good *ET* predictions in the eddy covariance sites, also if some differences arose depending on the type and density of vegetation

Keywords: Evapotranspiration, energy balance, remote sensing, SEBAL, eddy covariance

1. INTRODUCTION

The last report of the *Intergovernmental Panel on Climate Change* (Houghton et al., 2001) showed that in the 20th century, due to the excessive release of CO_2 in the atmosphere produced by the increased human activities, the average temperature of the Earth raised about 0.6 °C. This increase determined a generalized rise in evapotranspiration that, in turn produced in the Northern hemisphere a growth in precipitations of about 10–20%.

From a hydrological point of view the global warming, supplying more energy to the atmosphere, has led to a drastic increase in extreme events. Brunetti et al. (2004) observed a reduction in the number of rainy days especially in southern Italy, while pluviometric analyses carried out by Xie et al. (2003) in the Mediterranean

G. Rossi et al. (eds.), Methods and Tools for Drought Analysis and Management, 125–147.

basin during the period 1986–2002 have shown a great variability of precipitation, compared to the average values of the period, evidencing that areas with higher local values of standard deviation (among these are the southern Italian regions) are also the more affected by recurrent, sudden and long droughts.

Drought is a phenomenon not easy to define (Dracup et al. 1980; Wilhite and Glantz, 1985), even though it is the extreme climatic event that affects the largest number of people and the largest territory worldwide, and causes the greatest economic damages (about 5 to 7 thousand million euros every year; Keyantash and Dracup, 2002). Reliable knowledge of hydrological cycle is essential for drought monitoring and in order to guarantee equilibrium between water availability and uses, but the estimate of the main water balance variables has been always poor. Specially for evapotranspiration (ET), the problem is emphasized because of low availability and intrinsic limits of instrumental measurements, and for limited reliability of indirect estimate methods.

Conventional and more recent techniques supplying punctual measures (more or less precise) are just representative of *ET* at a local scale. In fact, surface heterogeneity and dynamic nature of heat transfer processes do not allow to consider the recorded data representative of wider areas, for which the most widespread method is the recommended FAO56 (Allen et al., 1998). This method consists in the estimate of evapotranspiration for a particular crop canopy ET_c by multiplying the crop coefficient K_c for the reference evapotranspiration ET_r, obtained through the Penman-Monteith method for a standard crop with a constant value of the surface resistance. Nonetheless, this resistance can vary according to the day and to the meteorological conditions, therefore also the spatial and temporal estimate of K_c is questionable (Neale et al., 2005).

In the last years the possibility of using long data series coming from remote sensing opened new and promising perspectives to the distributed estimate of *ET* over different scales, ranging from a single field or little basin to regional scale up to continental scale. Various methods and procedures have been developed based on satellite images (a detailed review is presented by Kustas and Norman, 1996). Courault et al. (2005) proposed to group these methods in four main categories. The former is composed by the *empirical direct methods*, characterized by the *ET* estimate with semi-empirical relationships between net radiation R_n and cumulative temperature difference $(T_s - T_a)$, where the surface temperature T_s is obtained by satellite images, while the air temperature T_a is given by ground based observations. The second category is described by *vegetation indices* or *inference methods*, based on the use of remote sensing for the computation of reduction factors (like the crop coefficient K_c) for the estimate of *ET*, once ET_r is known through ground based measurements. The third group is made up of methods that use remote sensing directly to calculate input parameters and, combining empirical relationships to physical models, to estimate the energy budget components, except *ET* that is computed as the residual of the energy budget equation (for this reason these methods are usually referred to as *residual methods*). The last category is that of the *deterministic methods*, usually based on complex soil-vegetation-atmosphere transfer models (SVAT), directly computing all the components of the energy budget, where remote sensing data can be used either as input parameters to characterize the different surfaces, or in data assimilation procedures for the calibration or the retrieving of particular parameters for the *ET* estimate.

All the remote sensing models are characterized by several approximations, and need detailed experimental validations. Precise ground based measurements of evaporative fluxes are necessary for this purpose, and the eddy covariance technique is generally considered one of the most reliable approaches for the analysis of the phenomenon (Kanda et al., 2004). The eddy covariance method allows the direct measure of turbulent flux density of a scalar along horizontal wind streamlines (Swinbank, 1951, 1955). The technique in the past years spread all over the world for the study of energetic and CO_2 fluxes (Baldocchi and Meyers, 1988; Baldocchi et al., 1996; Wilson et al., 2002; Valentini, 2003), due to several important advantages concerning the possibility of having direct measures without parameters calibration, spatially distributed flux measures over a wide area (footprint), and finally simultaneous measures of vapor, heat and CO_2 fluxes.

After a brief theoretical discussion in the next section on the energy budget, aimed at showing relations between *ET* and the variables from which it depends, in the third section an introduction to the four listed categories of remote sensing methods is presented. We will particularly linger over the third approach, very interesting for operational applications, and the fourth approach that, thanks to the data assimilation techniques, seems to represent the most promising way to the high resolution distributed estimate of *ET* and other energy balance components. Finally, on the fourth section we will present some examples of application of a residual method, the Surface Energy Balance Algorithm for Land (SEBAL) scheme, proposed by Bastiaanssen et al. (1998) and Bastiaanssen (2000). SEBAL has been applied starting from images of the Moderate Resolution Imaging Spectroradiometer (MODIS) in the visible, near and thermal infrared bands, to determine the spatially distributed *ET* in southern Italy. Distributed estimates of turbulent fluxes (latent flux λE and sensible flux H) and available energy ($R_n - G$, where G is the soil heat flux) have been locally verified for different days from summer 2004 to summer 2006 with data recorded by eddy covariance stations located in three areas characterized by different physiographic and vegetative conditions (sparse vegetation, crop canopy and high mountain vegetation).

Obtained results appear to be acceptable, even though some differences in the estimate of the single components of the energy balance arose, specially due to the approximations of the method and to the different scales between the remote sensing estimated values (1 km pixel resolution) and the observed values (depending on the footprint). To overcome this problem, and to allow *ET* estimates with higher spatial resolution, a downscaling procedure is proposed using also images of the Advanced Spaceborne Thermal Emission and Reflection radiometer (ASTER). Finally, in the chapter it is shown that to obtain reliable estimates in some specific areas it is not appropriate to apply the procedure on the whole southern Italian territory, because the application of SEBAL requires relatively constant atmospheric conditions over the image.

2. ENERGY BUDGET AND EVAPOTRANSPIRATION ESTIMATES

Different terms are involved in the soil surface energy balance, depending on the finite volume examined. Considering an infinitesimal soil layer without vegetation in the atmosphere-surface interface, conservation of energy, both instantaneously and averaged in time, is given by the following equation:

$$R_n - G = \lambda E + H \tag{1}$$

Even though in many practical situations the energy balance is fully described by equation (1), whose terms are expressed in Wm^{-2}, sometimes other additional terms are not negligible (e.g. a term representing the energy storage per unit area in the canopy should be considered in presence of tall vegetation, while introducing an advection term can be very important in some particular cases). Furthermore, both radiant and convective fluxes can be described considering the observed surface as a *single-layer*, or considering *two-sources*, i.e. bare soil and vegetation, also taking in account more vegetation layers (*multi-layer* approaches).

Net radiation R_n is obtained through the balance of three fluxes, the global shortwave radiative flux R_t, the downward longwave flux L_{in} and the upward longwave flux L_{out}:

$$R_n = (1-\alpha)\, R_t + \varepsilon_s L_{in} - L_{out} = (1-\alpha)\, R_t + + \varepsilon_s L_{in} - \varepsilon_s\, \sigma\, T_s^4 \tag{2}$$

where α is the shortwave albedo of surface, ε_s the longwave emissivity, $\sigma = 5.67 \times 10^{-8} Wm^{-2}K^{-4}$ is the Stefan-Boltzmann constant, T_s is given in Kelvin [K]. The upward longwave flux L_{out} is directly derived from the Plank's law of black body radiation, approximating Earth surface as a grey body.

Values of the soil energy flux at the interface or at a shallow depth depend on many factors concerning intensity of solar radiation, soil type and properties and soil moisture content (Garratt, 1992). Generally, the soil heat flux G at any level z' can be described by discretizing Fourier's law for heat conduction in a homogeneous body:

$$G(z') = -k_s \partial T_s / \partial z' \cong k_s \frac{T_s - T'}{\Delta z'} \tag{3}$$

where k_s is the thermal conductivity and temperature T' is related to level z'.

In a single-layer approach, sensible heat flux H can be obtained by the following bulk aerodynamic equation:

$$H = \frac{\rho c_p (T_0 - T_a)}{r_a + r_{ex}} = \frac{\rho c_p (T_0 - T_a)}{r_{ah}} \tag{4}$$

where $\rho [kg\, m^{-3}]$ is the air density, c_p $[J\, kg^{-1} K^{-1}]$ the specific heat at constant pressure, T_0 the so-called "aerodynamic temperature" of the surface (very similar to T_s),

r_{ah} [s m^{-1}] the bulk aerodynamic resistance to sensible heat exchange, r_{ex}[s m^{-1}] an excess resistance associated with heat transport, and r_a[s m^{-1}] the aerodynamic resistance between the surface and the reference height z_a in the lower atmosphere, which can be expressed by the following equation (Brutsaert, 1982):

$$r_a = \frac{\left[\ln\left(\frac{z_a - d_0}{z_{0m}}\right) - \Psi_m\right]\left[\ln\left(\frac{z_a - d_0}{z_{0m}}\right) - \Psi_h\right]}{k^2 u_a} \tag{5}$$

In equation (5) d_0 is the displacement height, u_a is the wind speed measured at height z_a, k is the von Karman constant ($\approx$0.4), Ψ_m and Ψ_h are the Monin-Obukhov stability functions for momentum and heat respectively, z_{0m} is the roughness length for momentum transport. The excess resistance is often related to a roughness length for heat:

$$r_{ex} = \frac{\ln\left(z_{0m}/z_{0h}\right)}{ku^*} \tag{6}$$

where z_{oh} (according to Kustas et al., 1989 $\approx$ 1/10 to 1/5 of z_{0m}) is the roughness length for heat transport and u^* is the friction velocity:

$$u^* = \frac{u_a k}{\ln\left(\frac{z_a - d_0}{z_{0m}}\right) - \Psi_m} \tag{7}$$

If the other three components of the energy budget are known, the latent heat flux λE (hence evapotranspiration) can be obtained with a residual method in a form like this:

$$\lambda E = R_n - G - \rho c_p (T_0 - T_a)/r_{ah} \tag{8}$$

Equation (8) is widely used to estimate the instantaneous value of λE. If this value is calculated around midday, it can be useful to determine the water stress of the vegetation. Considering that during the day the evaporative fraction Λ:

$$\Lambda = \frac{\lambda E}{R_n - G} \tag{9}$$

is almost constant, several methods estimate through instantaneous measures of Λ and λE the daily value of evapotranspiration ET_{24}. For longer periods, the use of ground-based ET measures is necessary to make temporal interpolations.

In a single-layer approach, also latent heat flux λE can be obtained by a bulk equation. For saturated surfaces it is equal to the potential term λE_p:

$$\lambda E \equiv \lambda E_p = \frac{\rho\lambda\left(q^*(T_0) - q_a\right)}{r_{av}} \tag{10}$$

where λ[J kg^{-1}] is the latent heat of vaporization of water, $q^*(T_0)$ the saturation specific humidity for T_0, q_a the specific humidity at the reference height z_a, and r_{av} [s m^{-1}] ($\approx r_{ah}$) the bulk aerodynamic resistance to latent heat exchange.

Evaporation is reduced if the surface humidity is less than the saturation value calculated at the existing surface temperature. In the case of a leaf surface, a surface resistance to water vapor transfer through foliage stomata exists whether the vegetation is under negligible or severe water stress (Garratt, 1992). This surface resistance r_s for a partly or fully vegetated surface can be defined by:

$$r_s = \frac{\rho\left(q^*\left(T_{eff}\right) - q_0\right)}{E} \tag{11}$$

where T_{eff} is the effective temperature of the surface (in a sparse canopy, it is an intermediate value between the leaf and soil-surface temperatures), and q_0 is the unknown surface specific humidity, that can be eliminated, obtaining:

$$\lambda E = \frac{\rho\lambda\left(q^*\left(T_{eff}\right) - q_a\right)}{r_{av} + r_s} \tag{12}$$

The resistance r_s is not readily interpreted physically, since E is not only due to transpiration but also to soil evaporation. If the surface is wholly and uniformly vegetated, the whole canopy can be represented as a single hypothetical "leaf", and the "big-leaf" model can be introduced, that uses the bulk stomatal resistance of the canopy r_{st}:

$$r_{st} = \frac{\rho\left(q^*\left(T_f\right) - q_0\right)}{E} \tag{13}$$

where T_f is the foliage temperature.

The temperature T_0 introduced in equation (4) is very similar to T_s because it is defined by extrapolating air temperature profile down to the level z_{0h}, that is very close to the surface level. Neither T_0 nor T_s can be measured, so they are often replaced by observation of radiometric surface temperature T_R, obtained by remotely sensed images. However, differences between T_0 and T_R can be greater than 10°C for sparse canopies (Kustas et al., 2004). This has forced many users of single-source approaches to adjust z_{0h} or the ratio $\ln(z_{0m}/z_{0h})$, to obtain good agreement with measured values of energy balance components, and at the same time encouraged the development of two-sources models, to obtain more realistic expressions for the vertical fluxes from the ground foliage system to the atmosphere. In these models the soil surface temperature T_g is allowed to differ from the foliage temperature T_f, resulting in fluxes from both underlying ground (E_g, H_g) and foliage (E_f, H_f) to the atmosphere (Deardorff , 1978). The total fluxes are given by:

$$H = H_f + H_g = \left[\rho c_p\left(T_0 - T_a\right)\right]/r_{ah} \tag{14}$$

$$E = E_f + E_g = \left[\rho\left(q_0 - q_a\right)\right]/r_{av} \tag{15}$$

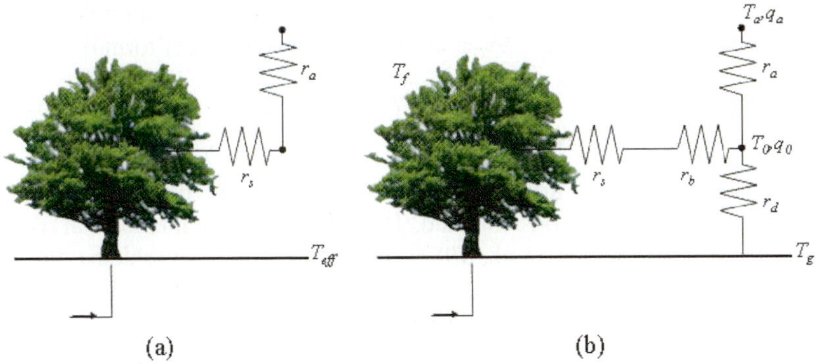

Figure 1. Schematic representation of the main elements of (a) a single-component and (b) a two-component canopy model (modified from Garratt, 1992)

where the individual fluxes can be described as:

$$H_f = \rho c_p (T_f - T_0)/r_b \tag{16}$$

$$H_g = \rho c_p (T_g - T_0)/r_d \tag{17}$$

$$E_f = \rho (q_f - q_0)/r_b \tag{18}$$

$$E_g = \rho (q_g - q_0)/r_d \tag{19}$$

with r_b like a boundary-layer resistance between the leaves and foliage air space, and r_d a resistance to transfer between the ground and canopy air space. Both r_b and r_d can be expected to be on the order of, or greater than r_a (Deardorff , 1978). Figures 1a and 1b show schematic representations of the main elements of a single-component and a two-component canopy model.

3. REMOTE SENSING METHODS FOR *ET* ESTIMATE

In this section a brief review of remote sensing methods for *ET* estimate is presented, subdividing them according to the classification proposed by Courault et al. (2005).

3.1 Empirical Direct Methods

These methods are essentially simplified relationships based on the theoretical assumption that ratio H/R_n is constant during the day and that the daily value of G

is null. The relationships, proposed by various authors (e.g. Jackson et al., 1977; Lagouarde, 1991; Courault et al., 1994), have the following typical formulation:

$$ET_{24} = R_{n24} + A - B\,(T_s - T_a) \tag{20}$$

where ET_{24} and R_{n24} are the daily values of evapotranspiration and net radiation, $(T_s - T_a)$ is the instantaneous difference in temperature measured around midday, and A and B are parameters to be calibrated. The accuracy of the method can reach 10–15% at local scale (Seguin et al., 1982), but apart from the problems due to calibration, issues linked to the spatial interpolation of the quantities have to be considered. In fact, also assuming that solar radiation can be spatially estimated through remote sensing techniques, geostatistical models used to interpolate T_a reduce the accuracy around 20 to 30%.

Since vegetation indices and surface temperature are related (higher ET values are usually associated with lower surface temperatures), Carlson et al. (1995) and Moran et al. (1994) explored the relationship between the cumulated temperature difference $(T_s - T_a)$, also named *stress degree day* (SDD), and the Normalized Difference Vegetation Index (NDVI), drawing up a trapezoidal scheme by which a classification of the different soil moisture conditions can be obtained.

3.2 Inference Methods

Reflectance values in the red and in the near infrared bands allow to estimate various vegetation indices. These indices are strictly correlated to parameters as fraction of vegetation cover or Leaf Area Index (LAI), that also affect the behavior of the crop coefficient K_c. Starting from these considerations, Heilman et al. (1982) investigated the relationships between percent cover and reflectance-based Perpendicular Vegetation Index (PVI) for alfalfa. Afterwards many studies have been conducted especially considering the NDVI (Neale et al., 1989; Choudhury et al., 1994; Bausch, 1995; Allen et al., 2005) or similar indices as the Weighted Difference Vegetation Index (Consoli et al., 2006), and the Soil Adjusted Vegetation Index (Garatuza-Payan et al., 1998; Neale et al., 2005), retrieving various empirical equations to link these indices to the crop coefficient. However, specially because of irrigation effects on soil moisture and water stress conditions, it has been verified that relationship between K_c and vegetation indices is not univocal (Allen et al., 2005).

Equations linking K_c to vegetation indices could be very useful for irrigation planning, particularly for K_c estimate in relatively dry soils, but results till now obtained are empirical, and it is necessary to work hard in order to develop more general relationships.

3.3 Residual Methods

Equation (20) essentially derives from a simplified approach to the energy budget equation, leading to the evapotranspirative term as the residual term of the equation.

More complex, but also more reliable residual methods, not simply empirical and widely adopted in operational applications, are the SEBAL, SEBI, S-SEBI, SEBS and T-SEB schemes, that use the spatial variability of radiance and reflectance of satellite images.

The Surface Energy Balance Algorithm for Land (SEBAL) scheme, proposed by Bastiaanssen et al. (1998) and Bastiaanssen (2000), is based on an intermediate approach, using both empirical relationships and physical parameterizations. This method was realized for a very flexible use, because it allows a regional scale estimate of energy fluxes with very few ground-based data, and it has been adopted in several applications (Bastiaanssen, 2000; Jacob et al., 2002; Bastiaanssen and Chandrapala, 2003; Mohamed et al., 2004; French et al., 2005; Patel et al., 2006).

SEBAL is based on equation (8) and on the hypothesis that evaporative fraction Λ is constant. Net radiation is estimated through remotely sensed images (supplying information about albedo), and the same is for soil heat flux, that is a function of R_n and NDVI. The aerodynamic resistance is estimated as a function of friction velocity u^* (that can be simply obtained with one wind speed measurement in the examined area), and is corrected for atmospheric stability through an iterative procedure. Also the roughness length is linked to vegetation indices through empirical functions. Distinctive of SEBAL scheme is the estimate of the spatial distribution of (T_s-T_a), determined considering two particular pixels inside the analyzed area, the "dry" pixel and the "wet" pixel. On the first pixel the latent heat flux λE can be considered null, so that the available energy (R_n-G) is completely transformed in sensible heat flux H, and the ΔT difference on the same pixel can be obtained inverting equation (4). On the second pixel the sensible heat flux is hypothesized as equal to zero, so that surface and air temperature coincide (ΔT=0). Once the couples of values (T_s, ΔT) are known on the two pixels, they are used to estimate a linear relation between T_s and ΔT by which, obtained the spatial distribution of T_s from remote sensing images, the spatial distribution of ΔT (hence of H) is estimated.

Also based on the contrast between wet and dry areas, the S-SEBI scheme (Simplified Surface Energy Balance Index; Roerink et al., 2000) determines a reflectance dependant maximum temperature for dry conditions and a reflectance dependant minimum temperature for wet conditions, partitioning the sensible and latent heat fluxes according to the actual surface temperature. The major advantages of this scheme are that no additional meteorological data is needed to calculate the fluxes if the surface hydrological extremes are present, and that the extreme temperatures for the wet and dry conditions vary according to changing reflectance values.

S-SEBI is a simplified version of SEBI scheme (Menenti and Choudhury, 1993). The latter obtains the extreme temperatures for dry and wet conditions from an external data source, and allows to operate even though in the analyzed area these temperatures are not present (typical examples where SEBI can be used, but not S-SEBI, are the images of England, where there are not dry areas; images of Sahara desert, where there are not wet areas; larger areas like Europe, because atmospheric conditions are not constant).

The Surface Energy Balance System (SEBS; Su, 2002) requires as inputs three sets of information. The first set consists of land surface albedo, emissivity, temperature, fractional vegetation coverage and LAI, and the height of the vegetation (or roughness height), that can be derived from remote sensing data in conjunction with other information about the concerned surface. The second set includes air pressure, temperature, humidity, and wind speed at a reference height (i.e. the height measurement for point application and the height of the Planetary Boundary Layer (PBL) for regional application). The third data set includes downward solar radiation, and downward longwave radiation which can be obtained through either direct measurements, model output or parameterization. Unlike SEBAL, in SEBS the sensible heat flux H associated to the wet pixel is not null, but can be derived by a combination equation similar to that suggested by Penman-Monteith. The evaporative fraction is not directly obtained, but as a function of the relative evaporation Λ_r, that can be evaluated as:

$$\Lambda_r = \frac{\lambda E}{\lambda E_{wet}} = 1 - \frac{H - H_{wet}}{H_{dry} - H_{wet}} \tag{21}$$

Λ_r can be directly linked to the soil degree of saturation θ/θ_s (Su et al., 2003), and can be used as a soil moisture index. Also Λ is related to θ/θ_s, but not directly, needing the calibration of some parameters (Scott et al., 2003).

While the methods so far presented consider the observed surface as a single-layer, TSEB scheme is a two-source approach based on land surface separation into two distinct, but linked components (soil surfaces and vegetation canopies), aimed at modeling in a more physically meaningful way surface resistance characteristics. TSEB derives energy flux estimates from modeling the land as a resistance network, between energy sources from soil, vegetation, and the overlying atmosphere. Two main variants of TSEB exist, one strictly applicable at local scales (as described in Norman et al., 1995), while the other, known as DisAlexi (Anderson et al., 1997; Mecikalski et al., 1999) is also useful at regional scales since it also models energy exchange at the atmospheric boundary layer. TSEB has three key assumptions: turbulent fluxes are constant within the near surface layer (Monin–Obukhov similarity is used for stability correction), radiometric temperature can be repartitioned into soil and vegetation components, and Priestley–Taylor transpiration (Priestley and Taylor, 1972) is applied for unstressed vegetation.

Relatively few comparisons have been made among the various residual methods. In the comparison carried out by Timmermans et al. (2005) at the SPARC2004 site it is shown that patterns in all four flux components are similar for SEBAL, SEBS and TSEB, although over drier areas the TSEB scheme shows higher values for sensible heat. French et al. (2005) found that both TSEB and SEBAL showed systematic agreement and responded to spatially varying surface temperatures and vegetation densities, also if direct comparison against ground eddy covariance data suggests that the TSEB approach is more helpful over sparsely vegetated terrain. Melesse and Nangia (2005), finally, used a hybrid model, where net radiation is determined using SEBAL, and the surface temperature together with the latent and sensible heat fluxes are partitioned applying TSEB.

Apart from the differences between single-source and two-source models, some common issues for residual models have to be considered. The first question arises about the evaporative fraction: is it really constant during the day? Crago (1996) concluded that a complicated combination of weather conditions, soil moisture, topography and biophysical conditions contribute to the conservation of Λ on the same day, also if variably cloudy weather and proximity to surface discontinuities or fronts may cause significant variability.

Another important issue is the estimate accuracy of surface temperature, depending on the performances of the retrieval algorithms. It has been calculated (Norman et al., 1995) that an error of 1°C in the estimate of (T_s-T_a) can lead to errors in the estimate of H varying from $8\,W\,m^{-2}C^{-1}$ (with canopy height equal to 1 m and wind speed equal to $1\,m\,s^{-1}$) to $87\,W\,m^{-2}C^{-1}$ (with canopy height equal to 10 m and wind speed equal to $5\,m\,s^{-1}$).

Spatial and temporal resolution required for *ET* estimates influence the choice of the source for the remote sensing images. Table 1 shows some of the main characteristics of the most used satellites for NDVI and T_s estimates. Some attempts to propose disaggregation procedures for low spatial resolution data are coming out in the last years (e.g. Kustas et al., 2003).

Spatial interpolation of T_a values locally measured could represent a source of error for the residual methods. Sometimes T_a is estimated in a distributed way through models simulating the evolution of the Planetary Boundary Layer (Carlson et al., 1995). Models that do not use directly T_a measurements (SEBAL, S-SEBI) need in the image the dry and the wet pixels to be present.

Finally it has to be considered that residual models are quite sensitive to the roughness length, a distributed parameter that is not estimable with great accuracy for wide and heterogeneous areas using only empirical functions of NDVI. Nowadays the most promising remote sensing methods for roughness length estimates are based on LIDAR techniques.

3.4 Deterministic Methods

Residual methods are able to supply *ET* estimates only if remote sensing images for the analyzed area are available. For the days without images (e.g. cloudy days)

Table 1. Main characteristics of the most used satellites for NDVI and T_s estimates

Satellite	Repeat cycle (day)	NDVI pixel resolution (m)	T_s pixel resolution (m)
ASTER	16	15	90
AVHRR	2 im/day	1100	1100
LANDSAT 5	16	30	120
MODIS	2 im/day	250	1000

temporal interpolations are needed, that can be obtained locally from ground-based measurements. Deterministic models (SVAT models) can operate also without having remote sensing data, because they use these data just as auxiliary input parameters or in data assimilation procedures. Therefore SVAT models are suitable for the estimate of energetic exchanges during the periods without remote sensing images. Furthermore, through SVAT models it is possible to obtain, together with the components of the energy balance, the estimates of many intermediate variables (e.g. LAI, soil moisture) strictly correlated to physiological and hydrological processes.

SVAT models are explicitly based on the principles of energy and mass balance. They can use purely "big-leaf" schemes (e.g., the 0-dimensional models based on Monteith, 1965; Priestley and Taylor, 1972; Shuttleworth and Wallace, 1985); intermediate "big-leaf" schemes (Raupach and Finnigan, 1986) with a vegetation layer considering implicitly a vertical dimension, through the parameterization of solar radiation transfer and turbulent exchanges (e.g. Sellers et al., 1996); or multi-layer schemes (e.g. Baldocchi and Wilson, 2001). Many SVAT models combine physical with biophysical processes to model photosynthesis, transpiration and decay of vegetation, and they are used in ecosystem dynamic models, hydrological applications and Land Surface Schemes of climatic models.

Big-leaf models rely more on lumped parameters than the others. They are chosen specially for long-term or large-scale applications, when computation time becomes an issue. Multi-layer models allow a deeper analysis of the energy exchange processes, but they require a larger number of a priori model parameters. The complexity of the models is usually related to an increase in the number of parameters. For example, the ISBA model (Interactions Soil Biosphere Atmosphere, Noilhan and Mahfouf, 1996), used in operative contexts by the French Meteorological Service, has about ten parameters and can be easily applied in different spatial scales, while the SiSPAT scheme (*Simple Soil-Plant-Atmosphere Transfers*, Braud et al., 1995), developed as a research instrument based on numerous physical and biophysical processes, needs about 60 parameters and variables to be initialized. Many of these parameters vary in space and time, and are often evaluated through in situ experiments. The definition of all the parameters greatly affects the behavior of the SVAT scheme, showing that it needs to be calibrated or regularly corrected. Due to the errors in the structure of the model and to the uncertainties of parameters, a single "optimal" set of calibrated parameters is often unavailable, thus multi-objective approaches have been developed (Yapo et al., 1998; Gupta et al., 1998; Demarty et al., 2004), aimed at determining sets of parameters that can not be distinguished just in terms of performances of the model.

Part of the parameters necessary to SVAT models can be estimated from remote sensing images, and assumed in the models by data assimilation techniques. Many methodologies and mathematical tools (e.g. Kalman filters) have been adopted to realize data assimilation procedures, but essentially they can be grouped in three categories (Olioso et al., 2005): (a) methods forcing the model input directly with the remote sensing measurements (for example, with these methods one can renounce to model a vegetation parameter, importing it directly from remote sensing images); (b) methods correcting the course of the variables each time remote sensing data are available (sequential assimilation); (c) method re-initializing or varying unknown parameters using data sets acquired during continuous periods of several days or weeks, and adopting either iterative or stochastic algorithms (variational assimilation).

Summarizing, Olioso et al. (2005) highlight various advantages in using SVAT models in combination with remote sensing data to derive *ET*: 1) they allow a continuous *ET* monitoring; 2) they do not need thermal infrared remote sensing data because, differently from residual methods, T_s is not necessary as an input, but can be estimated by solving the energy balance equation; 3) they allow to estimate the evolution of many auxiliary parameters, like vegetation indices or soil moisture.

Of course, some drawbacks have to be considered: 1) an high number of distributed parameters is needed, not all of them obtainable through data assimilation; 2) a continuous input of meteorological data is needed, because SVAT schemes have a very reduced time step (even some minutes); 3) information about soil hydraulic properties and plant physiology is needed, seldom available for wide areas; 4) if SVAT models are applied over large areas, they need a big computational expense (this disadvantage can be partially overcome using parallel computing; Mendicino et al., 2006).

Although the drawbacks listed are sometimes relevant, it should be emphasized that SVAT modeling according to assimilation procedures is still at its beginning, and the opportunity that these methods promise are very attractive.

4. APPLICATION OF SEBAL SCHEME IN SOUTHERN ITALY

SEBAL has been applied starting from images of the Moderate Resolution Imaging Spectroradiometer (MODIS) in the visible, near and thermal infrared bands, to determine distributed estimates of turbulent fluxes λE and H and available energy $R_n - G$, and finally the spatially distributed *ET* in southern Italy.

The performances of SEBAL have been locally verified for different days from summer 2004 to summer 2006 with data recorded by eddy covariance stations located in three areas characterized by different physiographic and vegetative conditions. Specifically, an eddy covariance station was sited in a mountain forest, in the territory of the town of Longobucco-CS (for further details, see Marino et al., 2005), while another station was placed in the years 2004–2005 in a plain characterized by a sparse vegetation near the town of Sibari-CS (Mendicino and Senatore, 2005), and in the year 2006 it was moved to an alfalfa field in Paglialonga-Bisignano-CS (Figure 2).

The eddy covariance stations provide continuous measurements of the main components of the energy balance averaged every 30 minutes, besides measuring other micrometeorological variables (such as wind speed and direction, air humidity, surface, air and soil temperature, soil water content, and also the content of CO_2 in the atmosphere). The recorded data of R_n, G, λE, H and cumulated daily *ET* have been used to analyze the performance of SEBAL.

Figure 2. Location of eddy covariance stations in southern Italy (Calabria). Three-dimensional elaboration from Landsat 7 images, view from north

A first comparison was carried out using both the measurements recorded by the Sibari and Longobucco stations during the period August 18–20, 2004. The differences in measured and estimated λE and H values are shown in Figures 3(a) and 3(b). Even though the analyzed period is very short, some general aspects can be pointed out. Specifically, it can be observed a more reliable estimate in the agricultural area (Sibari) than that obtained in the mountain area (Longobucco). The difference is mainly due to the correct definition of the aerodynamic resistance of equation (5), which for tall forests becomes very difficult to determine. In this case sudden changes in vegetation height and density strongly affect the satellite

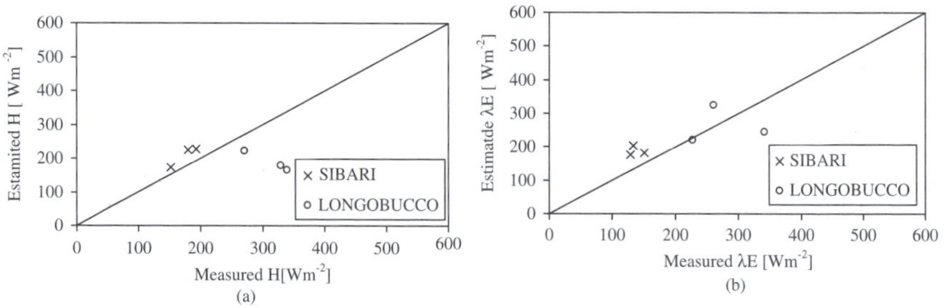

Figure 3. Differences in measured and estimated values of (a) λE and (b) H for the Sibari and Longobucco stations during the period August 18–20, 2004

estimate, based on a representative mean value. Therefore, in addition to the detailed knowledge of the actual density and phenologic characteristics, also a suitable spatial resolution of satellite images is required to capture the heterogeneity of vegetation. With the aim of increasing the spatial information content of a satellite image, a downscaling procedure can be applied through a multi-sensors approach. This procedure has been tested only for the area of the Sibari station, because just for it were available, for the same period, images recorded by diverse sensors typified by different spatial resolutions.

For the period June 22-July 08, 2005 the 1 km spatial resolution MODIS images were integrated with a 90 m spatial resolution image of the Advanced Spaceborne Thermal Emission and Reflection radiometer (ASTER). A disaggregation procedure has been applied to every pixel of the MODIS images, subdividing the single 1 km^2 pixel in a 90 m regular sub-grid composed by 121 cells, whose values for the input variables have been obtained through the following equations:

$$T_{s,MODIS-90m}(i,j) = T_{s,MODIS-1km} \cdot \frac{T_{s,ASTER}(i,j)}{\bar{T}_{s,ASTER}} \tag{22}$$

$$\alpha_{MODIS-90m}(i,j) = \alpha_{MODIS-1km} \cdot \frac{\alpha_{ASTER}(i,j)}{\bar{\alpha}_{ASTER}} \tag{23}$$

$$NDVI_{MODIS-90m}(i,j) = NDVI_{MODIS-1km} \cdot \frac{NDVI_{ASTER}(i,j)}{\overline{NDVI}_{ASTER}} \tag{24}$$

where the over-signed terms are the average values of the variables estimated starting from the ASTER sub-grid covering the same elementary area interested by the MODIS pixel.

For June 22, 2005, Figure 4 shows the comparison between the different daily *ET* spatial distributions obtained considering MODIS, ASTER and the downscaled

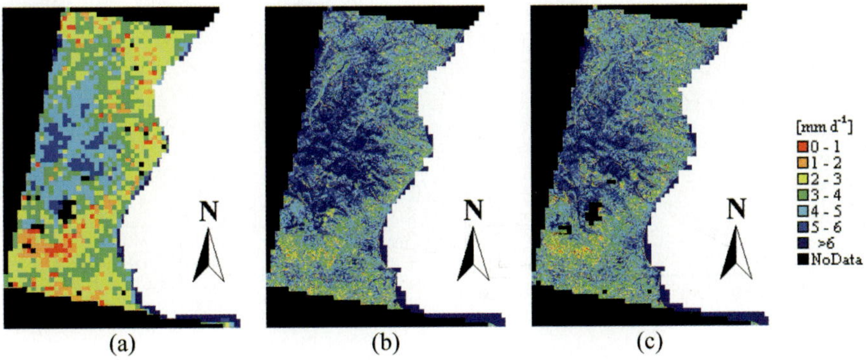

Figure 4. Comparison of daily *ET* spatial distributions for June 22, 2005, estimated using (a) MODIS, (b) ASTER and (c) MODIS downscaled images

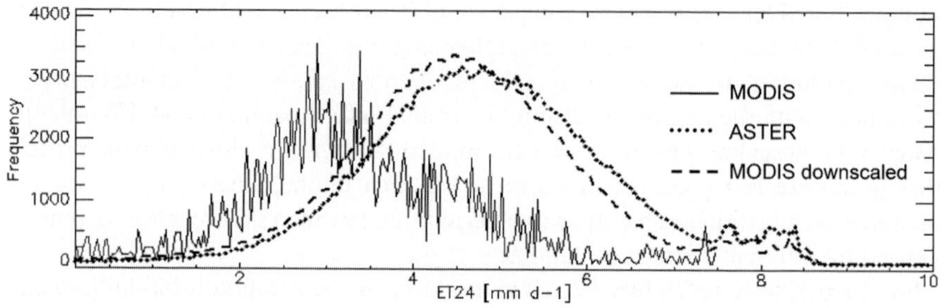

Figure 5. Comparison among *ET* frequency distributions obtained from MODIS, ASTER and downscaled MODIS images for the same area and period (in order to obtain a significant comparison, pixels of MODIS image were simply disaggregated to get the same number of pixels of the other two images, without applying the downscaling procedure)

MODIS images. The positive downscaling effects in the *ET* estimate are mainly evident in the mountain zones of the analyzed area (higher *ET* values); they can be also pointed out by considering the frequency distributions of the same images (Fig. 5), where for the downscaled MODIS image a more realistic *ET* overestimate is observed compared with the original MODIS image, also confirmed by on-site measurements.

In the case of the single energy balance components, the effects produced by the disaggregation procedure are shown from Figure 6(a) to Figure 6(d). These effects, for the valley site of the Sibari station are not very evident, even though the RMSE

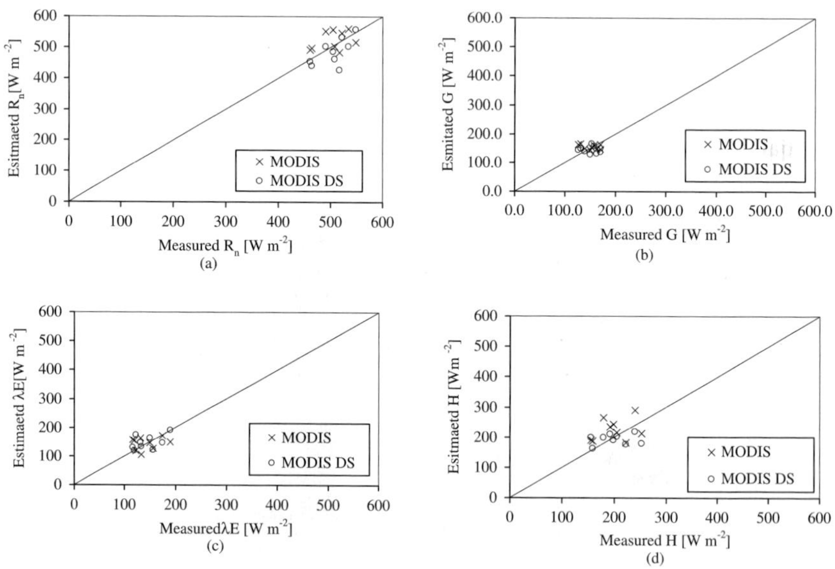

Figure 6. Differences in measured and estimated (a) R_n, (b) G, (c) λE and (d) H using MODIS and downscaled MODIS images during the period June 22-July 08 2005 (Sibari)

values obtained with the downscaling procedure are usually better than those derived by the original MODIS image (RMSE for the original MODIS: $R_n = 36.4\,\text{Wm}^{-2}$, $G = 20.4\,\text{Wm}^{-2}$, $\lambda E = 28.6\,\text{Wm}^{-2}$, $H = 45.9\,\text{Wm}^{-2}$; RMSE for the downscaled MODIS: $R_n = 37.5\,\text{Wm}^{-2}$, $G = 20.8\,\text{Wm}^{-2}$, $\lambda E = 24.1\,\text{Wm}^{-2}$, $H = 34.4\,\text{Wm}^{-2}$).

The disaggregation procedure herein described if, on one hand, allows a more realistic description of the *ET* phenomenon, on the other, can be only applied at the basin scale or on limited areas where multi-sensor images are available. Therefore, regional *ET* estimates have to be based on a greater scale equal to that represented by the 1 km spatial resolution MODIS images. A typical MODIS-based regional spatial distribution of the daily actual evapotranspiration is shown in Figure 7,

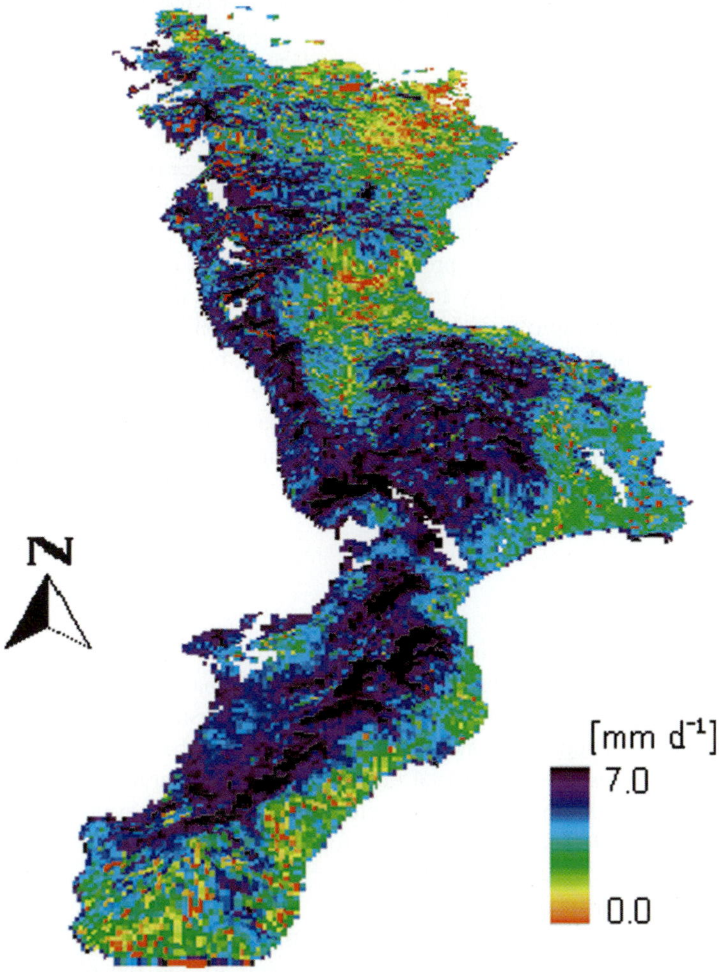

Figure 7. Spatial distribution of daily *ET* obtained using MODIS image of July 08, 2005. White areas are cloudy areas

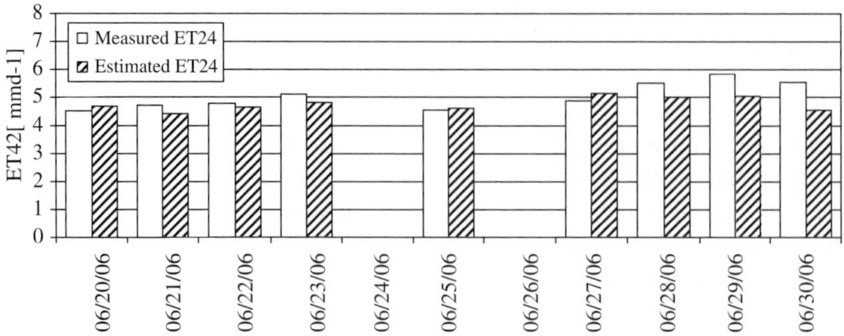

Figure 8. Comparison between measured and estimated ET values during the period June 20–30, 2006 for the Paglialonga station

where for the period June 22-July 08, 2005, the latter day was selected for the almost total lack of clouds (whose presence negatively affect the *ET* estimates).

Finally, it is interesting to show the more recent comparison carried out during the period June 20–30, 2006, on the third site (Paglialonga-Bisignano-CS), characterized not only by a different soil use (alfalfa field), but above all by a good availability of soil water. This aspect produced eddy measures characterized by higher λE values, and consequently by lower H values. Such a condition implies for the SEBAL procedure a more careful selection of the "wet" pixel, which can not be indistinctly chosen for the whole southern Italian territory, but has to be selected among those characterized by higher surface temperature values. This selection avoids an underestimate of λE, which in some cases can assume unrealistic values on wide areas of the analyzed region. In our case the correct choice of the "wet" pixel determined good results for the daily *ET* (Figure 8), providing for the analyzed period a RMSE equal to 0.38 mm.

5. CONCLUSION

After a brief theoretical discussion on the energy budget and an introduction to the main remote sensing methods for *ET* estimate, the performances given by the Surface Energy Balance Algorithm for Land (SEBAL) model using images of the Moderate Resolution Imaging Spectroradiometer (MODIS) have been analyzed by means of ground *ET* measurements carried out through eddy covariance systems at three different sites in southern Italy, characterized by different physiographic and vegetative conditions (sparse vegetation, crop canopy and high mountain vegetation).

Results obtained appear to be acceptable, even though some differences arose, specially due to approximations of the method and to the different scales between the remote sensing estimated values (1 km pixel resolution) and the observed values (depending on the footprint). To overcome this problem, and to allow *ET* estimates with higher spatial resolution, a downscaling procedure has been proposed using also images of the Advanced Spaceborne Thermal Emission and Reflection radiometer (ASTER).

Furthermore, analyses carried out on surfaces characterized by a good availability of soil water have pointed out that simplified methods like SEBAL work correctly when surface and atmospheric conditions are relatively constant over the satellite image, instead heterogeneous regions require higher levels of detail in the definition of the aerodynamic parameters and in the choice of the surface hydrological extremes ("wet" and "dry" pixels).

Nevertheless, in the cases analyzed in southern Italy the differences between estimated and recorded *ET* values are very reduced, confirming the validity of the procedure adopted for the *ET* estimate.

ACKNOWLEDGMENTS

The research presented in this chapter was supported by the EU Program INTERREG III B MEDOCC – <<Sécheresse et Désertification dans le bassin Méditerranée II>> (SEDEMED II). The authors would like to thank Ing. Raffaele Niccoli and Ing. Giuseppe Iiritano (CFS-MIDMAR) for the meteorological data at regional scale, Dr. Caterina Marino and Dr. Giorgio Matteucci (CNR-ISAFoM) for the data of the Longobucco eddy covariance station and Eng. Giacomo Martirano (EPSILON Italia srl) for ASTER images.

REFERENCES

Allen, R., Pereira, L.S., Raes, D. and Smith, M. (1998) Crop Evapotranspiration – Guidelines for Computing Crop Water Requirements. *FAO Irrigation and Drainage Paper* 56, FAO, Rome.

Allen, R., Tasumi, M., Morse, A. and Trezza, R. (2005) A Landsat-based energy balance and evapo-transpiration model in Western US water rights regulation and planning. *Irrigation and Drainage Systems* 19: 251–268.

Anderson, M., Norman, J., Diak, G., Kustas,W. and Mecikalski, J. (1997) A two-source time-integrated model for estimating surface fluxes from thermal infrared satellite observations. *Remote Sensing of Environment* 60: 195–216.

Baldocchi, D.D. and Meyers, T.D. (1988) A Spectral and Lag-correlation Analysis of Turbulence in a Deciduous Forest Canopy. *Boundary-Layer Meteorology* 45: 31–58.

Baldocchi, D.D. and Wilson, K.B. (2001) Modeling CO_2 and water vapor exchange of a temperate broadleaved forest across hourly to decadal time scales. *Ecological Modelling* 142: 155–184.

Baldocchi, D.D., Valentini, R., Running, S., Oechel, W., and Dahlman, R. (1996) Strategies for measuring and modelling carbon dioxide and water vapour fluxes over terrestrial ecosystem. *Global Change Biology* 2: 159–168.

Bastiaanssen, W.G.M. and Chandrapala, L. (2003) Water balance variability across Sri Lanka for assessing agricultural and environmental water use. *Agricultural Water Management* 58 (2): 171–192.

Bastiaanssen, W.G.M. (2000) SEBAL-based sensible and latent heat fluxes in the irrigated Gediz Basin, Turkey. *Journal of Hydrology* 229: 87–100.

Bastiaanssen, W.G.M., Menenti, M., Feddes, R.A. and Holtslag, A.A.M. (1998) A remote sensing surface energy balance algorithm for land (SEBAL). 1. Formulation. *Journal of Hydrology* 212–213: 198–212.

Bausch,W.C. (1995) Remote sensing of crop coefficients for improving the irrigation scheduling of corn. *Agricultural Water Management* 27: 55–68.

Braud, I., Dantas-Antonino, A.C., Vauclin, M., Thoony, J.L. and Ruelle, P. (1995) A simple soil-plant-atmosphere transfer model (SiSPAT) development and field verification. *Journal of Hydrology* 166: 213–250.

Brunetti, M., Buffoni, L., Mangianti, F., Maugeri, M. and Nanni, T. (2004) Temperature, precipitation and extreme events during the last century in Italy. *Global and Planetary Change* 40: 141–149.

Brutsaert, W. (1982) *Evaporation into the atmosphere: Theory, history, and applications*, D. Reidel Publishing Company, Dordrecht, Holland, 299 pp.

Carlson, T.N., Gillies, R.R. and Schmugge, T.J. (1995) An interpretation of methodologies for indirect measurement of soil water content. *Agricultural and Forest Meteorology* 77: 191–205.

Choudhury, B.J., Ahmed, N.U., Idso, S.B., Reginato, R.J. and Daughtry, C.S.T. (1994) Relations between evaporation coefficients and vegetation indices studies by model simulations. *Remote Sensing of Environment* 50: 1–17.

Consoli, S., D'Urso, G. and Toscano, A. (2006) Remote sensing to estimate ET-fluxes and the performance of an irrigation district in southern Italy. *Agricultural Water Management* 81: 295–314.

Courault, D., Clastre, P., Guinot, J.P. and Seguin, B. (1994) Analyse des sécheresses de 1988 à 1990 en France à partir de l'analyse combine de données satellitaires NOAA-AVHRR et d'un modèle agrométéorologique. *Agronomie* 14: 41–56.

Courault, D., Seguin, B. and Olioso, A. (2005) Review on estimation of evapotranspiration from remote sensing data: from empirical to numerical modeling approaches. *Irrigation and Drainage Systems* 19(3/4): 223–249.

Crago, R.D. (1996) Conservation and variability of the evaporative fraction during the daytime. *Journal of Hydrology* 180: 173–194.

Deardorff, J.W. (1978) Efficient prediction of ground surface temperature and moisture, with inclusion of a layer of vegetation. *Journal of Geophysical Research* 83(C4): 1889–1903.

Demarty, J., Ottlé, C., Braud, I., Olioso, A., Frangi, J.P., Bastidas, L.A. and Gupta, H.V. (2004) Using a multiobjective approach to retrieve information on surface properties used in a SVAT model. *Journal of Hydrology* 287: 214–236.

Dracup, J.A., Lee, K.S. and Paulson, E.G. (1980) On the definition of droughts. *Water Resources Research* 16(2): 297–302.

French, A.N., Jacob, F., Anderson, M.C., Kustas, W.P., Timmermans, W., Gieske, A., Su, Z., Su, H., McCabe, M.F., Li, F., Prueger, J. and Brunsell, N. (2005) Surface energy fluxes with the Advanced Spaceborne Thermal Emission and Reflection radiometer (ASTER) at the Iowa 2002 SMACEX site (USA). *Remote Sensing of Environment* 99(1–2): 55–65.

Garatuza-Payan, J., Shuttleworth, W.J., Encinas, D., McNeil, D.D., Stewart, J.B., deBruin, H. and Watts, C. (1998) Measurement and modelling evaporation for irrigated crops in Northwest Mexico. *Hydrological Processes* 12: 1397–1418.

Garratt, J.R. (1992) *The atmospheric boundary layer*, Cambridge University Press, Cambridge, UK.

Gupta, H.V., Sorooshian, S. and Yapo, P.O. (1998) Toward improved calibration of hydrologic models: multiple and non-commensurable measures of information. *Water Resources Research* 34(3): 751–763.

Heilman, J.L., Heilman, W.E. and Moore, D.G. (1982). Evaluating the crop coefficient using spectral reflectance. *Agronomy Journal* 74: 967–971.

Houghton, J.T., Ding, Y., Griggs, D.J., Noguer, M., Van der Linden, P.J. and Xiaosu, D. (2001) *Climate change 2001: the scientific basis, Contribution of Working Group 1 to the Third Assessment Report of the Intergovernmental Panel on Climate Change (IPCC)*, Cambridge University Press, Cambridge, 944 pp.

Jackson, R.D., Reginato, R.J. and Idso, S.B. (1977) Wheat canopy temperature: a practical tool for evaluating water requirements. *Water Resources Research* 13: 651–656.

Jacob, F., Olioso, A., Gu, X.F., Su and Seguin, B. (2002) Mapping surface fluxes using airborne visible, near infrared, thermal infrared remote sensing data and a spatialized surface energy balance model. *Agronomie* 22(5): 669–680.

Kanda, M., Inagaki, A., Letzel, M.O., Raasch, S. and Watanabe, T. (2004) LES study of the energy imbalance problem with eddy covariance fluxes. *Boundary-Layer Meteorology* 110: 381–404.

Keyantash, J. and Dracup, J.A. (2002) The Quantification of Drought: an Evaluation of Drought Indices. *Bulletin of the American Meteorological Society* 83(8), August: 1167–1180.

Kustas, W.P., Jackson, R.D. and Asrar, G. (1989) Estimating surface energy-balance components from remotely sensed data. In: G. Asrar (Ed.) *Theory and Applications of Optical Remote Sensing*, John Wiley, New York, 743 pp.

Kustas, W.P., Norman, J., Anderson, M.C. and French, A.N. (2003) Estimating subpixel surface temperatures and energy fluxes from the vegetation index radiometric temperature relationship. *Remote Sensing of Environment* 85: 429–440.

Kustas, W.P., Norman, J.M., Schmugge, T.J. and Anderson, M. (2004) Mapping surface energy fluxes with radiometric temperature. In: D.A. Quattrochi and J.C. Luvall (Eds.) *Thermal Remote Sensing in Land Surface Processes*, CRC Press, Boca Raton, Florida, pp. 205–253.

Kustas, W.P and Norman, J.M. (1996) Use of remote sensing for evapotranspiration monitoring over land surfaces. *Hydrological Sciences Journal* 41(3): 495–516.

Lagouarde, J.P. (1991) Use of NOAA-AVHRR data combined with an agrometeorological model for evaporation mapping. *International Journal of Remote Sensing* 12: 1853–1864.

Marino, C., Manca, G., Matteucci, G. and Scarascia Mugnozza, G. (2005) Cambiamenti climatici nel mediterraneo: un caso di studio sul ciclo del carbonio in una pineta della Sila, Calabria. *Forest@* 2(1): 52–65.

Mecikalski, J., Diak, G., Anderson, M. and Norman, J. (1999) Estimating fluxes on continental scales using remotely sensed data in an atmospheric land exchange model. *Journal of Applied Meteorology* 38(8): 1352–1369.

Melesse, A.M. and Nangia, V. (2005) Estimation of spatially distributed surface energy fluxes using remotely-sensed data for agricultural fields. *Hydrological processes* 19: 2653–2670.

Mendicino, G. and Senatore, A. (2005) Analysis of mass and energy fluxes in southern Italy. *ICID 21st Europ. Reg. Conf.* 15–19 May 2005, Frankfurt (Oder) and Slubice-Germany and Poland.

Mendicino, G., Senatore, A. and Versace, P. (2006) Un modello per la stima dei processi di trasferimento di energia e massa tra suolo, vegetazione ed atmosfera. *XXX Convegno di Idraulica e Costruzioni Idrauliche*, Rome, 10–15 September, 2006.

Menenti, M. and Choudhury, B.J. (1993) Parameterization of land surface evapotranspiration using a location dependent potential evapotranspiration and surface temperature range. In: H.J. Bolle et al. (Eds.), *Exchange Processes at the Land Surface for a Range of Space and Time Scales*, IAHS Publication 212, Wallingford, UK, 561–568.

Mohamed, Y.A., Bastiaanssen, W.G.M. and Savenije, H.H.G. (2004) Spatial variability of evaporation and moisture storage in the swamps of the upper Nile studied by remote sensing techniques. *Journal of Hydrology* 289 (1–4): 145–164.

Monteith, J.L. (1965) Evaporation and environment. In: G.E. Fogg (Ed.) *The State and Movement of Water in Living Organisms. Proceedings of the XIX Symposium of the Society of Experimental Biology.* Cambridge University Press, Cambridge, pp. 205–234.

Moran, M.S., Clarke, T.R., Inoue, Y. and Vidal, A. (1994) Estimating crop water management using the relation between surface-air temperature and spectral vegetation index. *Remote Sensing of Environment* 49: 246–263.

Neale, C., Jayanthi, H. and Wright, J.L. (2005) Irrigation water management using high resolution airborne remote sensing. *Irrigation and Drainage Systems* 19(3/4): 321–336.

Neale, C., Bausch, W.C. and Heerman, D.F. (1989) Development of reflectance-based crop coefficients for corn. *Transactions of ASAE* 32(5): 1891–1899.

Noilhan, J. and Mahfouf, J.-F. (1996) The ISBA land surface parameterisation scheme. *Global and Planetary Change* 13(1–4): 145–159.

Norman, J., Kustas, W. and Humes, K. (1995) A two-source approach for estimating soil and vegetation energy fluxes from observations of directional radiometric surface temperature. *Agricultural and Forest Meteorology* 77: 263–293.

Olioso, A., Inoue, Y., Ortega-Farias, S., Demarty, J., Wigneron, J.-P., Braud, I., Jacob, F., Lecharpentier, P., Ottlé, C., Calvet, J.-C. and Brisson, N. (2005) Future directions for advanced evapotranspiration modeling: assimilation of remote sensing data into crop simulation models and SVAT models. *Irrigation and Drainage Systems* 19: 377–412.

Patel, N.R., Rakhesh, D. and Mohammed, A.J. (2006) Mapping of regional evapotranspiration in wheat using Terra/MODIS satellite data. *Hydrological Science Journal* 51(2): 325–335.

Priestley, C. and Taylor, R. (1972) On the assessment of surface heat flux and evaporation using large-scale parameters. *Monthly Weather Review* 100: 81–92.

Raupach, M.R. and Finnigan, J.J. (1986) Single-layer models of evaporation from plant canopy are incorrect but useful, whereas multilayer models are correct but useless: discuss. *Australian Journal of Plant Physiology* 15: 705–716.

Roerink, G.J., Su, B. and Menenti, M. (2000) S-SEBI A simple remote sensing algorithm to estimate the surface energy balance. *Physics and Chemistry of the Earth (B)* 25(2): 147–157.

Scott, C.A., Bastiaanssen, W.G.M. and Ahmad, M. (2003) Mapping root zone soil moisture using remotely sensed optical imagery. *Journal of Irrigation and Drainage Engineering* 129(4): 326–335.

Seguin, B., Baelz, S., Monget, J.M. and Petit, V. (1982) Utilisation de la thermographie IR pour l'estimation de l'évaporation régionale. II: Résultats obtenus à partir de données satellites. *Agronomie* 2(2): 113–118.

Sellers, P.J., Randall, D.A., Collatz, G.J., Berry, J.A., Field, C.B., Dazlich, D.A., Zhang, C., Collelo, G.D. and Bounoua, L. (1996) A revised land surface parameterization (SiB2) for atmospheric GCMs. Part I. Model formulation. *Journal of Climate* 9: 676–705.

Shuttleworth, W.J. and Wallace, J.S. (1985) Evaporation from sparse crops – an energy combination theory. *Quarterly Journal of the Royal Meteorological Society* 111: 839–855.

Su, Z. (2002) The surface energy balance system (SEBS) for estimation of turbulent heat fluxes. *Hydrology and Earth System Sciences* 6(1): 85–99.

Su, Z., Yacob, A., Wen, J., Roerink, G., He, Y., Gao, B., Boogaard, H. and van Diepen, C. (2003) Assessing relative soil moisture with remote sensing data: theory, experimental validation, and application to drought monitoring over the North China Plain. *Physics and Chemistry of the Earth (B)* 28(1–3): 89–101.

Swinbank, W.C. (1951) Measurement of Vertical Transfer of Heat and Water Vapour by Eddies in the Lower Atmosphere. *Journal of Meteorology* 8: 135–145.

Swinbank, W.C. (1955) *Eddy Transports in the Lower Atmosphere*, Tech. Paper No. 2, Division of Meteorological Physics, Commonwealth Scientific and Industrial Research Organization, Melbourne, Australia, 30 pp.

Timmermans, W.J., van der Kwast, J., Gieske, A.S.M., Su, Z., Olioso, A., Jia, L. and Elbers, J. (2005) Intercomparison of energy flux models using ASTER imagery at the SPARC 2004 site (Barrax, Spain). *SPARC final Workshop*, Enschede, The Netherlands, 4–5 July, 2005.

Valentini, R. (2003), Ed. *Fluxes of Carbon, Water and Energy of European Forests*, Ecological Studies 163, Springer-Verlag, Berlin Heidelberg, 450 pp.

Wilhite, D.A. and Glantz, M.H. (1985) Understanding the drought phenomenon: the role of definitions. *Water International* 10(3): 111–120.

Wilson, K., Goldstein, A., Falge, E., Aubinet, M., Baldocchi, D., Berbigier, P., Bernhofer, C., Ceulemans, R., Dolman, H., Field, C., Grelle, A., Ibrom, A., Law, B.E., Kowalski, A., Meyers, T., Moncrieff, J., Monson, R., Oechel, W., Tenhunen, J., Valentini, R. and Verma, S. (2002) Energy balance closure at FLUXNET sites. *Agricultural and Forest Meteorology* 113: 223–243.

Xie, P., Janowiak, J.E., Arkin, P.A., Adler, R., Gruber, A., Ferraro, R., Huffman, G.J. and Curtis, S. (2003) GPCP pentad precipitation analyses: an experimental dataset based on gauge observations and satellite estimates. *Journal of Climate* 16: 2197–2214.

Yapo, P.O., Gupta, H.V. and Sorooshian, S. (1998) Multiobjective global optimization for hydrologic models. *Journal of Hydrology* 204: 83–97.

CHAPTER 8

TESTING A MODIFICATION OF THE PALMER DROUGHT SEVERITY INDEX FOR MEDITERRANEAN ENVIRONMENTS

L. S. PEREIRA, R. D. ROSA AND A. A. PAULO

Institute of Agronomy, Technical University of Lisbon, Portugal

Abstract: The Palmer drought severity index (PDSI) has been developed for the Great Plains of USA, considering the environmental conditions and the crops in the area. After testing the applicability of the PDSI for Southern Portugal, a modification of the water balance was attempted assuming dryland olives as the drought-reference crop and using the FAO methodology to compute the crop evapotranspiration. All basic procedures of the Palmer method were kept, mainly those relative to the anomaly index and the PDSI. Results show that the modified PDSI behaves coherently with the original PDSI but its behaviour responds better to the observed drought conditions, characterized with the SPI. Particularly, its response to different soil water holding capacities is more coherent than the one of the original PDSI

Keywords: Drought Indices, PDSI, Drought-reference crop, Water balance, Olive crop

1. INTRODUCTION

The Palmer drought severity index (PDSI) has been developed for the Great Plains of USA considering the environmental conditions and the crops in the area, mainly assuming the winter wheat as the drought reference crop. It is computed from a generalized sequential soil water balance where crop evapotranspiration (ET) is estimated from the Thornthwaite equation (Palmer, 1965). The PDSI is computed in many regions of the world as originally proposed or just changing the ET equation, thus not questioning the base assumptions of the water balance calculation. However, a few authors questioned the validity of the method for different environments, including for the Mediterranean region (Cancelliere *et al.*, 1996). The need to improve the PDSI methodology in order to adapt it to the Mediterranean conditions has been identified in previous studies (Paulo *et al.*, 2003) despite the fact that the results of PDSI are coherent with those of SPI (Paulo and Pereira, 2006). This led to a first approach testing an alternative soil water balance applied to the olive crop (Pereira *et al.*, 2005).

The olive crop has been selected because it is a typical perennial crop of the Mediterranean environment and its demand for water is well studied

G. Rossi et al. (eds.), Methods and Tools for Drought Analysis and Management, 149–167.

(Moreno *et al.*, 1996; Orgaz and Fereres, 1997; Villalobos *et al.*, 2000; Palomo *et al.*, 2002; Nuberg and Yunusa, 2003), also under rainfed and deficit irrigation conditions (Fernández *et al.*, 2003; Moriana *et al.*, 2003). The FAO crop evapotranspiration methodology (Allen *et al.*, 1998) was adopted because it is nowadays recognized as the most appropriate when crop specific models are not available (Burt *et al.*, 2005). The water balance method was changed because, after Palmer developed his methodology, progresses in modelling provided more powerful simulation tools, such as the ISAREG model (Teixeira and Pereira, 1992; Pereira *et al.*, 2003). In addition, the results for some water balance terms obtained with the Palmer method are worse if compared with those obtained in Mediterranean for several crops (e.g. studies reported in Rossi *et al.*, 2003).

In this paper data relative to the meteorological station of Évora (38.57° N, altitude 309 m), southern Portugal, are used for the calibration and testing of the modified PDSI. The region may be assumed as representative of the Mediterranean conditions and is well studied for what concerns droughts (Paulo *et al.*, 2003 and 2005; Paulo and Pereira, 2006). In this paper, also the ET methodology has been improved applying the dual K_c methodology (Allen *et al.*, 1998; 2005) which allows estimating the contribution of evaporation from the soil surface in addition to crop transpiration (Pereira , 2004). This approach leads to a more reliable soil water balance, which is of great importance since its results are the basis for the PDSI calculation (Alley, 1984), thus influencing the final index values. A comparative analysis was conducted in terms of evapotranspiration, soil water content, soil water depletion, soil water recharge and runoff, between the values obtained with the ISAREG applied to the olive crop, and the corresponding values obtained with the standard Palmer procedure. The values of the moisture anomaly index (Z_i) computed with the modified PDSI and with the standard Palmer method (Palmer, 1965; Alley, 1984) were compared, as well as the indices resulting from the modified and original PDSI methods. The impact of the total available soil water (TAW) on both original and modified PDSI values was also tested. In addition, the modified and original PDSI were compared and with the Standardized Precipitation Index (SPI) computed for the same locations relative to 9- and 12-month time scales.

2. SOIL WATER BALANCE COMPUTATION: COMPARING THE MODIFIED AND ORIGINAL PDSI CALCULATION PROCEDURES

2.1 The Water Balance Model

As aforementioned, the water balance of the olive crop has been computed with the ISAREG model, which is a water balance simulation model aimed at the study of crop water and irrigation requirements (Teixeira and Pereira, 1992; Pereira *et al.*, 2003). Whereas the model is utilized mainly for irrigation planning, in this study it was only used to execute the soil water balance under rainfed conditions. To simulate the soil water balance, the model uses meteorological, crop and soil data that are stored in several files.

The generic equation of the soil water balance used by the ISAREG, not considering both irrigation and the upward flux of water from a water table, is:

$$\Delta R = (P - ET_a - E_s - Dr)\ \Delta t \tag{1}$$

where ΔR is the variation of the soil water reserve (mm) P is precipitation (mm); E_s is runoff due to non-infiltrated precipitation (mm), ET_a is actual crop evapotranspiration (mm), Dr is deep percolation (mm), and Δt is the time step of the computation (day), varying from the day to the month. Considering the effective precipitation $Pe = P - E_s$, Eq. 1 simplifies into:

$$\Delta R = (Pe - ET_a - Dr)\ \Delta t \tag{2}$$

All the terms in Eq. 2, except the effective precipitation, behave differently depending upon which storage "zone" the soil water is (Pereira *et al.*, 2003):

1. *Deep percolation zone:* the excess water zone above field capacity, corresponding to gravitational water, which is not immediately available for the plants.
2. *Non-stress zone:* between the field capacity, θ_{FC}, and the lower limit where the water can still be easily extracted by the plant without any stress, θ_p.
3. *Water stress zone:* between θ_p and the wilting point, θ_{WP}, where water is still available to the plants but requires increased energy for extraction by the roots.

Deep percolation is considered when the soil water content on day j is above the field capacity, and precipitation on day $j+1$ exceeds the maximum evapotranspiration on this same day. Thus

$$\Delta R = (Pe - ET_m - Dr)\ \Delta t \tag{3}$$

with $ET_a = ET_m$. The maximum crop evapotranspiration is $ET_m = K_c\ ET_o$, where K_c is the crop coefficient and ET_o is the reference evapotranspiration, currently computed with the Penman-Monteith equation (Allen *et al.*, 1998). When soil water is in the optimal yield zone, deep percolation is not considered and $ET_a = ET_m$, therefore

$$\Delta R = (Pe - ET_m)\ \Delta t \tag{4}$$

When water stress occurs, i.e. when the soil water content is below θ_p, $ET_a < ET_m$, decreasing progressively while the soil is loosing water, thus

$$ET_{a\,(i+1)} = \frac{ET_{m\,(i+1)}}{R_{\min}} \cdot R_{(i)} \tag{5}$$

and

$$\Delta R = \left[Pe_{(i+1)} - \frac{ET_{m\,(i+1)}}{R_{\min}} \cdot R_{(i)} \right] \cdot \Delta t \tag{6}$$

where $ET_{a(i+1)}$ is actual evapotranspiration on day $i+1$, R_{min} is water storage corresponding to the lower limit of the optimal yield zone, and R(i) is water storage at the end of day i.

2.2 Crop Coefficients For the Olive Crop

The crop coefficients (K_c) adopted for estimating the evapotranspiration of the olives crop used in the first studies concerning the modification of the PDSI (Pereira *et al.*, 2005) were those obtained by Orgaz and Fereres (1997) for a mature and dense olive orchard (up to 60% ground cover) in Cordoba, south of Spain, where the climate is similar to that of Alentejo region. However, results from field research have shown that for computing ET of olive crops it is important to consider separately both components of ET, transpiration and soil water evaporation, because the latter may contribute for about 30% of total crop ET due to the important fraction of soil not covered by the trees and so exposed to radiation (Bonachela *et al.*, 1999; 2001). Hence, the dual crop coefficient approach (Allen *et al.*, 1998; 2005) has been adopted to estimate ET in the soil water balance of the olive crop.

The dual crop coefficient approach consists in adopting a basal crop coefficient K_{cb} to represent crop transpiration and in computing a soil evaporation coefficient K_e to represent soil evaporation. Then the (potential) maximum crop evapotranspiration is given by

$$ET_m = (K_{cb} + K_e)\ ET_o \tag{7}$$

The K_e values are computed with a daily water balance relative to the 0.15 m soil evaporation layer, which also requires that a daily time step is adopted for ET_0. Monthly K_{cb} values were estimated from literature (Orgaz and Fereres, 1997; Testi *et al.*, 2006). Computations refer to the period 1965 to 2000.

The monthly averages of the K_e daily values were computed for every month and then added to the K_{cb} values of the corresponding month to produce a single month value of K_c that was used by the ISAREG to compute the monthly crop evapotranspiration. This change from daily to monthly values is justified by the fact that water balance calculations for PDSI are performed on a monthly time step.

Table 1 presents the monthly average of K_{cb}, K_e and K_c relative to the period 1965–2000. It can be seen that during the rainy months, when soil evaporation is non-negligible, higher values for K_e produce larger K_c values. Final K_c values follow a trend similar to those presented by Testi *et al.* (2006).

The methodology above described requires daily precipitation and reference evapotranspiration data to compute the evaporation coefficient (K_e). However, because such daily data are not available in many locations or for longer data sets, this required a procedure to estimate K_c using monthly data only. The procedure consists in defining a monthly precipitation threshold above which the soil wetting is assumed to be enough to produce an evaporation rate corresponding to the average K_e of that month (Table 1). When the precipitation of any month is below that upper boundary soil evaporation, hence K_e values are reduced. Thus arbitrary classes of precipitation for the previous and current month were created by dividing the difference between that threshold value and the null precipitation and assigning to each class a single value of K_c.

To test this procedure, the values for olive ET computed with monthly data and crop coefficients corrected as indicated above were compared with those calculated

Table 1. Monthly average values for the basal crop coefficient (K_{cb}), the soil evaporation coefficient (K_e) and the time averaged crop coefficient (K_c) (averages refer to 1965–2000)

Month	K_{cb}	K_e	K_c
Jan	0.50	0.31	0.81
Feb	0.50	0.30	0.80
Mar	0.65	0.19	0.84
Apr	0.60	0.19	0.79
May	0.55	0.14	0.69
Jun	0.55	0.07	0.62
Jul	0.50	0.01	0.51
Aug	0.50	0.01	0.51
Sep	0.55	0.07	0.62
Oct	0.60	0.21	0.81
Nov	0.65	0.27	0.92
Dec	0.50	0.31	0.81

with daily data, i.e. where K_e daily values were used. Results (Fig. 1) show that monthly estimates of olive ET using the corrected values for K_e are close to those computed for the same months using daily weather data: the coefficients of the regression forced to the origin (b) are close to 1.0 and the determination coefficients (R^2) are above 0.96.

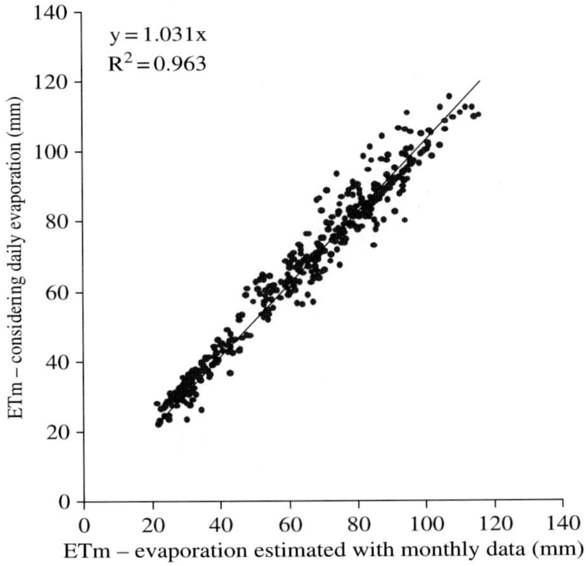

Figure 1. Comparison of the monthly values of olives evapotranspiration computed with daily and monthly data when the K_e coefficient is corrected according to soil wetting by rain (Évora)

2.3 Comparing the Water Balance Terms Computed with the Original and the Modified PDSI Methods

Differences in methodologies used for computing the soil water balance, following that one proposed by Palmer (Palmer, 1965; Alley, 1984) and that described above for the olive crop, necessarily produce different results for the water balance terms despite the same monthly weather data were used and soil water characteristics were the same (TAW = 150 mm). It is then worthwhile to analyze such differences in results. The case of Évora is used as an example.

The actual evapotranspiration (ET_a) computed with the ISAREG model for the olive crop (ISAREG-Olive) is generally smaller than that computed with the Palmer method for the periods with high soil water availability and slightly higher than the latter for the periods when the soil is dry (Fig. 2). The FAO-PM reference evapotranspiration (Allen *et al.*, 1998) was used with the Palmer method replacing the original Thornthwaite equation. This reflects a better use of soil water by the olive crop, which is extremely well adapted to the Mediterranean climate, compared to the reference crop adopted by Palmer, that was supposed to represent the US High Plains conditions.

The soil water reserve at the end of each month (SWR) computed by both methods is presented in Fig. 3. Results show a higher SWR when computations refer to the olive crop, particularly for 1980–83, 1991–93, 1995, and 1998–2000 when major droughts were observed. This may be explained by the fact that potential ET in the original Palmer is replaced by the potential ET of the olive crop, and percolation and runoff are differently computed with the ISAREG and the Palmer method. Again, results reflect the good adaptation of olives to the local climate, differently from results of the Palmer water balance.

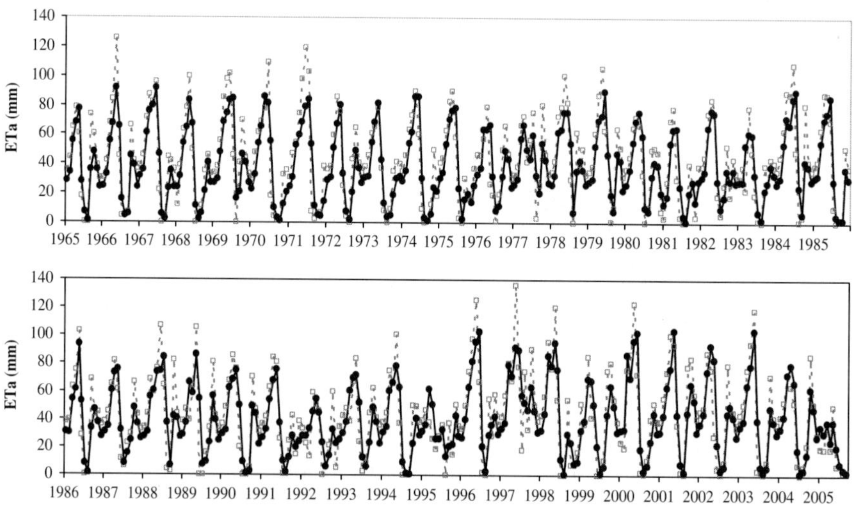

Figure 2. Time series of monthly actual evapotranspiration, ET_a, (mm) computed with the original Palmer (----) and with ISAREG for the olive crop (—•—), Évora

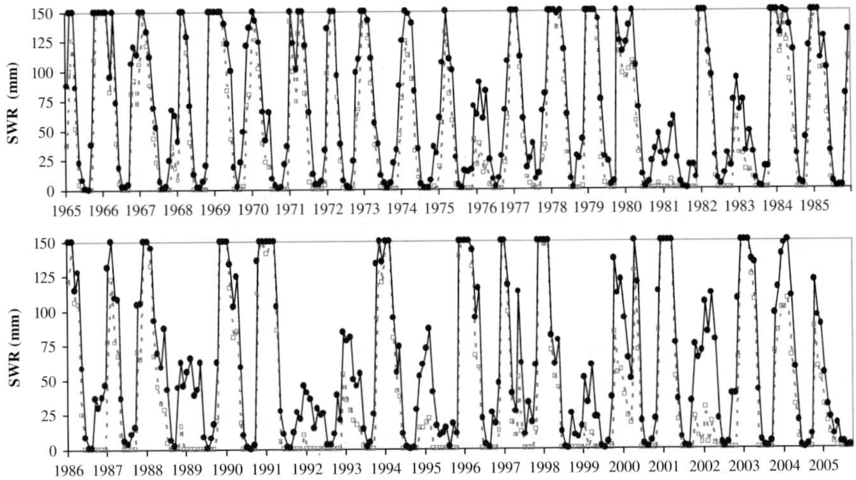

Figure 3. Time series of the soil water reserve at the end of each month (SWR) computed with the original Palmer (----) and with ISAREG for the olive crop (—•—), Évora

The soil water recharge computed by both methods shows evident differences between the two methods (Fig. 4), namely for the peaks in the recharge and for the number of months in which soil water recharge occurs. These differences are inherent to the soil water balance methods. A clear trend favouring one or another could not be noticed.

The actual soil water depletion (Fig. 5), which represents the water use by the vegetation that is not originated from the rainfall during the current month, shows that the peaks of soil water depletion occur first for the original Palmer method,

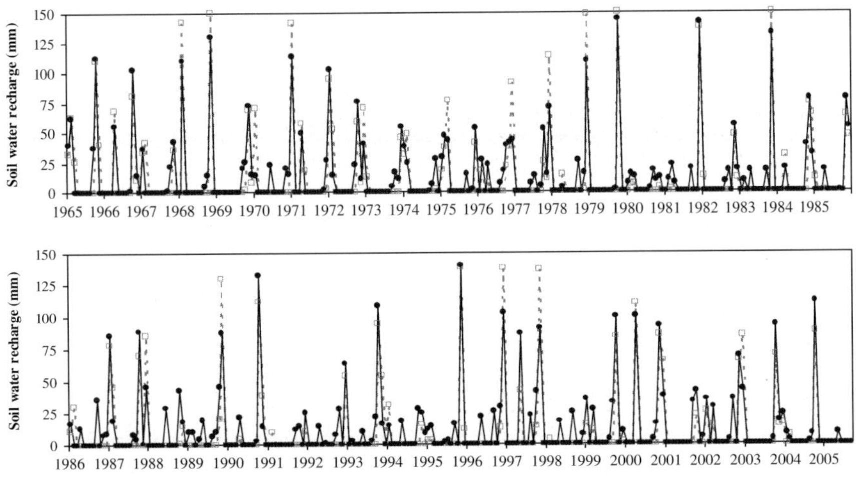

Figure 4. Time series of the monthly soil water recharge (mm) computed with the original Palmer (----) and the ISAREG model for the olive crop (—•—), Évora

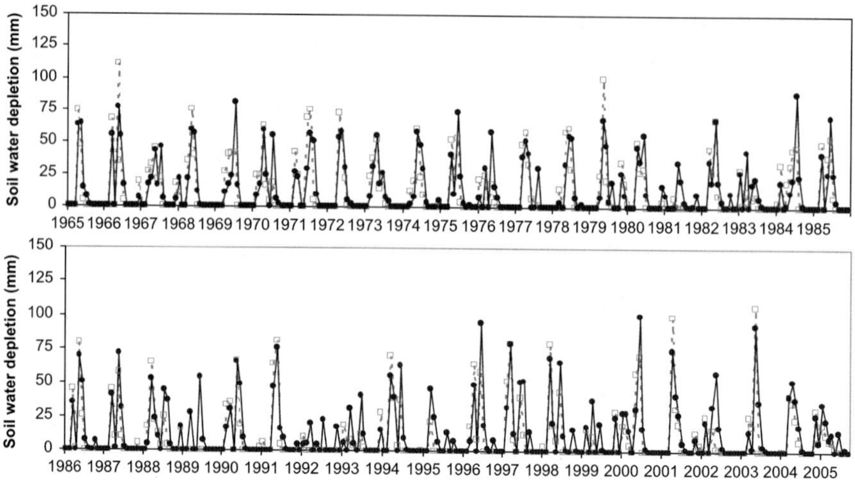

Figure 5. Time series of monthly soil water depletion (mm) computed with the original Palmer (----) and the ISAREG model for the olive crop (—•—), Évora

mainly during the spring season, and later for the ISAREG simulations relative to the olive crop. This is partly explained by the noticed differences in ET_a and in the SWR at the end of the previous month, as well as by the procedures of computation, which are quite simplified in the Palmer method and better elaborated in ISAREG.

Relative to the runoff values, Fig. 6 shows that peaks are generally coincident for the same periods but runoff is more often computed with ISAREG and peaks

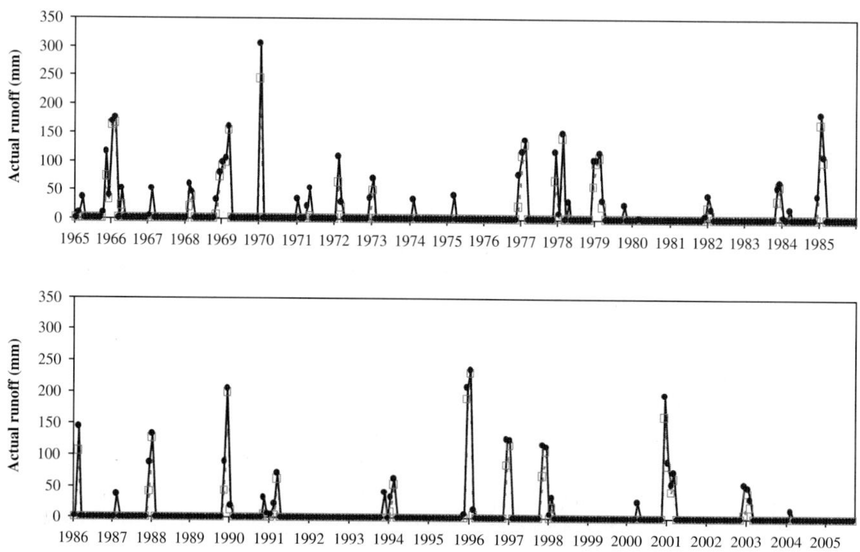

Figure 6. Time series of monthly runoff values (mm) computed with the original Palmer (----) and the ISAREG model for the olive crop (—•—), Évora

tend to be higher. Figure 6 shows both higher values and frequency of occurrence for the ISAREG model applied to the olive crop.

3. MOISTURE ANOMALY INDEX

Besides using an alternative soil water balance procedure applied to the olive crop as characteristic crop of the Mediterranean region, a modification was introduced to compute the potential runoff (PRO). PRO is used in the Palmer equation to compute the moisture departure (d_i), that is the deviation between actual precipitation, P_i, and the precipitation expected to occur for the average conditions of the climate, P'_i, which is computed with a monthly time step as:

$$d_i = P_i - P'_i = P_i - (\alpha_i \cdot ETP_i + \beta_i \cdot PR_i + \gamma_i \cdot PRO_i - \delta_i \cdot PL_i) \quad (8)$$

where ETP_i is the potential evapotranspiration (mm), PR_i is the potential soil moisture recharge (mm), PRO_i is potential runoff (mm) and PL_i is the potential soil moisture depletion (mm), all relative to the time interval i. The coefficients α_i, β_i, γ_i, and δ_i are the ratios between the average of each of the actual values for ET, soil moisture recharge, runoff, and soil moisture depletion to the average of the corresponding potential values (ETP, PR, PRO, and PL) over a calibration period. The subscript j refers to the month of the year.

The potential runoff, as defined by Palmer, corresponds to the difference between an arbitrary potential precipitation (PP) and potential recharge. Palmer adopted originally for PP the value of the total available soil water (TAW) whereas, as herecognized later, that choice wasn't particularly "elegant" since it could suggest a correlation between precipitation and TAW. Thus, for the modified PDSI, like proposed later by Palmer (Alley, 1984), the value adopted for PP is the triple of the average precipitation in each month during the period under analysis.

In Table 2 are displayed the monthly values of the climatic coefficients, α_i, β_i, δ_i and γ_i, defined above in Eq. (8), which have to be calibrated to compute the moisture departure (d_i). These coefficients were calibrated for both the modified and original PDSI, the latter using ET_o calculated by the FAO-PM method for the water balance computations. It was observed that α_i and β_i are generally higher when the soil water balance is computed with ISAREG for the olive crop, with considerable differences for the spring and summer months. Considering that the coefficient α_i is the ratio between the average of the actual value for ET and the average of the corresponding potential ET (estimated from ET_m) over the calibration period, an higher value for this coefficient means less variability of the actual values. This may indicate that the modified method may be closer to the reality than the original one. On the contrary, δ_i (relative to soil water depletion) is always higher for the original Palmer method except on July. The parameter γ_i is also always higher for the original Palmer method because the potential runoff estimated for the modified method, which is in the denominator of the ratio γ_i, is generally higher than that in the original Palmer as said above.

Table 2. Climatic coefficients for Alentejo computed with the original Palmer and the modified method with ISAREG simulations for the olive crop

	From ISAREG computations				From the original Palmer			
	α_i	β_i	δ_i	γ_i	α_i	β_i	δ_i	γ_i
Jan	0.983	0.377	0.092	0.208	0.927	0.273	0.144	0.386
Feb	0.987	0.289	0.156	0.198	0.916	0.249	0.250	0.275
Mar	0.982	0.188	0.304	0.137	0.870	0.118	0.353	0.196
Apr	0.955	0.158	0.348	0.037	0.829	0.084	0.482	0.014
May	0.913	0.081	0.526	0.039	0.661	0.017	0.639	0.025
Jun	0.666	0.019	0.744	0	0.341	0	0.785	0
Jul	0.300	0.005	0.899	0	0.077	0	0.805	0
Aug	0.115	0.009	0.873	0	0.030	0	0.921	0
Sep	0.233	0.082	0.209	0	0.224	0.004	1.000	0
Oct	0.563	0.26	0.169	0.012	0.577	0.147	0.994	0.061
Nov	0.808	0.359	0.096	0.089	0.811	0.289	0.130	0.255
Dec	0.935	0.449	0.062	0.181	0.939	0.374	0.131	0.413

Whereas sensible differences on the water balance terms between the two methods have been founded, with consequent differences in the climatic coefficients, the moisture anomaly index (Z_i) obtained indicates a similar general behaviour. It is computed as

$$Z_i = k_j d_i \tag{9}$$

where k_j is the climatic characteristic of month j (no-dimensions) and d_i is the moisture departure (mm) defined in Eq. (8). Results for Évora in Figure 7 show that there is a trend to estimate higher anomaly values with the modified method, which overestimation, compared to original Palmer.

4. CALIBRATION OF THE PDSI TO THE REGIONAL CLIMATIC CONDITIONS

Both the original and the modified PDSI were calibrated for the climatic conditions of southern Portugal as proposed by Heddinghaus and Sabol (1991), by determining the relationship between the cumulative values of the moisture anomaly index (Z_i) concerning the most severe drought events and their respective durations (Fig. 8). The 7 most severe droughts that occurred during the period from 1942 to 2005 sites in four sites Alentejo, Évora, Beja, Elvas, and Alvalade (the latter relative to 1942 until 2000) were selected. Then a regression line was adjusted to the values corresponding to PDSI < -4.

The resulting regression equation is then transformed in the equation that computes the final values of the PDSI:

$$X_i = p\, X_{i-1} + q\, Z_{(i)} \tag{10}$$

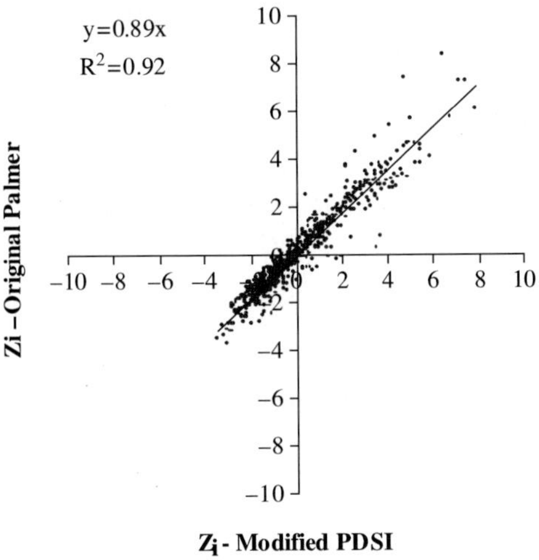

Figure 7. Linear regression to compare the moisture anomaly index when computed with the original Palmer and the modified method using ISAREG for the olive crop

where p is the coefficient for the previous value of the index in order to maintain the drought severity and q is the coefficient for the moisture anomaly index.

If m and b are respectively the slope and the ordinate at the origin in the respective regression line, the following equations are used to obtain p and q:

$$p = 1 - \frac{m}{m+b} \tag{11}$$

$$q = \frac{C}{m+b} \tag{12}$$

where C = −4.00.

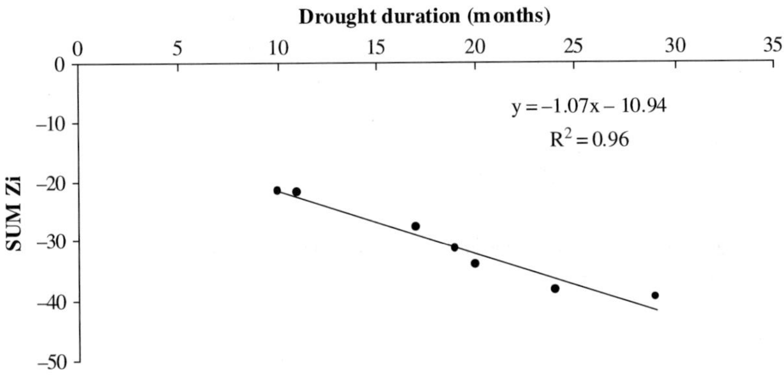

Figure 8. Relationship between the cumulative values of the moisture anomaly index, Zi, for the most severe drought events and their respective duration

The equation obtained for the modified PDSI for the Portuguese conditions is:

$$X_i = 0.911\, X_{i-1} + 0.33\, Z_i \tag{13}$$

and that obtained by Palmer for the USA climatic conditions is:

$$Xi = 0.897 X_{i-1} + Zi/3 \tag{14}$$

Because Eq. 13 and 14 are very close, the equation obtained by Palmer was used for the drought analyses presented in this study.

5. COMPARING THE MODIFIED AND THE ORIGINAL PDSI

The Fig. 9 shows for the Évora site a similar behaviour of the PDSI series obtained with the original and modified Palmer methods. Both clearly identify the same major droughts but they show relatively large differences on the number of months per class of severity (Table 3) with a clear trend for the modified PDSI to have

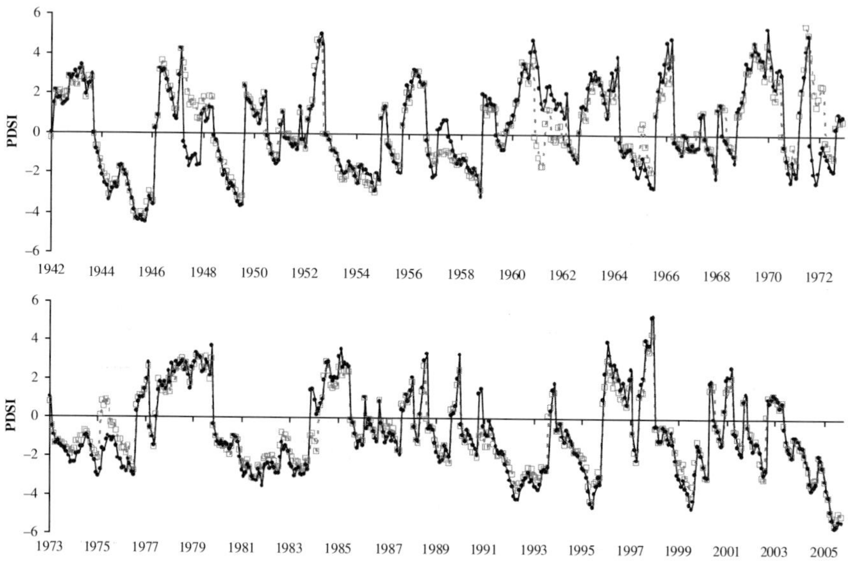

Figure 9. PDSI time series obtained by the modified (----) and the original Palmer (—•—) methods for Évora

Table 3. Number of months in each drought severity class when the PDSI is computed with the modified and the original Palmer methods, with alternative ET equations

Site	Drought severity	Modified PDSI	Original PDSI with ET_0 FAO-PM	Original PDSI with ET_0 Thornthwaite
Évora	Mild	172	174	179
	Moderate	103	91	75
	Severe	51	40	41
	Extreme	20	17	13

more negative values. This is coherent with the trend identified for the anomaly index as analyzed above.

Results in Fig. 9 indicate that the behaviours of the modified and original PDSI are coherent for all the study period. Results for other sites in Alentejo are similar to those for Évora.

Comparing results through a regression forced to the origin (Fig. 10), it can be confirmed that both methods behave similarly. However, as for the moisture anomaly, the modified method tends to overestimate the severity index by about 15%, more evidently for the more severe droughts. The determination coefficients are high (> 0.85) and the dispersion around the regression line is larger for the positive values, i.e. for non-drought conditions.

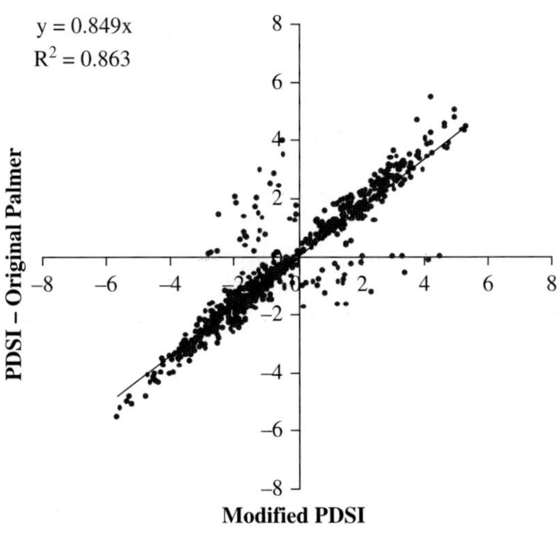

Figure 10. Linear regression between the PDSI values obtained through the modified and the original Palmer methods, Évora

6. SENSITIVITY OF THE PDSI TO DIFFERENT VALUES OF TOTAL AVAILABLE SOIL WATER

In a former study (Paulo *et al.*, 2003) it was found that the frequency of moderate, severe and extreme droughts increased when the total available soil water (TAW) increased. This has no explanation since soils with larger available soil water are not less able to respond to drought than soils with lower TAW. Thus, PDSI computations using the modified and the original procedure were performed for several TAW values: 100, 150, 200 and 250 mm (Fig. 11).

Results of the modified method are better than the ones obtained with the original Palmer, particularly when the FAO-PM equation is used. Results for other sites in the region are similar. Results from the modified method (Fig. 11 c) show a slight increase of drought frequency with TAW, particularly for the severe and extremely severe droughts. On the contrary, those frequencies sharply increase for the more severe drought classes when higher values for TAW are adopted to compute the PDSI with the original Palmer method, either with ET_o computed by the FAO-PM equation (Fig. 11 a) or with the Thornthwaite equation (Fig. 11 b).

7. COMPARING THE MODIFIED PDSI WITH THE SPI

In former studies, the PDSI (with ET_o computed with the FAO-PM equation) has been compared with the Standardized Precipitation Index, SPI (Paulo *et al.*, 2003; Paulo and Pereira, 2006). It was found that a reasonable relationship exists between the PDSI and the SPI 12-month, both indices behave similarly, so that they produce coherent results. However, the SPI values have shown a trend to classify droughts as

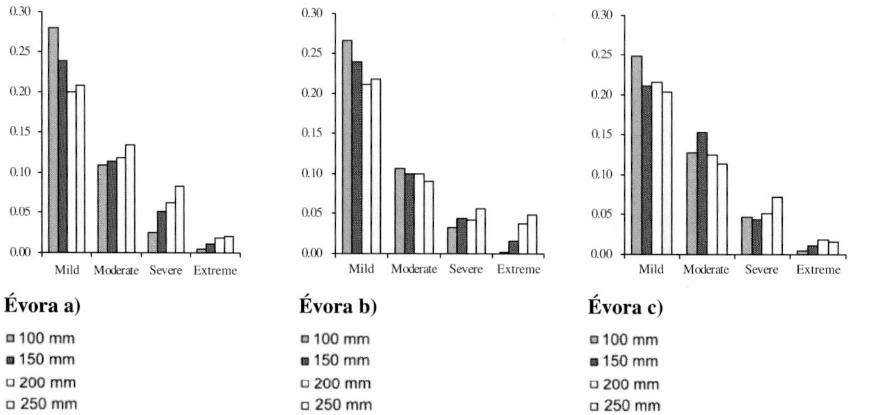

Figure 11. Frequency of drought months in selected drought severity classes for several TAW values: a) computed by the Palmer method with ET_o given by the FAO-PM equation; b) computed by the original Palmer method with ET_o given by the Thornthwaite's equation; c) computed by the modified PDSI method

more severe than with the Palmer method (Paulo and Pereira, 2006). Despite PDSI and SPI have different theoretical basis, it is important that both indices behave coherently because it is important that drought studies are performed by more than one index (Guttman, 1998). In this study it is also important to verify whether the modified PDSI is acceptably comparable with SPI and may add information to results obtained with this one. Therefore, results of the modified PDSI were compared with the SPI for 9- and 12-month time scales. Results are presented in Figs. 12 and 13 and in Table 4.

Results in Fig. 12 show that a reasonable correspondence exists between the modified PDSI and the SPI for a 9- and 12-month time scale. Since the SPI severity scale is about half of the PDSI severity scale, the values for SPI were doubled for an easier comparison with the PDSI. It is common to have a considerable difference between the two indices, with some months classified as drought for one index and non-drought for the other, which is evidenced through the dispersion around the regression line. Results in Table 4 indicate that this dispersion is slightly smaller with the modified PDSI than with the original one.

There is a slight trend for the modified PDSI to classify droughts as more severe then the SPI, both for the 9 and 12-month time scale, which was observed also for other stations in the region.. Moreover, the modified PDSI tends to identify the initiation of drought events slightly before the SPI (Fig. 13). This behaviour is improved with respect to the original PDSI. This aspect is clearly exemplified for the location of Évora, where the modified PDSI shows an evident trend to be more negative than the SPI-12 month for the major droughts recorded.

When the comparison is performed with the SPI for the 9-month time scale, it can be seen for the location of Évora, that the modified PDSI tends to classify the droughts that occurred in the second half of the analyzed period more severely; on the contrary, the SPI–9 month classifies as more severe the ones occurred in the first half (Figure 12). Results for other studied sites are similar.

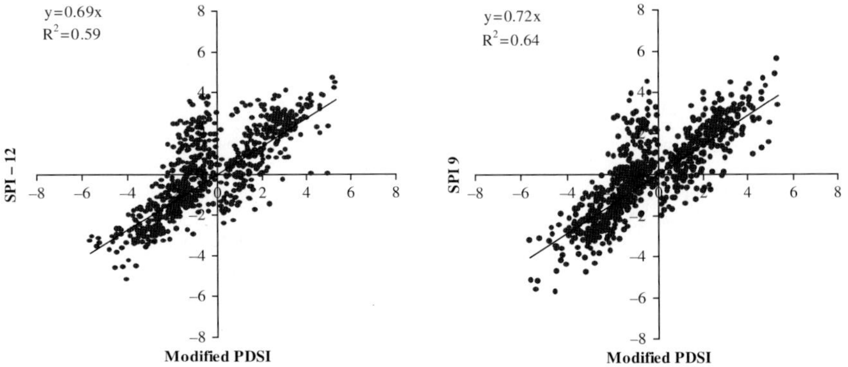

Figure 12. Linear regression between the modified PDSI (ISAREG applied to the olive crop) and the SPI 9- and 12-month time scale for Évora (SPI values are doubled)

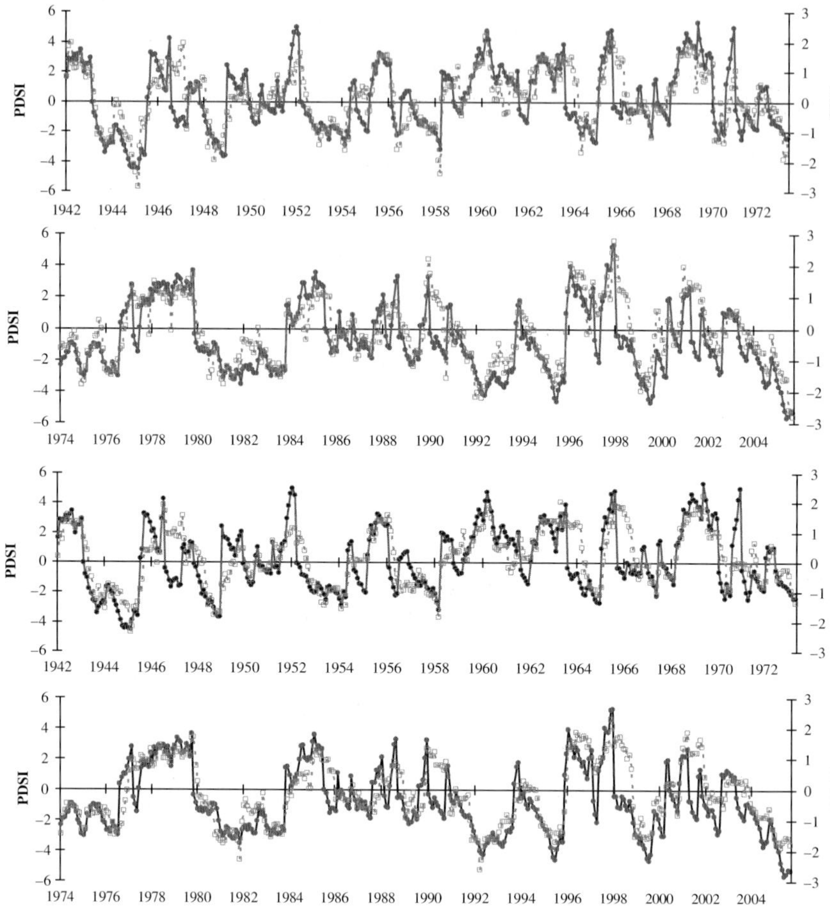

Figure 13. Comparison between the modified PDSI values (—•—) and the SPI (----) computed with 9- and 12-month time scale, Évora

Table 4. Results of the linear regression through the origin between the modified PDSI values and the SPI computed with 9- and 12- month time scales

Method	SPI 9-month		SPI 12-month	
	Regression coef	Determination coef	Regression coef	Determination coef
Modified PDSI	0.72	0.64	0.69	0.59
Original PDSI	0.77	0.61	0.73	0.56

8. CONCLUSIONS

This paper shows that using the modified PDSI to identify and characterize droughts may be appropriate for Mediterranean environments. The modified PDSI uses an improved soil water balance compared with the original method, it refers to a perennial Mediterranean and well studied crop, the olive crop, and it uses the basic features of drought severity classification as proposed by Palmer.

Results show that the water balance terms when considering the olive crop as drought reference crop instead of the original cereal crop in the Central Plains are closer to reality. Moreover, it is possible to parameterize better the crop evapotranspiration and through the water balance procedures to estimate better the respective actual values. The procedure for adopting a crop coefficient that takes into account soil evaporation and its decrease when soil water recharge is smaller than an average threshold, constitutes an important improvement for the computations. In fact, the calibrated monthly parameters for the terms of the water balance – evapotranspiration, soil water depletion, soil water recharge and runoff – show a smaller variation compared with those computed with the original Palmer method.

The modified PDSI method produces more coherent results than the original Palmer method when we compare the frequency of severe or extremely severe droughts as influenced by the total available soil water, since the original method tends to produce more frequent severe droughts when the total available soil water increases, which is not reasonable if compared with reality. Therefore, this more coherent behaviour of the modified PDSI could be considered a non-negligible improvement of the Palmer method.

The changes in the water balance procedures and drought reference crop make the modified PDSI method produce higher moisture anomaly values than the original method and, consequently, higher PDSI values. The resulting frequency of drought months in the moderate, severe and extreme classes of drought severity become closer to those identified with the SPI 12-month. This allows an easy and coherent joint use of both indices.

When comparing the modified PDSI for a soil with TAW of 150 mm with the SPI 12-month it shows to better follow this one than the original PDSI because the frequency of months with moderate, severe and extreme droughts increase relative to the original PDSI, so approaching the corresponding frequency of SPI identified droughts. Comparing the modified PDSI with several time scales of SPI, best results concern the 9 and 12 months time scale. Further improvements may be expected when testing the proposed modified method in locations with different climates where the crop and soil parameterization could be carefully done when performing the base water balance.

ACKNOWLEDGEMENTS

This research was funded by the INTERREG project SEDEMED I and II. The support of the Agricultural Engineering Research Center is also acknowledged.

REFERENCES

Alley, W.M. (1984) The Palmer Drought Severity Index: limitations and assumptions. *J. Climate and Applied Meteorology*, 23: 1100–1109.

Allen, R.G., Pereira, L.S., Raes, D., Smith, M. (1998) *Crop Evapotranspiration. Guidelines for Computing Crop Water Requirements*. FAO Irrig. Drain. Pap. 56, FAO, Rome, 300p.

Allen, R.G., Pereira, L.S., Smith, M., Raes, D., Wright, J.L. (2005) FAO-56 Dual Crop Coefficient Method for Estimating Evaporation from Soil and Application Extensions. *J. Irrig. Drain. Engng.* 131(1): 2–13.

Bonachela, S., Orgaz, F., Villalobos, F.J., Fereres, E. (1999) Measurement and simulation of evaporation from soil in olive orchards. *Irrig. Sci.* 18: 205–211.

Bonachela, S., Orgaz, F., Villalobos, F.J., Fereres, E. (2001) Soil evaporation from drip-irrigated olive orchards. *Irrig. Sci.* 20: 65–71.

Burt, C.M., Mutziger, A.J., Allen, R.G., Howell, T.A. (2005) Evaporation research: review and interpretation. *J. Irrig. Drain. Engn.*, 131(1): 37–58.

Cancelliere, A., Rossi, G., Ancarani, A. (1996) Use of Palmer index as a drought indicator in Mediterranean regions. In: *IAHR Congress From Floods to Drought*, Cape Town, South Africa.

Fernández, J.E., Palomo, M.J., Díaz-Espejo, A., Girón, I.F. (2003) Influence of partial soil wetting on water relation parameters of the olive tree. *Agronomie* 23 545–552.

Guttman, N.B. (1998) Comparing the Palmer drought index and the standardised precipitation index. *J. Am. Water Res. Assoc.* 34(1): 113–121.

Heddinghaus, T.R., Sabol, R.M. (1991) A review on the Palmer Drought Severity Index and where do we go from here? In: *7th Conference on Applied Climatology*, Am Meteorology Soc., pp. 242–246.

Moriana, A., Orgaz, F., Pastor, M., Fereres, E. (2003) Yield responses of mature olive orchard to water deficits. *J. Am. Soc. Horticultural Science* 123(3): 425–431.

Moreno, F., Fernandez, E., Clothier, B.E. and Green, S.R. (1996) Transpiration and root water uptake by olive trees. *Plant Soil* 184: 85–96.

Nuberg, I., Yunusa, I. (2003) *Olive water use and yield – monitoring the relationship*. RIRDC Publication No 03/048, Barton, Austrália.

Orgaz, F., Fereres, E. (1997) Riego. In: *El Cultivo del Olivo*. Mundi Prensa, Madrid, pp 251–272.

Palmer, W. (1965) *Meteorological Drought*. U.S. Weather Bureau, Res. Paper N^0 45, Washington.

Palomo, M.J., Moreno, F., Fernández, J.E., Díaz-Espejo, A., Girón, I.F. (2002) Determining water consumption in olive orchards using the water balance approach. *Agric. Water Manage.* 55(1): 15–35.

Paulo, A.A., Pereira, L.S. (2006) Drought concepts and characterization. Comparing drought indices applied at local and regional scales.*Water International* 31(1): 37–49.

Paulo, A.A., Pereira L.S., Ferreira, E. (2003) O índice de Palmer e o índice normalizado de precipitação na identificação de períodos secos. In: JPL Ferreira *et al.* (Eds.) *6° Simpósio de Hidráulica e Recursos Hídricos dos Países de Língua Oficial Portuguesa* (APRH/ABRH/AMCT/INGRH, Praia, Cabo Verde), APRH, Lisboa, pp. 293–307.

Paulo, A.A., Ferreira, E., Coelho, C., Pereira, L.S. (2005) Drought class transition analysis through Markov and Loglinear models, an approach to early warning. *Agric. Water Manage.* 77: 59–81.

Pereira, L.S. (2004) *Necessidades de Água e Métodos de Rega*. Publ. Europa-América, Lisboa, 313 p.

Pereira, L.S., Teodoro, P.R., Rodrigues, P.N., Teixeira, J.L. (2003) Irrigation scheduling simulation: the model ISAREG. In: Rossi, G., Cancelliere, A., Pereira, L.S., Oweis, T., Shatanawi, M., Zairi, A. (Eds.) *Tools for drought mitigation in Mediterranean Regions*. Kluwer, Dordrecht, pp. 161–180.

Pereira, L.S., Paulo, A.A., Rosa, R.D. (2005) A modification of the Palmer drought stress index for Mediterranean environments. In: R. Dannowski (Ed) *Integrated Land and Water Management: towards Sustainable Rural Development* (Proc. ICID 21st European Reg. Conf., Frankfurt (Oder), Germany, and Slubice, Poland), ICID German Nat. Com., Münchberg, CD-ROM.

Rossi, G., Cancelliere, A., Pereira, L.S., Oweis, T., Shatanawi, M., Zairi, A., Editors (2003) *Tools for Drought Mitigation in Mediterranean Regions*. Kluwer, Dordrecht, 257 p.

Teixeira, J.L., Pereira, L.S. (1992) ISAREG, an irrigation scheduling model. *ICID Bulletin* 41(2): 29–48.

Testi, L., Villalobos, F.J., Orgaz, F., Fereres, E. (2006) Water requirements of olive orchards: I simulation of daily evapotranspiration for scenario analysis. *Irrig. Sci.* 24: 69–76.

Villalobos, F.J., Orgaz, F., Testi, L., Fereres, E. (2000) Measurement and modelling of evapotranspiration of olive (*Olea europaea* L.) orchards. *Eur. J. Agron.* 13: 155–163.

CHAPTER 9

REGIONAL DROUGHT IDENTIFICATION AND ASSESSMENT. CASE STUDY IN CRETE

G. TSAKIRIS, D. TIGKAS, H. VANGELIS AND D. PANGALOU

National Technical University of Athens, Laboratory of Reclamation Works & Water Resources Management, Athens, Greece

Abstract: Two basic characteristics of meteorological drought were analysed in this study. Namely the severity and the areal extent of drought. Severity was represented by two general drought indices the Standardised Precipitation Index and the Reconnaissance Drought Index. A spatial distribution model, based on the reciprocal of the square distance, was employed to model the spatial extent of drought. Annual drought maps based on discretized platforms were produced for the selected case study in Eastern Crete. Eastern Crete is a drought prone area of Greece consisting of the prefectures of Heraklio and Lassithi. It was shown that an effective way to estimate the areal extent of each level of drought is to use the cumulative "or more" curves, in which the level of drought is plotted versus the area affected by this or by a higher level of drought

Keywords: regional drought, drought assessment, drought indices, SPI, RDI

1. INTRODUCTION

Drought is usually considered as a phenomenon that occurs when a significant reduction in precipitation with respect to the normal value is observed for a given period of time and a significant areal extent. It is difficult to identify when a drought starts and when it terminates. In fact, we realise that a drought event is occurring long after it has started. Therefore, some authors characterise drought as a creeping phenomenon (Wilhite, 2000).

In more general terms, drought is caused by lack of precipitation. However, the hardship caused by drought could be due to other factors such as overexploitation of existing resources or unexpected increase in consumption. Thus, a distinction can be made between meteorological drought and socio-economic drought. Other types of drought are the hydrological and the agricultural drought.

This paper is devoted to the identification and assessment of meteorological drought in Eastern Crete. Since drought is a regional phenomenon, special attention was paid to the spatial extension of its occurrence (Rossi *et al.*, 1992).

The identification of drought was realised by using drought indices. Two indices were used in this study. Namely the SPI, which is based on precipitation, and the

G. Rossi et al. (eds.), Methods and Tools for Drought Analysis and Management, 169–191.

RDI, which is based on both precipitation and potential evapotranspiration. For the quick and efficient calculation of the drought indices, a newly developed software package, DrinC, was used.

Regarding the assessment of the area, two ways of presentation were adopted based on the detailed evaluation of the drought indices on all the squares of the grid system produced by the discretized GIS platform. Illustrative pseudo-3D maps of the area were produced for each year of the historical record. Also, cumulative "or more" curves were produced for the direct assessment of the drought extent under the various levels of drought.

The analysis showed that Eastern Crete is vulnerable to droughts and susceptible to meteorological or climatic changes.

2. DROUGHT INDICES

Two indices were used for the detection of drought in the study area: the Standardised Precipitation Index (SPI) and the Reconnaissance Drought Index (RDI). The fundamental properties of these indices are described in this chapter. These indices were estimated by DrinC. This is a software package developed within the framework of the project SEDEMED II. The main features of DrinC are presented at the end of this chapter.

2.1 Standardized Precipitation Index (SPI)

The Standardized Precipitation Index (SPI) was developed for the definition and monitoring of drought (McKee et al., 1993).

The SPI estimation for any location is based on the long-term precipitation record during a period of time. This long-term record is fitted to a probability distribution, which is then transformed into a normal distribution so that the mean SPI for the location and the desired period of time is zero (Edwards and McKee, 1997). Positive SPI values indicate precipitation greater than the median precipitation, and negative values indicate precipitation less than the median precipitation. The SPI is normalized; therefore wet and dry climates can be represented in the same way. Wet periods can also be monitored using the SPI.

Thom (1958) found that the gamma distribution fits well precipitation time series. The gamma distribution is defined by its probability density function:

$$g(x) = \frac{1}{\beta^{\alpha}\Gamma(\alpha)} x^{\alpha-1} e^{-x/\beta}, \quad \text{for } x > 0 \qquad (1)$$

where α and β are the shape and scale parameters respectively, x is the precipitation amount and $\Gamma(\alpha)$ is the gamma function. Computation of the SPI involves fitting a gamma probability density function to a given frequency distribution of precipitation totals for a station. The alpha and beta parameters of the gamma probability density function are estimated for each station, for each time scale of interest (1, 3, 6, 9, 12 months, etc.), and for each month of the year. Maximum likelihood solutions are used to optimally estimate α and β:

$$\alpha = \frac{1}{4A}\left(1+\sqrt{1+\frac{4A}{3}}\right), \ \beta = \frac{\bar{x}}{\alpha}, \text{ where } A = \ln(\bar{x}) - \frac{\sum \ln(x)}{n} \tag{2}$$

and n = number of observations

The resulting parameters are then used to find the cumulative probability of an observed precipitation event for the given month and time scale for the station in question. Since the gamma function is undefined for $x = 0$ and a precipitation distribution may contain zeros, the cumulative probability becomes:

$$H(x) = q + (1-q)\,G(x) \tag{3}$$

where q is the probability of a zero and $G(x)$ the cumulative probability of the incomplete gamma function. If m is the number of zeros in a precipitation time series, then q can be estimated by m/n. The cumulative probability, $H(x)$, is then transformed to the standard normal random variable z with mean zero and variance of one, which is the value of the SPI. Once standardized the strength of the anomaly is classified as it is shown in Table 1. Table 1 also contains the corresponding probabilities of occurrence for each level of severity arising naturally from the normal probability density function. Thus, at a given location for an individual month, moderate droughts (SPI ≤ -1) have a probability of occurrence 15.9%, whereas extreme droughts (SPI ≤ -2) have a probability of 2.3%. The SPI will have extreme values, by definition, with the same frequency at all locations.

The SPI can be used to track drought events on multiple time-scales. The U.S. National Drought Mitigation Center (NDMC) computes the SPI with five running time intervals, i.e. 1-, 3-, 6-, 9-, and 12-months, but the index is flexible with respect to the chosen time interval. This powerful feature can provide an overwhelming amount of information, unless the exact desired intervals are known by the researchers.

Table 1. Drought classification by SPI value and corresponding event probabilities

SPI value	Category	Probability (%)
2.00 or more	Extremely wet	2.3
1.50 to 1.99	Severely wet	4.4
1.00 to 1.49	Moderately wet	9.2
0 to 0.99	Mildly wet	34.1
0 to −0.99	Mild drought	34.1
−1.00 to −1.49	Moderate drought	9.2
−1.50 to −1.99	Severe drought	4.4
−2 or less	Extreme drought	2.3

The method of estimation includes the following steps:

- Data preparation. Selection of a time series of the precipitation value of interest. At least 30 years of data are required.
- Estimation of the parameters of the probability distribution that statistically fits the available precipitation series.
- Computation of the non-exceedence probabilities of aggregated precipitation values.
- Calculation of the standard normal quantiles corresponding to the non-exceedence probabilities previously computed. Such normal quantiles correspond to SPI values.

2.2 Reconnaissance Drought Index (RDI)

The Reconnaissance Drought Index (RDI) was recently introduced. A detailed presentation of the RDI can be found in Tsakiris & Vangelis (2005) and Tsakiris *et al.* (2006). Two mathematical forms for the annual RDI are available:

The Normalised RDI:

$$RDI_n^{(i)} = \frac{a_0^{(i)}}{\bar{a}_0} - 1 \tag{4}$$

and the Standardised RDI:

$$RDI_{st}^{(i)} = \frac{y_i - \bar{y}}{\hat{\sigma}} \tag{5}$$

where $a_0^{(i)}$ is the initial value of RDI for the ith year [i = 1(1)N] , $\bar{a}_0$ is the mean of a_0 for N years, y_i is the $\ln(a_0^{(i)})$, $\bar{y}$ is the arithmetic mean and $\hat{\sigma}$ is the standard deviation of y_i.

The initial value for calculation of the RDI $a_0^{(i)}$ for the ith year [i = 1(1)N] is:

$$\alpha_0^{(i)} = \frac{\sum_{j=1}^{12} P_{ij}}{\sum_{j=1}^{12} PET_{ij}} \tag{6}$$

where P_{ij} and PET_{ij} are the precipitation and potential evapotranspiration of the jth month of the ith year, starting from October as it is customary for the Mediterranean countries and N is the number of years of the data.

The above formulation is based on the assumption that a_0 values follow a lognormal distribution. The Standardised RDI behaves as SPI and the interpretation of the results is similar. Therefore, Table 1 of SPI can be used for the standardised RDI.

It can be noted that RDI is based both on the precipitation and on the potential evapotranspiration. The mean of a_0 (i.e. $\bar{a}_0$) corresponds to the normal climatic conditions of an area and is equal to the Aridity Index proposed by FAO for assessing the aridity of an area.

RDI can be estimated for any period of time from one month to one year. For instance, if the effect of drought on agricultural production is under study, the RDI could be calculated for the period covering the growing season of the major crops grown in the area. This allows an effective linkage of the RDI with the expected rainfed crop production and therefore with the anticipated losses in the agricultural sector due to the occurrence of drought.

A more general expression of the initial value for the estimation of RDI, for each month of the year can be written as:

$$\alpha_0^{jk} = \frac{\sum_{j=1}^{k} P_{ij}}{\sum_{j=1}^{k} PET_{ij}} \tag{7}$$

where k is the number of the month starting preferably from October, in case of application in the Mediterranean Area, or any other month if it is appropriate.

The main advantages of RDI (mainly in comparison with SPI) are the following:

- It has a physical meaning given that the aggregated deficit between precipitation and the evaporative demand of the atmosphere is estimated.
- It can be estimated for any period of time (e.g. 1 month, 2 months).
- The estimated value is comprehensible.
- It can be compared with the FAO aridity index. Therefore, it can be directly linked to the climatic conditions of the region.
- It can be used for "climate instability" conditions, to examine the effect of various changes of climatic factors on drought and desertification.

Given the above, it is concluded that RDI is an advantageous tool for the reconnaissance assessment of drought severity for general use. Also, it provides comparable results within a large geographic area (e.g. Mediterranean).

Usually droughts are accompanied with high temperatures, resulting in higher evapotranspiration rates. Therefore, RDI is expected to be a more sensitive index than other indices that relate only to precipitation.

Previous studies have shown that the use of precipitation (and therefore SPI) as the variable corresponding to agricultural production is inadequate. However, the inclusion of Potential Evapotranspiration in the estimation of RDI (which could be considered as closely representative of atmospheric demand) confirms that RDI is a suitable indicator for the risk assessment of drought in the agricultural sector.

Similarly, Potential Evapotranspiration may also reflect the consumption of various activities apart from the agricultural sector. Thus, RDI is expected to be a suitable indicator for the risk assessment of water supply, energy production and recreational activities during a drought event. The application of RDI to sectors other than agriculture may require a modification to include the spatial characteristics of the sector. An example is the sector of tourism. In the areas where the water supply for tourism has priority over other uses, the Potential Evapotranspiration, used for the calculation of a_0, can be replaced by standardised values reflecting the water demand of tourism.

2.3 SOFTWARE DrinC

The estimation of the drought indices can be complex, considering the challenges involved in the assessment of the spatial distribution for each indicator. DrinC (Drought Indices Calculator) software was developed to facilitate the estimation of the drought indices (Deciles, SPI and RDI). Its development was carried out as part of the SEDEMED II project.

DrinC is a stand-alone PC software, operates on Windows platforms and has been programmed in Visual Basic 6. Input data for the estimation of Deciles and SPI is the annual or monthly precipitation, while for the estimation of RDI, potential evapotranspiration (PET) data are also required. Optionally, temperature data can be used to calculate PET, by implementing the Thornthwaite's method. The main window of the software interface appears in Figure 1.

The software is able to handle many data series. In order to improve the interface of the software the input and output files are in MS Excel worksheet format. For the estimation of the indices on annual basis, input data may be either annual or monthly, while for calculations on seasonal basis (monthly, 3-month, 6-month), monthly data is required.

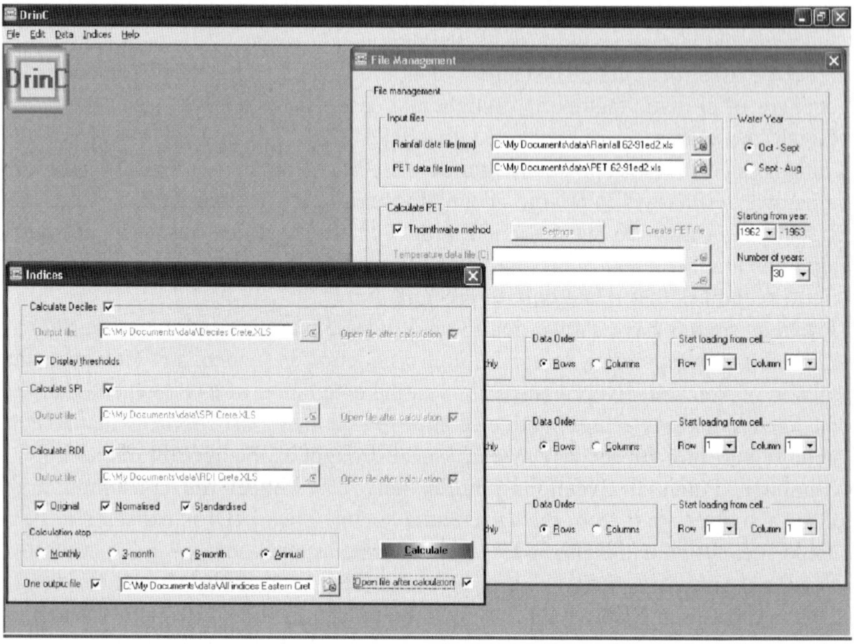

Figure 1. The main window of DrinC software

3. SPATIAL DISTRIBUTION MODEL

The spatial distribution model is based on the Geographical Information Systems (GIS) concept and comprises of two parts, the graphical environment and the database. Both parts are developed during the implementation of the model. A pre-developed package would not be applicable due to the nature of the involved tasks.

The graphical environment is divided in two parts. The first part comprises of linear and point elements that define the area under study or describe the main hydrological, geomorphological and/or other characteristics (hydrological basins, river and stream networks etc.). Such data are usually available in GIS or CAD formats. In this work, these elements had to be digitised from raster maps obtained from the Hellenic Military Geographical Service. The elements that were digitized include the coastal line, contour lines, the hydrographical network and the positions of the meteorological and hydrometric stations. Additionally, elements like the position of boreholes, springs and reservoirs, the fragile ecological sites and the protected areas were digitised.

An important issue in the digitisation procedure is the accurate geographical reference of the digitised elements. It is important to use a reference system that can be transformed in global reference systems. This allows the comparison of the produced results to those produced by other groups. In this case study the Hellenic Geographical Reference System EGSA of 1987 was used. It is the new reference system used in Greece for most mapping tasks and it is fully adaptable to global reference systems. Another important issue is the interval between the contour lines. The interval should be small enough to ensure high density of the contour lines and consequently high data resolution, and big enough to ensure time efficiency in the digitisation process. The data for the definition of the watersheds is obtained from the digitised elevation contour lines.

The second part of the graphical environment comprises of the elements that make the final cell grid. This grid is used as a base for the presentation of drought thematic maps and maps of drought prone areas. A graphical topology is created to link the database with the graphical environment. Topology can not be created for the whole grid but for each cell of the grid separately. Therefore, the cells are designed as squares. This doesn't allow the creation of the grid using horizontal and vertical lines. Instead, it is created using small squares attached to one another in a mesh. Overlapping square sides are deleted to correct the mesh and also because the topology can not be created otherwise. Each cell of the mesh was referenced by its unique code attached to its centre. Furthermore, the coordinates and height of each cell centre are derived from the graphical environment according to the Reference System that is used.

Although this procedure may appear to be complicated, the implementation model provides tools for the assessment of the procedure. In the end of this procedure, each grid cell has a unique reference code that links it to the database that is created at a next step. The coordinates and elevation of each cell must be known. The developer of the system decides on the size of the squares, which determines the spatial scale of the thematic map. The smaller are the squares, the denser is the grid and the better is the spatial representation. Each cell represents a record in the database. Consequently, the denser is the grid, the higher is the number of records (lines). This increase in the number of records is exponential relative to the increase of the grid density. Thus, the accuracy assessment of the database records may be difficult if the database is too large. In this case study, the optimal cell size, to achieve high grid density and obtain a sufficiently small database, was considered to be 2km × 2km.

The next step of the procedure is the creation of the database, which is the second main component of the spatial distribution model. The database can be considered as a large matrix, where the rows correspond to the records of each cell of the grid and the columns to the various elements that are represented in the thematic maps. The first column of the matrix has a special entry, which is important for the structure of the system. This entry is a "key", unique to each record and thus to each grid cell. It is the identification code of each grid cell in the matrix. The link between the database and the graphical environment can be achieved only through that code. In order to make this link possible, a point in the graphical environment is connected directly to the corresponding database "key". This point can be the centre of a grid cell. Besides, the system requires a link between the centres of the squares of the grid and the database, and a graphical topology between the squares of the grid and their centres. This complex task is once again performed by the model. As a result, each cell of the grid represents a value from a column per record (line). Unfortunately, the "keys" have to be assigned to the centres of the cells manually, because the model can not detect the sequence of the squares in the graphical environment. An important error may occur in this phase of implementation. An incorrect square code assignment may result in a blank spot. This can be avoided in only one way; to reassign the code. Assuming that the designed system fulfils the set requirements, the data can be inserted in the matrix in many different ways. Two of the most useful ones are the numeric representation and the chromatic representation of selected ranges of values.

Clearly, the above described model is worthless without the appropriate data. For this study, these data are the values of the drought indices or only meteorological data represented spatially. Sub-databases can also be useful. For example, the horizontal coordinates and the elevation of the square centres or other constant variables that may be used in the procedure can be stored in a separate sub-database. The data in the system are stored in cells. Therefore, it is essential that values of data acquired from point sources (e.g. data recorded in gauge stations) are assigned in all the cells.

This assignment is performed by estimating a weighted mean value of the acquired data. The "weight" that will be used for this data distribution is an important issue. The "weight" that was finally used in this study is the square of the distance between the grid cell and each gauge station. This is based on the assumption that each station affects its surrounding area. Clearly, a meteorological station closer to

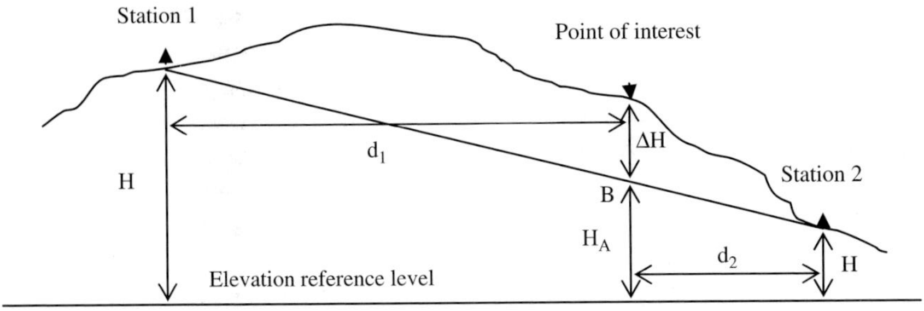

Figure 2. Data spatial distribution method

the point of interest will affect the meteorological conditions of that area more than a station further away. Nevertheless, a station further from the point of interest will also have an effect, to a degree, and this should be included in the estimation of the parameters to increase the accuracy of the results. The distance is a parameter that can be accurately measured (Tsakiris and Vangelis, 2004). Figure 2 is illustrative of the method, whereas detailed calculations are presented in the Appendix of this paper.

The meteorological data can be input in all the cells of the grid following the above procedure. After the spatial distribution of the meteorological data, the drought indices can be estimated by the developed software DrinC. The results of the estimation of the drought indices are stored in the database and can be accessed by the system to produce the thematic maps. Each colour on each thematic map represents a range of values (stored in the database) for the relevant property. This provides a comprehensive visual representation of the overall condition in the area (Figure 3).

Figure 3. Spatial distribution of precipitation

The thematic maps are produced for each year and in different time scales. The integration of these thematic maps allows not only the monitoring of the drought conditions but also the identification of drought prone regions in the study area. Thematic maps of drought prone regions of various severity levels can be created by implementing drought severity thresholds. Also, the use of data for fragile ecological sites and/or regions with extensive water consumption (e.g. irrigated regions) will further improve the assessment of the areas affected by drought.

4. CASE STUDY AREA: ISLAND OF CRETE

4.1 Overview of the Island

The island of Crete covers an area of 8336 km^2, the mean altitude is 460m and the total population is about 600,000 inhabitants. Crete is divided in four prefectures, namely from west to east: Chania, Rethymnon, Heraklio and Lassithi. The climate ranges between sub-humid Mediterranean and semi-arid. The average annual precipitation is estimated to be 750mm and the potential renewable water resources is 2650Mm^3. The actual water use is about 485Mm^3/year (Department of Industry, Energy & Technology, 1989). The main water use in Crete is irrigation, reaching 83.3% of the total consumption. The domestic use, including tourism, and the industrial use are 15.6% and 1% of the total consumption, respectively. In Crete, regional variations in water availability are significant. The eastern and southern parts are more arid than the west and northern parts. It is estimated that about 1/3 of the cultivated land is irrigated.

4.2 Agriculture

Agriculture is an important sector of the local economy in Crete. It contributes 13% to the GDP of the island, while services and tourism contribute 77% and the industry 10% to the GDP. Approximately 6.7% of the work force in the island is employed in the agricultural sector.

Olive oil production is the most important in the agricultural sector of the island. The most important crops cultivated in Crete are presented in Table 2, based on information from the Region of Crete.

Table 2. Cultivated crops in Crete

Crops	Cultivated Area (km^2)	Percentage (%) of total
Raw crops	320.0	9.9
Vegetable crops	80.0	2.7
Vineyards	309.5	9.6
Fruit crops	1850.2	57.4
Fallow fields	653.5	20.4
Total cultivated land	3223.2	100.0

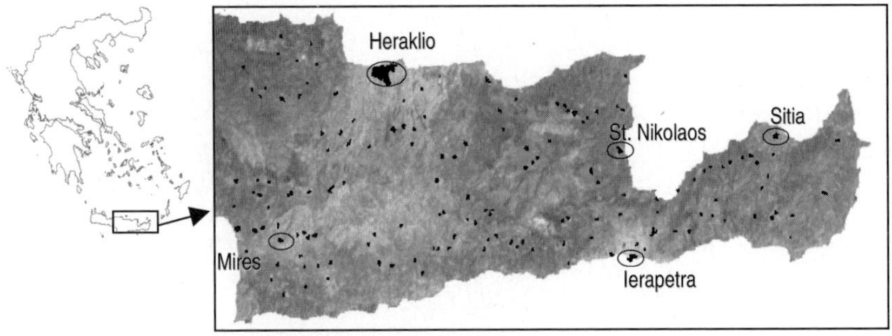

Figure 4. Study area (spots indicate the urban areas)

4.3 Selected Area for the Drought Identification Study

The selected area for the drought identification study is the eastern part of the island, which includes the prefectures of Heraklio and Lassithi (Figure 4).

The principal cities of this area and their population, according to the national census of 2001 are presented in Table 3.

5. METEOROLOGICAL STATIONS AND DATA PROCESSING

The meteorological network of Eastern Crete is rather dense when compared to other areas of Greece. It should be noted that some stations have large data gaps or quality problems in the time series due to malfunctions or other technical problems.

The meteorological stations of the area appear in Figure 5. The precipitation data used in this study were acquired from 21 stations, while temperature data are from 10 stations, which are shown circled on the map. The list of the used stations is presented in Table 4 and Table 5.

Figures 6 and 7 show the average annual precipitation in the Heraklio prefecture and the Lassithi prefecture respectively. Also, Figures 8 and 9 show the mean annual distribution of precipitation (in months) in the prefectures of Heraklio and Lassithi respectively.

Furthermore, the variation of the mean annual temperature and the monthly distribution of the mean temperature are shown in Figures 10 to 13.

Table 3. Population of the principal cities of eastern Crete

City	Population
Heraklio	116,000
St. Nicolaos	18,000
Ierapetra	22,500
Sitia	15,000
Mires	11,000

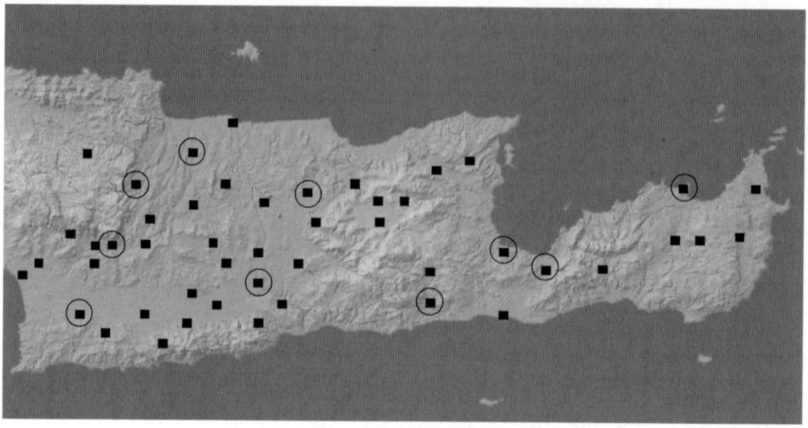

Figure 5. Meteorological stations of the study area (dots indicate availability of precipitation data, circles indicate availability of temperature data)

Table 4. Temperature stations of the study area

Temperature Station	Latitude	Longitude	Altitude (m)	Aver.PET (mm)
Finikia	599837	3904607	45	874.9
Protoria	604705	3876931	220	811.0
Pompia	578874	3874822	150	922.6
Kasteli	621157	3897473	350	817.0
Gergeri	584836	3887815	450	883.3
Sitia	690995	3898679	115	936.2
Pahia Ammos	665423	3883381	50	921.8
Mithoi	644236	3877473	200	997.3
Kalo Horio Lassithiou	657760	3886943	12	943.8

Table 5. Precipitation stations of the study area

Precipitation Station	Latitude	Longitude	Altitude (m)	Average Precipitation (mm)
Kroussona	589282	3898950	440	1062.1
Finikia	599837	3904607	40	681.4
Profitis Ilias	599939	3895364	340	749.6
Voni	613595	3895526	330	744.1
Ano Arhanes	605965	3899131	380	748.1
Metaxohori	603658	3888005	430	711.2
Tefeli	606139	3884343	360	708.9
Protoria	604705	3876931	220	577.7
Demati	616868	3877080	190	479.7
Kasanos	619811	3884513	320	601.4
Armaha	622749	3891948	500	869.7

Kastelli	621157	3897473	340	727.7
Avdou	630233	3899448	230	786.1
Exo Potamoi	639390	3895887	840	1375.5
Neapoli Lassithiou	645371	3901529	280	844.0
Mithi	644236	3877473	220	597.9
Malles	644148	3883018	590	805.6
Kalo Horio Lassithiou	657760	3886943	20	509.8
Stavrohori Lassithiou	676060	3883581	320	772.8
Maronias	689622	3889403	140	667.8
Katsidoniou	694178	3889500	480	875.5
Zakros Lassithiou	701772	3889666	200	588.4

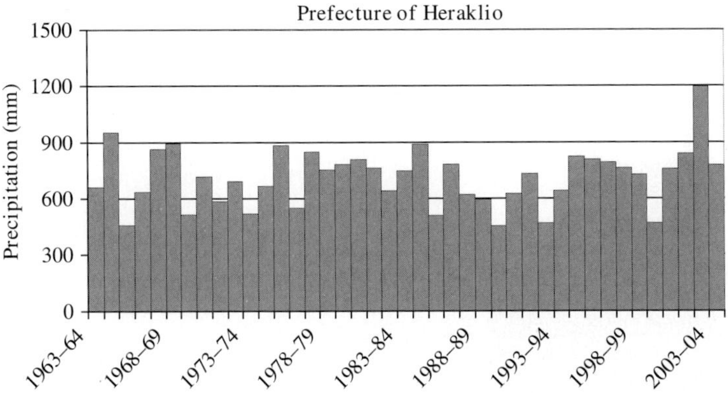

Figure 6. Annual precipitation for Heraklio prefecture (1963–2004)

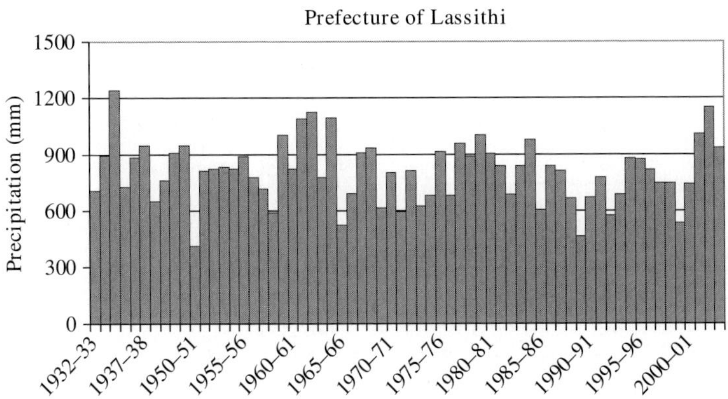

Figure 7. Annual precipitation for Lassithi prefecture (1932–2004)

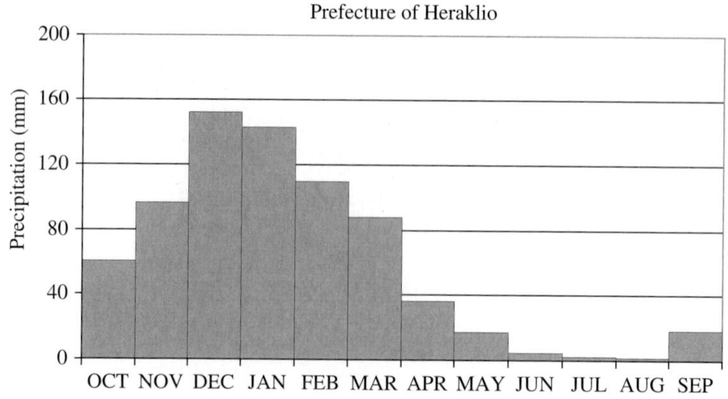

Figure 8. Mean annual distribution of precipitation for Heraklio prefecture

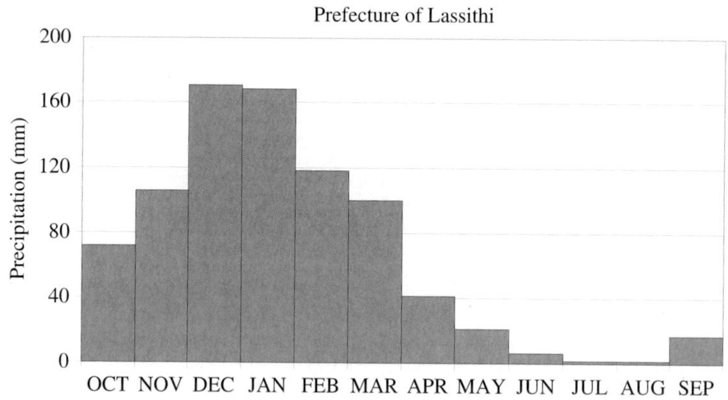

Figure 9. Mean annual distribution of precipitation for Lassithi prefecture

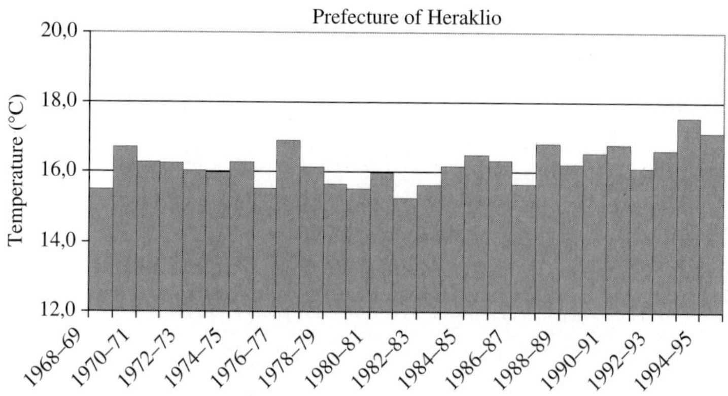

Figure 10. Mean annual temperature for Heraklio prefecture (1968–2004)

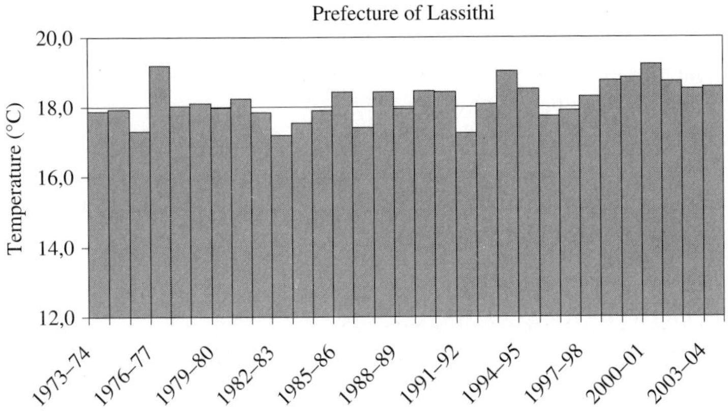

Figure 11. Mean annual temperature for Lassithi prefecture. (1973–2004)

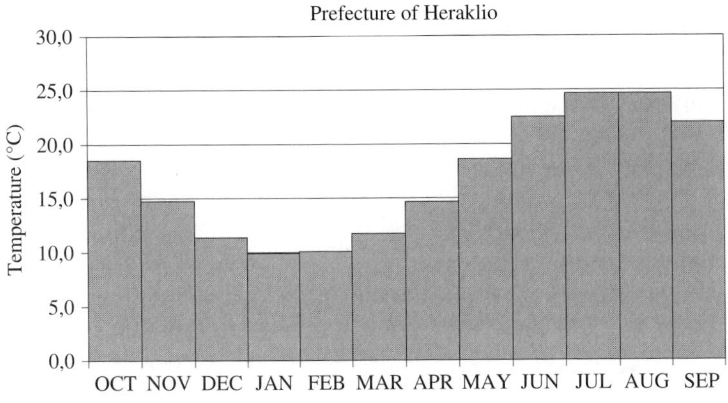

Figure 12. Monthly distribution of mean temperature for Heraklio prefecture

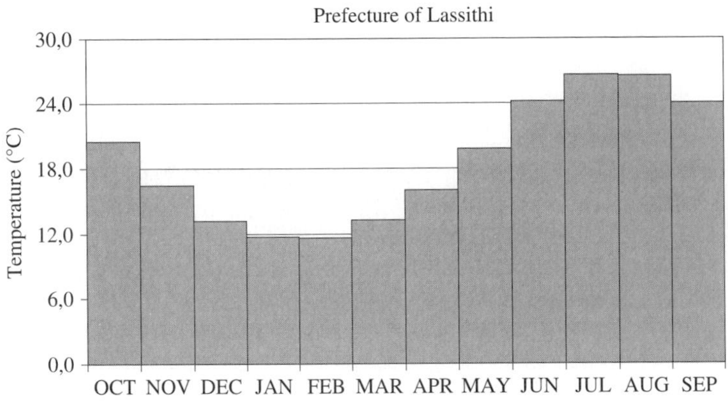

Figure 13. Monthly distribution of mean temperature for Lassithi prefecture

The figures mentioned above show that there are only small differences in the precipitation and temperature records between the two prefectures. This justifies the assumption that the entire area of the two prefectures can be analysed as a homogeneous unit as far as concerns drought.

Monthly data of precipitation and temperature have been used. Gaps in the time series were filled following standard hydrological methodologies. The stations were checked for consistency and the missing monthly values were estimated using a linear regression approach based on base stations.

6. SPATIAL AND TEMPORAL ANALYSIS

The drought identification includes the spatial and temporal analysis of drought in the study area. The spatial distribution model based on the discretized GIS platform was applied for each hydrological year (October-September). The basic objectives of this analysis were the identification of the drought periods and the spatial extent of the drought.

6.1 Identification of Drought Events

Annual values of SPI and RDI were estimated for Heraklio and Lassithi prefectures to provide an overview of drought occurrences. Figure 14 and Figure 15 show that a significantly persistent drought occurred from 1987 to 1994, while distinct drought events occurred in the years 1973–74, 1976–77, 1985–86 and 1999–00. To summarise, the values of 11 out of 30 years were below the baseline, and only minor differences can be observed between the two selected indices.

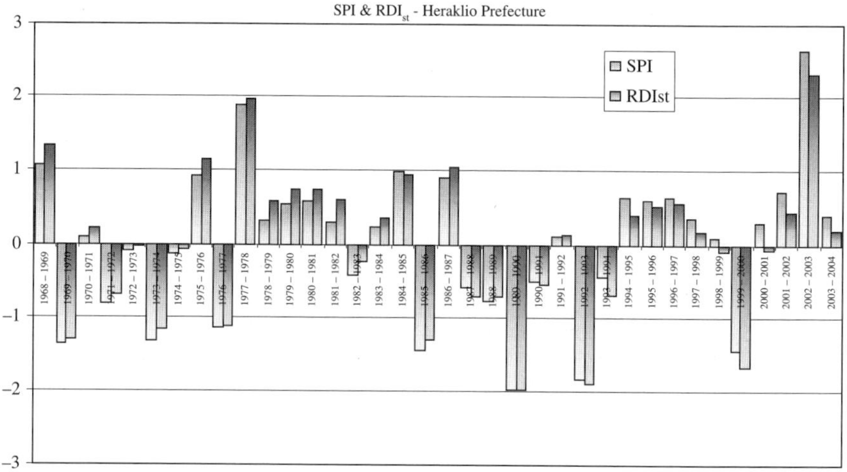

Figure 14. Annual SPI & RDI_{st} for each hydrological year (Oct–Sep) for Heraklio prefecture

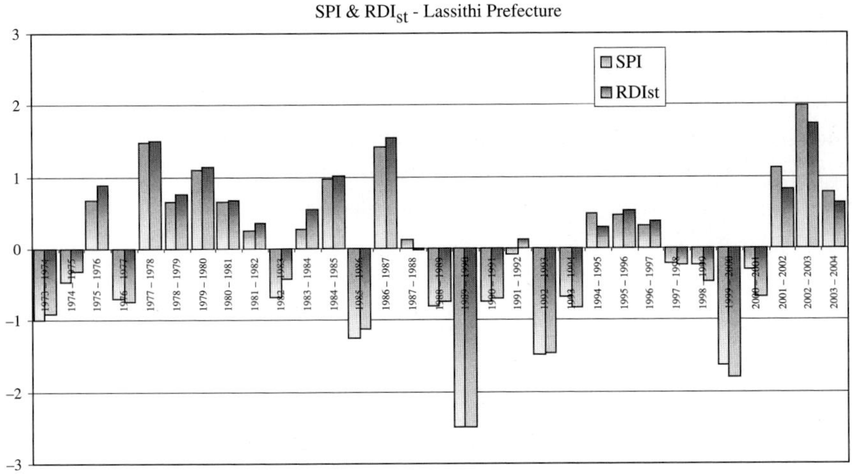

Figure 15. Annual SPI & RDI_{st} for each hydrological year (Oct-Sep) for Lassithi prefecture

6.2 Drought Maps

The area was divided based on the discretisation procedure, which is presented in the Appendix. Each grid cell is 2×2 km. Drought maps were produced using the estimated annual values of SPI and RDI. Typical samples of these maps are presented in Figures 16, 17 and 18.

In most cases, the results from both indices are similar. However, in some cases, depending on the local conditions, the results from RDI are more sensitive.

6.3 Cumulative "Or More" Curves

A better representation of the spatial extent of the drought can be achieved using a type of curves known as cumulative 'or more' curves (ogives). These curves can be produced by plotting the severity of drought (y-axis) versus the percentage of the affected area (x-axis). The severity of drought is presented by a drought index and the area refers to that affected by at least the corresponding severity. This type of graphs can be used not only for the characterisation of drought and the determination of its areal extent, but also for comparisons with the critical area percentage (related to severity) directly. Clearly, more than one thresholds referring to the percentage of critical area can be used to define levels of severity.

For example, a type of drought can be characterised as severe for a region if more than 50% of the area is under drought and more than 30% is under severe drought.

Figure 19 presents the cumulative 'or more' curve for the hydrological year 1969-70 based on the SPI, whereas in Figure 20 the same curve and year are presented based on the RDI. Figure 19 shows that 19% of the area is under extreme

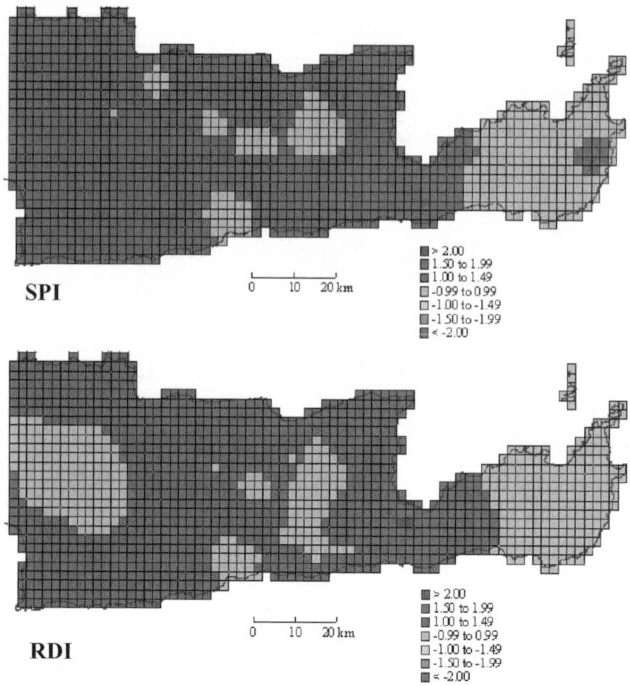

Figure 16. SPI & RDI_{st} for a wet year (1984–85)

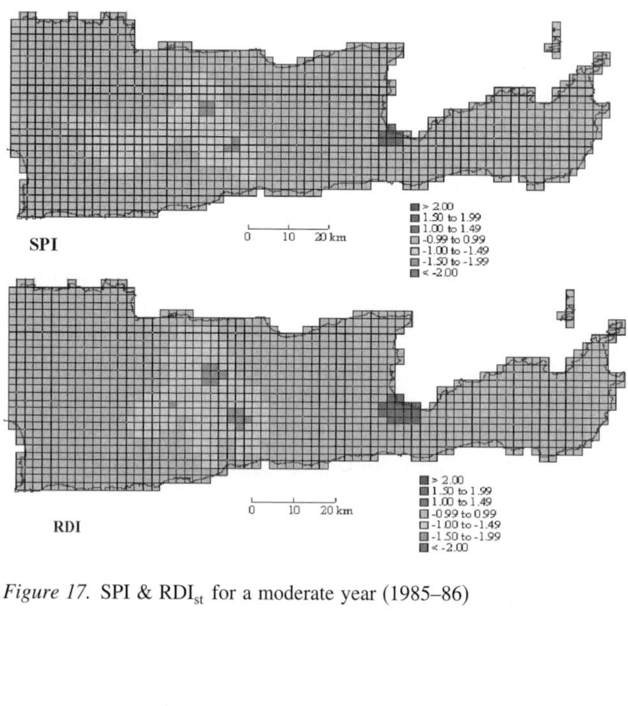

Figure 17. SPI & RDI_{st} for a moderate year (1985–86)

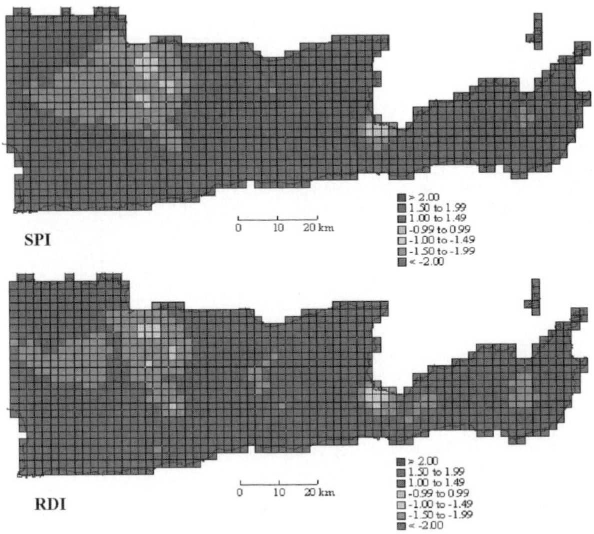

Figure 18. SPI and RDI_{st} for a dry year (1989–90)

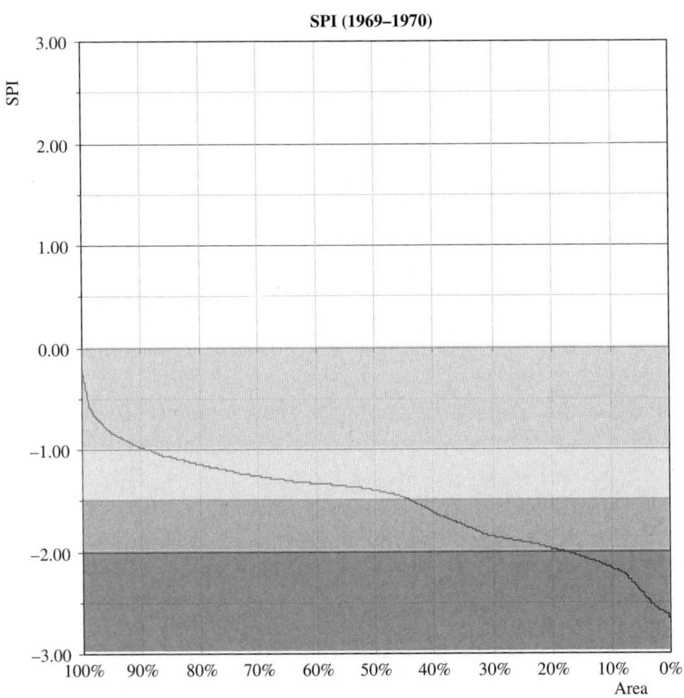

Figure 19. 'Or more' curve for the annual SPI (1969–70)

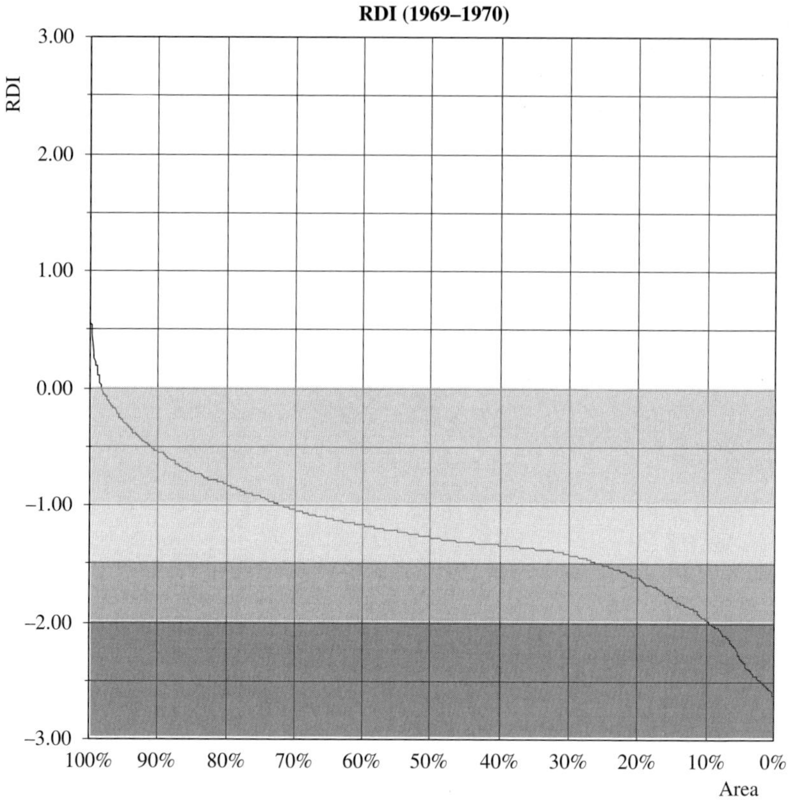

Figure 20. 'Or more' curve for the annual RDI (1969–70)

drought, 44% under severe drought and 89% under moderate drought conditions. Similarly, in Figure 20 the corresponding area percentages are 10%, 26% and 72%, respectively. It may be noticed that estimations are different, depending on the chosen index.

It can be concluded that the cumulative "or more" curve represents the areal extent of drought in a very convenient way for estimating the critical area percentage directly.

7. CONCLUDING REMARKS

The area of eastern Crete, one of the most drought-prone regions of Greece, was used as a case study to present the drought identification analysis. The applied methodology was based on a spatial distribution model utilising a grid analysis. Two drought indices, the SPI and the RDI, were estimated for a historical record of thirty years, using the recently developed software package DrinC.

Drought maps based on discretized platforms were produced. The spatial extent was estimated by the cumulative 'or more' curves for each year of the historical record. The use of these curves allows the direct assessment of the areal extent of drought episodes by setting thresholds and finding the corresponding critical area percentage. Both indices used, provide similar results. However, some discrepancies can be observed mainly in a regional scale. The proposed methodology is easy to implement and can become a practical tool for the assessment of regional drought phenomena.

ACKNOWLEDGEMENTS

The work presented in this paper was supported by the project SEDEMED II (Interreg Community Initiative Framework). The constructive comments of the anonymous referees are kindly acknowledged.

REFERENCES

Department of Industry, Energy & Technology (1989) *Water Resources in Greece*. Athens, Greece (in Greek).

Edwards, D. C. and McKee, T. B. (1997) Characteristics of 20th Century Drought in the United States at Multiple Timescales. *Climatology Report No. 97-2,* Colorado State University, Fort Collins, pp. 155.

McKee, T.B., Doesken, N.J. and Kleist, J. (1993) The Relationship of Drought Frequency and Duration to Time Scales. *8th Conference on Applied Climatology*, Boston, American Meteorological Society.

Rossi, G., Benedini, M., Tsakiris, G. and Giakoumakis, S. (1992) On regional drought estimation and analysis, *Water Resources Management* 6: 249–277.

Thom, H.C.S. (1958) A note on the gamma distribution. *Mon. Weath. Rev*. 86: 117–122.

Tsakiris, G. and Vangelis, H. (2004) Towards a Drought Watch System based on Spatial SPI, *Water Resources Management* 18:1–12.

Tsakiris, G. and Vangelis, H. (2005) Establishing a drought index incorporating evapotranspiration, *European Water* 9-10:1–9.

Tsakiris, G., Pangalou, D. and Vangelis, H. (2006) Regional Drought assessment based on the Reconnaissance Drought Index, *Water Resources Management,* Special Issue, Eds A. Iglesias and L. Garrote (in print).

Wilhite, D. (2000) Drought preparedness in the U.S. In J. V. Vogt and F. Somma (Eds.) *Drought and Drought Mitigation in Europe*, Kluwer, The Netherlands, pp. 119–132.

APPENDIX – THE SPATIAL DISTRIBUTION MODEL

Let us consider Figure 2, which shows the transfer of data from two stations to our point of interest. In this case, the required data to a point of interest in space are represented by a cell of the grid with code "A". The data from the two stations are considered of equal reliability regarding their infrastructure, so their distance from the cell “A” is the only parameter that affects the final result. The parameter P (e.g. precipitation) is estimated by the following equation:

$$P_A = \frac{P_1 \frac{1}{d_1^2} + P_2 \frac{1}{d_2^2}}{\frac{1}{d_1^2} + \frac{1}{d_2^2}} \tag{A1}$$

where P_A is the estimated parameter, P_1 and P_2 are the parameters in each station, and d_1 and d_2 are the horizontal distances from each of the two stations to the point of interest (location of the cell). Equation A1 is linear in respect to the input parameters P_1 and P_2. It is obvious that the linear approach will be misleading because the calculated value is referring to an imaginary point different from the real surface point. Station 2 is closer to the point of interest than Station 1. Thus, the contribution of the data from Station 2 to the estimated value P_A is higher than that of the data from Station 1. However, the elevation difference between the point of interest and Station 1 is smaller than the elevation difference with Station 2. Therefore, it is expected that the meteorological conditions at the point of interest have more similarities with those at Station 1 than with those at Station 2. This can be resolved by adding or subtracting a value, which includes a “parameter gradient”, from the estimated parameter P_A. In this way, the adjusted parameter (P_A^F) represents more realistically the meteorological conditions at the point of interest. The “parameter gradient” is derived from the data from all the available stations in the area or part of the area, if necessary. The corrected parameter is estimated by the following equation:

$$P_A^F = P_A + \Delta H \cdot \beta \tag{A2}$$

where P_A and P_A^F are the parameters before and after the correction respectively, β is the “parameter gradient” and ΔH is the elevation difference between the point of interest and a simulated elevation estimated by the following equation:

$$H_A = \frac{H_1 \frac{1}{d_1^2} + H_2 \frac{1}{d_2^2}}{\frac{1}{d_1^2} + \frac{1}{d_2^2}} \tag{A3}$$

where H_A is the simulated elevation of the point of interest, H_1 and H_2 are the elevations of the two stations, d_1 and d_2 are the distances between the point of interest and Station 1 and Station 2 respectively.

The equations presented above, can be extended to include data from all the stations in the study area, in the following form:

$$H_i = \frac{H_1 \frac{1}{d_1^2} + H_2 \frac{1}{d_2^2} + \ldots + H_n \frac{1}{d_n^2}}{\frac{1}{d_1^2} + \frac{1}{d_2^2} + \ldots + H \frac{1}{d_n^2}} \tag{A4}$$

where H_i is the simulated elevation of a point of interest referring to cell i, H_1, $H_2, \ldots, H_n$ are the elevations of Station 1, Station 2, …, Station n and d_1 to d_n are the distances between the point of interest and each of the n stations.

$$P_i = \frac{P_1 \frac{1}{d_1^2} + P_2 \frac{1}{d_2^2} + \ldots + P_n \frac{1}{d_n^2}}{\frac{1}{d_1^2} + \frac{1}{d_2^2} + \ldots + H \frac{1}{d_n^2}} \tag{A5}$$

where P_i is the estimated parameter, P_1 to P_n are the parameters in each of the n stations and d_1 to d_n are the distances between the point of interest and each of the n stations.

Given Equation A2 the final parameter is estimated as follows:

$$P_i^F = P_i + \Delta H \cdot \beta \tag{A6}$$

PART III

WATER RESOURCES MANAGEMENT UNDER DROUGHT CONDITIONS

CHAPTER 10

DROUGHT MANAGEMENT DECISION SUPPORT SYSTEM BY MEANS OF RISK ANALYSIS MODELS

J. ANDREU[(1)], M. A. PÉREZ[(2)], J. FERRER[(1)], A. VILLALOBOS[(2)] AND J. PAREDES[(2)]

[(1)] *Júcar Basin agency, Environmental Ministry, Valencia, Spain*

[(2)] *Polytecnic University of Valencia, Spain*

Abstract: Droughts in arid and semiarid Mediterranean river basins have an increasing socioeconomic and environmental impact. The problem is especially complex in basins where droughts are very frequent and intense and where water resources are under a massive use. For that reason in the Mediterranean basins it is necessary to manage the water resources in constant alert in order to relieve the consequences of a drought. In this chapter a complete methodology for the mitigation of the droughts is defined. The methodology includes different tools: from the definition of an alert system for the different types of drought up to the simulation and optimization software for water system management. This methodology has been applied to the different hydrographic basins included in the Confederación Hidrográfica del Júcar (CHJ, the local water agency) and it has been currently used during the extreme drought occurred in the Júcar river basin since the hydrological years 2004/05. This chapter shows how the proposed methodology is able to forecast the occurrence of a drought and to realize a water resources management focused on the mitigation of the economic, social and environmental effects of water shortage

Keywords: Drought, water resources management, drought mitigation, simulation model

1. INTRODUCTION

The water resources management in the arid and semiarid regions of the Mediterranean area is an especially complex task, which includes multiple factors and a great number of different fields of study. Hydrological, climatic, environmental, social, economic and management factors must be considered in order to guarantee the protection of the environment and to satisfy adequately the water demands needed by different productive sectors, allowing to ensure suitable levels of quality of life.

Droughts are a frequent phenomenon in the Mediterranean area and they increase considerably the complexity of the water resources management in these basins. A suitable management of water resources in river basins where droughts take

G. Rossi et al. (eds.), Methods and Tools for Drought Analysis and Management, 195–216.

place with certain frequency can reduce considerably the negative impacts on the environment and on the regional economy. For this reason water authorities must detect the occurrence of droughts as soon as possible, and rely on the necessary tools to improve the management, as the decision support systems, in order to mitigate the impacts that drought produces. It is necessary to add to this situation the current uncertainties due to the possible existence of a global climatic change, since the river basins of semiarid regions are, once again, the most vulnerable to climate modifications. These circumstances have led to an important effort dedicated to the water resources management and to the analysis of the possible measures to apply both in situations of drought and in the previous phases. Measures for the environmental protection (water saving, improvement in the management or the search for alternative water supply) are analyzed by developing methodologies for the analysis of the efficiency of the measures that could be applied before the occurrence of a drought. The extreme current drought of the Júcar River started in the 2004/05 hydrological year and nowadays it has reached its major severity, that makes necessary to dedicate important efforts to the analysis and to the definition of mitigation measures. Definition of the mentioned measures, as well as their execution, and estimation of the efficiency that each of them could have, require the assistance of simulation models in order to predict the behaviour of the water resource systems in case of different management alternatives.

Management Simulation Models foresee the future behaviour of the water system and quantify the risk of failure expected from the system by means of the probabilistic analysis of the water management in case of multiple possible hydrological scenarios. This chapter shows some of the results that the Technical Office for Drought Management is performing for the Permanent Droughts Commission, constituted in the CHJ by the Royal Decree 1265/2005, October 21, 2005, to simplify the decision process and the definition of the measures to apply to the current drought.

2. METHODOLOGY

In arid and semiarid basins the implantation of a efficient system of drought alert and management is necessary in order, to foresee the possible occurrence of a drought and to define as soon as possible the most suitable measures to mitigate the drought effects.

Drought management system must be able to evaluate the efficiency of measures designed in order to mitigate the drought effects, so the most effective measures must be prioritised.

The experience acquired through the last episodes of extreme drought that occurred in the Mediterranean basins of the Iberian Peninsula led to a drought management system constituted by four phases of development, as appears in Figure 1.

Next the different stages that compose the methodology are described.

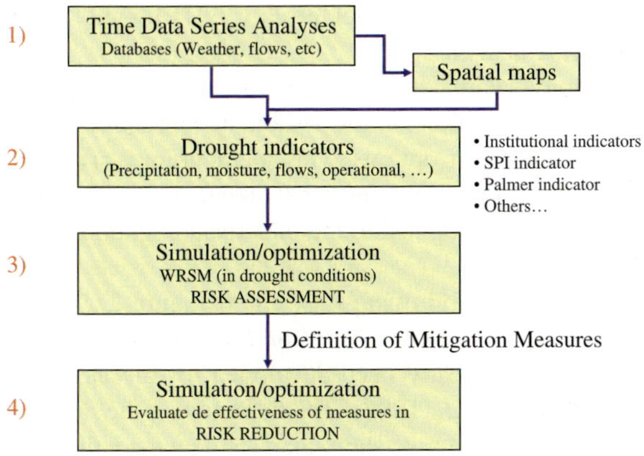

Figure 1. Conceptual drought management system

2.1 Stage 1: Information Analysis

The first stage of the drought management system consists on the spatial and temporal analysis of the meteorological and hydrological information and on the definition of the rainfall network dedicated to the drought forecasting. The meteorological information spatial and temporal variability requires a high number of stations and a huge amount of temporal information in order to realize a suitable hydrological and meteorological characterization.

Previously to the information analysis stage a basic network has to be defined as a set of measurement stations that provide long duration and with a suitable spatial coverage data series.

2.2 Stage 2: Drought Indicators

The second stage of the droughts management system consists in the definition of specific indicators for the different types of drought:

- Meteorological drought indicators. Based on the SPI index (McKee et al., 1993).
- Agricultural drought indicators. Based on the analysis of the soil moisture by means of a runoff simulation model of the river basin.
- Hydrological drought indicators. Based on the inflows statistical analysis.
- Operative drought indicators. Based on water system management simulation models and on the CHJ institutional indicators.

2.3 Stage 3: Water System Simulation. Risk Analysis

The third phase of the drought management system is based on the application of water system simulation models, which are considered the only capable tools to reproduce the whole complexity of river basins. The work consists on the application of the methodology developed by Sánchez et al. (2001) for the water system management by means of risk analysis.

The decision support methodology evaluates by means of statistical methods the probabilities associated with the evolution of the system as consequence of the adoption of each one of the management options. This methodology offers an objective criterion for the comparison of the consequences associated to each of the proposed decisions, which facilitates the dialogue and the understanding among the parts involved in a conflict.

The methodology can be resumed in the following flow chart (Figure 2).

The application of the management methodology based on risk analysis requires a number of informatics tools that have already been developed for the Decision support System (DSS) Aquatool (Andreu et al., 1996). The main tool is the Simrisk module for the multiple simulations of water system management and the risk estimation. The module Simrisk has been designed to be used as a support tool for the management of a Water Resources River Basin, and provides information about the risk of failure due to operative drought and the probabilities related to the water reserves in the future.

The probabilities related to the future water availability in the system calculated by the module Simrisk have to be correctly understood in order to take the most

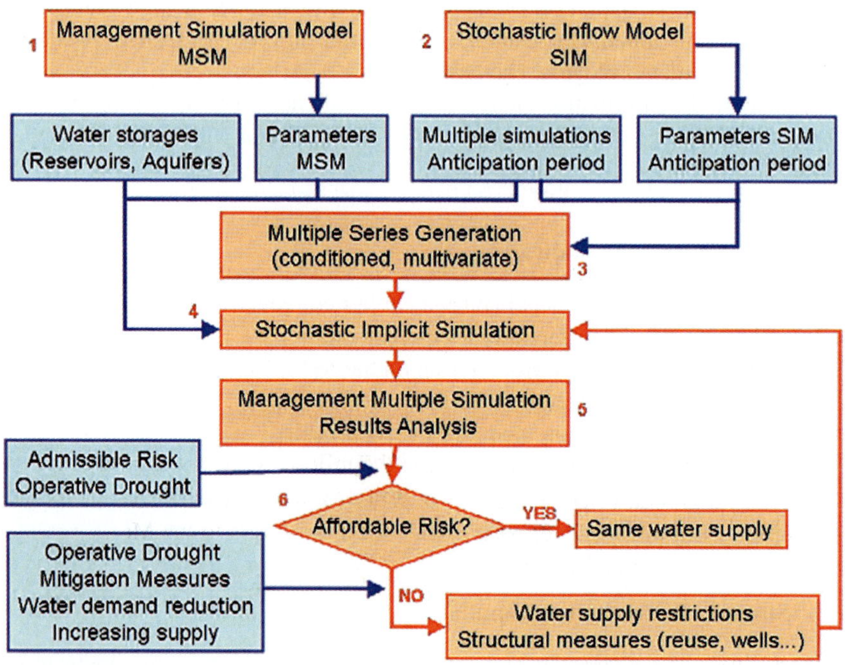

Figure 2. Decision support methodology for water systems management by means of risk analysis (Sánchez et al., 2001)

suitable decisions. For that reason it is necessary that the decision maker has the maximum information about the relationship between the probabilities estimated by the model and the expected situations in the future.

All the tools have been developed starting from the Simges simulation model (Andreu et al., 1992) and they are based on the multiple repetition of the Simges management simulations. It guarantees the validity of the results because the Simges module has been previously validated through the simulation of several real cases.

Besides, in order to integrate these tools in a SSD, others tools have been developed to make easier the input of information and the analysis of results.

The Simrisk module makes the implicit stochastic simulation of the water system management, which essentially consists on:

- multiple water system simulations through the same period, starting from the same initial situation and considering different and equiprobable inflow scenarios,
- available water probabilities estimation by means of the multiple simulation for every month of the simulation period.

This simulation is made on monthly scale and it begins at any month using the hydrological information of the same month for the same time interval as the anticipation period. Internally this module is based on the module Simges. This one has been modified to begin the simulation at any month and to make multiple simulations and risk evaluation.

The two following steps are executed by the Simrisk module and they consist on the multiple water system management simulation and the statistical analysis of the results in order to estimate the risk.

2.4 Stage 4: Mitigation Measures Definition and Effectiveness Estimation

Once the possible occurrence of a drought is detected, the last phase of the drought management system, the stage 4, starts. In this stage the most suitable mitigation measures are planned, and, by means of the simulation models, it is possible to analyse the efficiency of each one of the measures and of all of them.

3. CASE OF STUDY

The Júcar River is the main river in the territory of the CHJ (Figure 3). The CHJ is located in the east side of the Iberian Peninsula and is formed by the aggregation of different hydrographical river basins that spill their waters to the Mediterranean Sea: the Júcar River, the Turia River and the Mijares River. The total surface of the CHJ is 42,989km^2.

The northwest zone of the CHJ has a typical high mountain climate, and it is the place where the Júcar and Turia rivers are born, whereas the west zone of the basin is a plateau where the river Júcar crosses the La Mancha plain, and the east zone is formed by coastal plains where the rivers end to the sea.

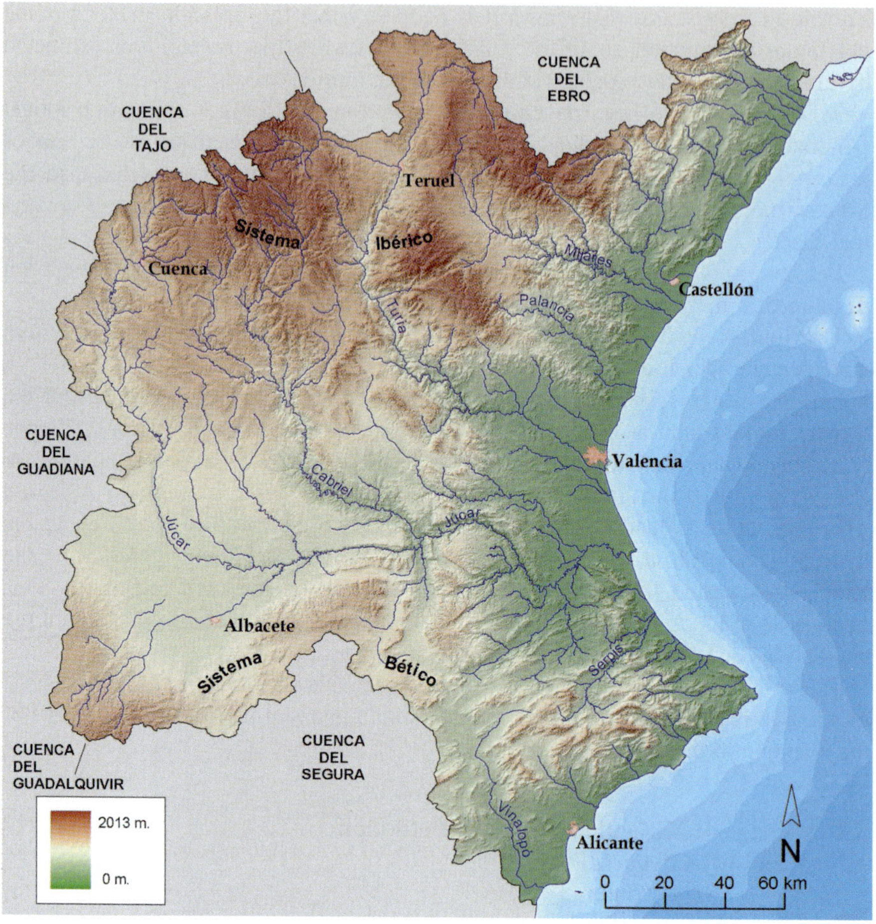

Figure 3. CHJ digital terrain model

The CHJ's area is composed of several principal rivers that form 9 hydrological systems (Figure 4). These systems are: 01-Cénia-Maestrazgo (1.875 km^2), 02-Mijares-Plana de Castellón (5.466 km^2), 03-Palancia-Los Valles (1.159 km^2), 04-Turia (6.913 km^2), 05-Júcar (22.378 km^2), 06-Serpis (990 km^2), 07-Marina Alta (839 km^2), 08-Marina Baja (583 km^2) and 09-Vinalopó-Alacantí (2.786 km^2).

CHJ's area natural resources guarantee the water supply, nowadays, to a 4 millions inhabitant population, and to a 300.000 has irrigation surface.

The Júcar system superficial water control is provided principally by Alarcón (maximum capacity of 1.112 hm3), Contreras (874 hm3) and Tous (340 hm3) reservoirs. The volume stored in the set of the three reservoirs constitutes the main superficial reserves of the system; therefore the analysis of the above mentioned reservoirs is considered the most relevant for the system.

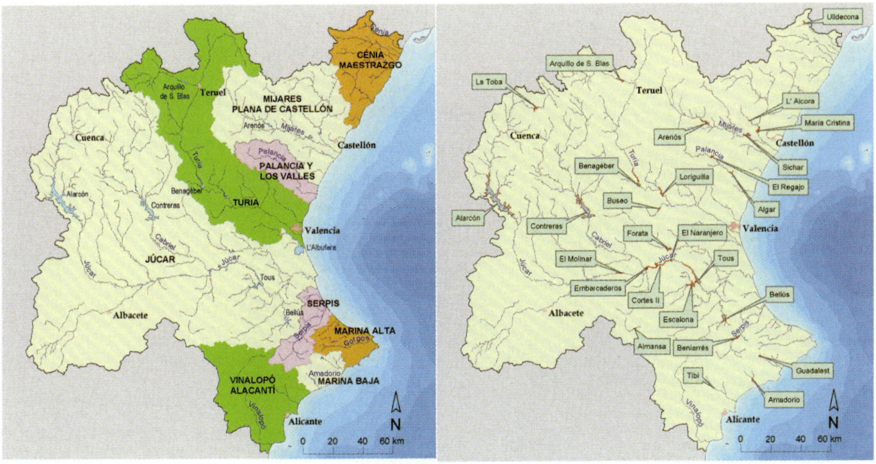

Figure 4. Water systems in the CHJ area and principal reservoirs

In the Júcar river there are also the Cortes (116 hm^3), Naranjero and Molinar reservoir. The Forata reservoir (37 hm^3) lies in the Magro River and the Bellús reservoir (69,2 hm^3) in the Albaida River, both tributaries of the Júcar.

Natural water resources in the area of the CHJ have an important spatial and temporary variability and they are generated by the rainfalls that take place in the continental area, which average annual value is estimated in 500 mm, with a minimum of 320 mm in the driest years and a maximum of almost 800 mm in the most humid years, as reported in Figure 5. Last year was especially dry with only 340 mm, it was the third worst year of the whole historical period with available information.

These average values present significant spatial differences, as in the southern regions the average annual rain is less than 300mm, while in other regions it exceeds 800 mm (Figure 6).

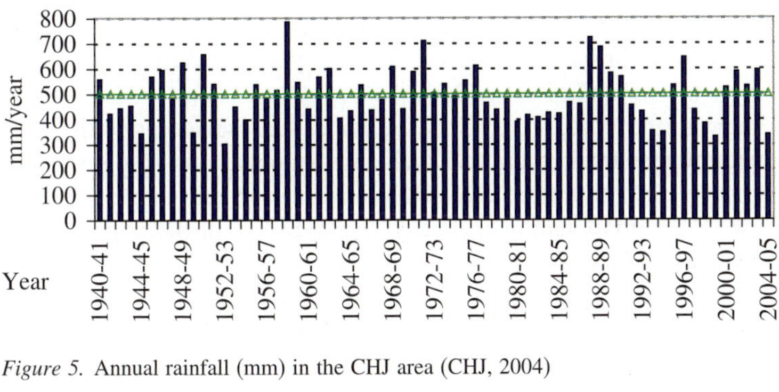

Figure 5. Annual rainfall (mm) in the CHJ area (CHJ, 2004)

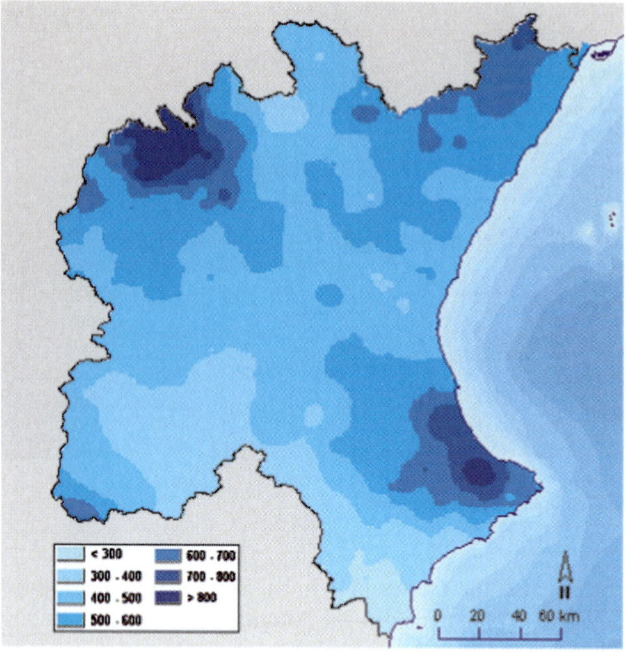

Figure 6. Average spatial rain distribution in the CHJ area (CHJ, 2004)

4. STAGE 1: TIME DATA SERIES ANALYSIS

The first phase consists in the analysis of the climatic and hydrological information, etc... available in the basin, in order to define the basic measurement network. The definition of a basic rainfall network has to include, at least, the following criteria:

1) Amount of available data (large series).
   The temporary variability of the climatic series leads to the use of long information series to characterize adequately the occurrence of a drought. Some international climatic organisations recommend that a climatic series, to be considered sufficient, must include at least 25 years of measurements. Figure 7 shows the list of rainfall stations that has been selected considering which is the amount of historical information available and if they are operative nowadays.
2) Spatial coverage of the rainfall network.
   Due to the great spatial variability of the rainfalls, the previous criterion has to be completed with a second one that includes the spatial distribution of the rainfall stations in order to ensure the spatial coverage of the whole territory. Figure 8 shows the rainfall stations network that keeps in all the points of the basin a minimal distance between any station.
3) Spatial coverage of the climatologic variables estimation.
   Due to the differences among the different zones of the hydrologic basin, where in a few regions predominate horizontal fronts and in other regions predominate

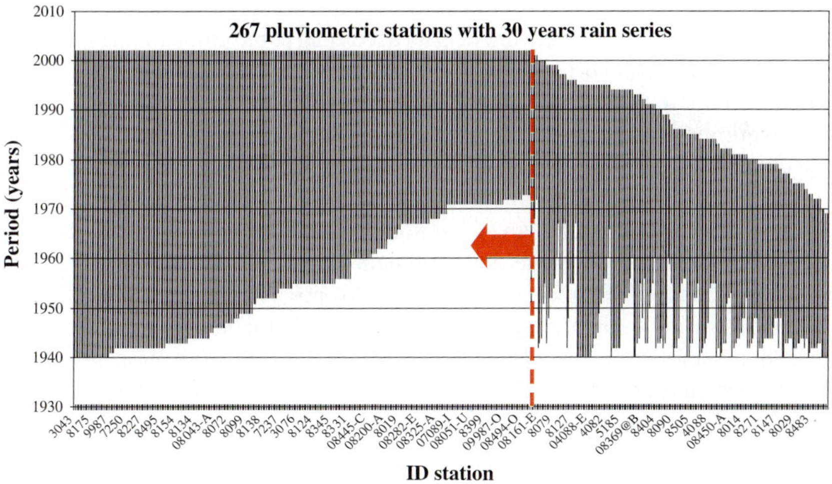

Figure 7. Pluviometric data available in the CHJ area, meteorological stations

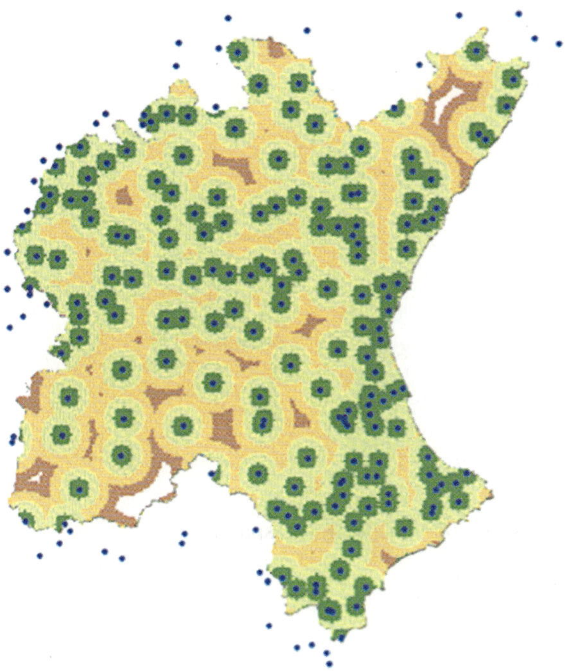

Figure 8. Basic rainfall network spatial coverage, minimal distance

convective storms, in order to minimize the rainfall estimation error, it is necessary to adapt the measurement network density according to the climatologic zone.

The regions where convective storms are present require more dense rainfall networks than the regions where predominate horizontal fronts. Figure 9 shows the basic network that minimizes the areal rainfall estimation error in the river basin.

The combination of the three previous criteria defines a climatologic basic network (Figure 10) for the rainfall estimation and for the early detection of droughts, which fulfils the following conditions: sufficient length of the information series, sufficient spatial coverage and minimum rainfall estimation error. On the other hand by means of the information analysis it is possible to identify the climatic characteristics of the basin and to define the different climatic regions.

The climatic information is completed by the hydrological information, registered on the gauge stations, reservoirs etc. that complements itself with the flows obtained by means of runoff simulation models.

The distributed conceptual model, "Patrical" (Pérez, 2005), calculate the hydrologic cycle variables in the whole extension of the basin: rainfall, soil moisture, natural river flows, outflows to the sea, etc. (Figure 11). That information is useful

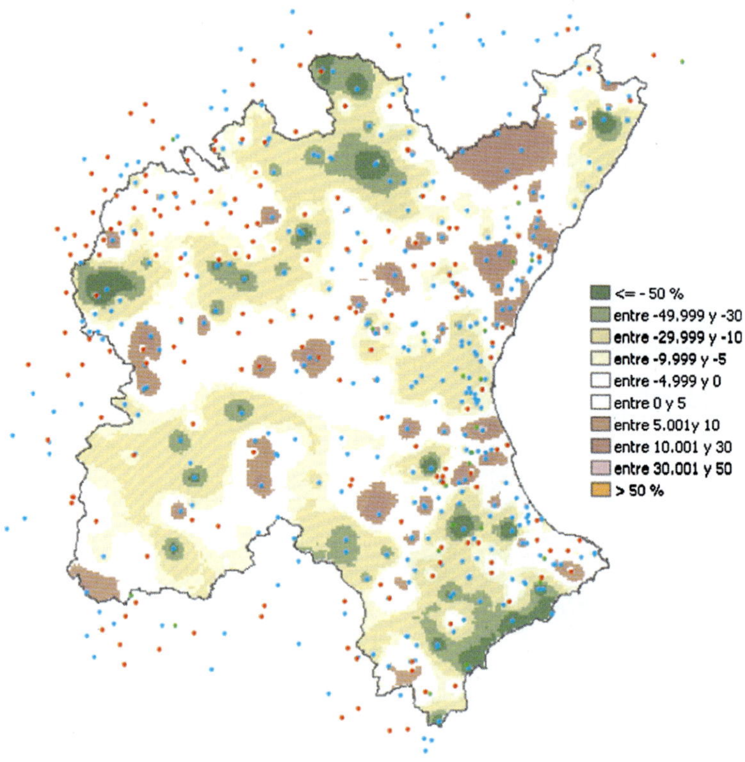

Figure 9. Basic network spatial coverage that minimizes the areal rainfall estimation error

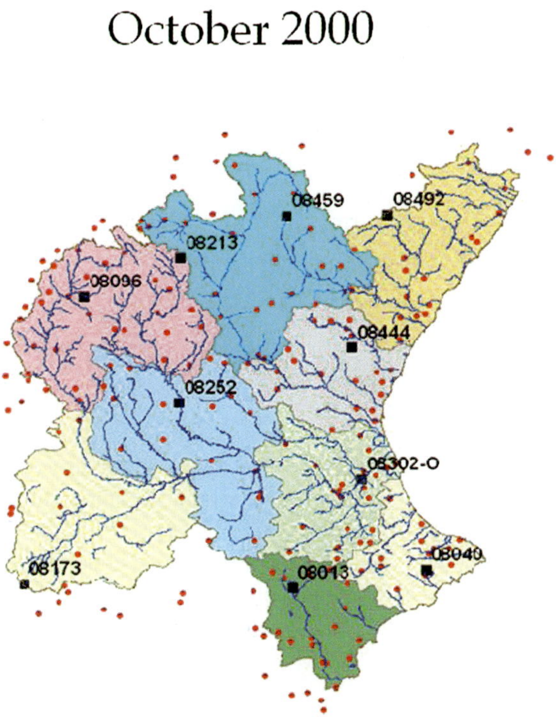

Figure 10. Basic rainfall network and climatic regions in the CHJ

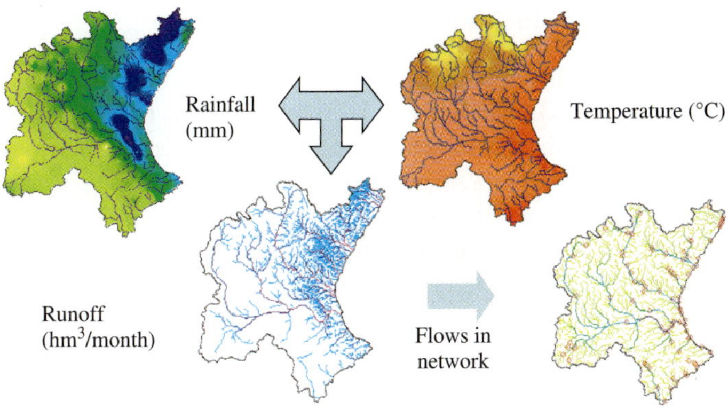

Figure 11. Determining the natural flows in the basin by means of the distributed runoff simulation model "Patrical" (Pérez, 2005)

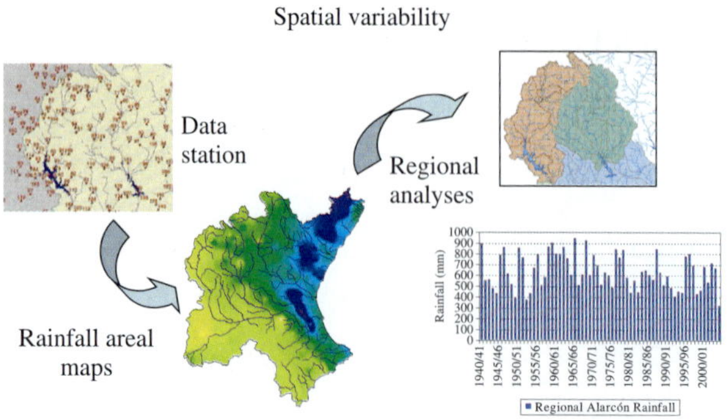

Figure 12. Areal rainfall definition process in a climatic region

to achieve a complete knowledge of the drought evolution across the hydrologic cycle and to determinate the indicators for the hydrological and the agricultural drought.

The great rainfall spatial variability and the need to achieve a regional-based climate characterization require first the areal determination of the rainfall by means of interpolation methodologies, and subsequently the regional analysis of the calculated areal rainfall, as it is shown in Figure 12.

5. STAGE 2: DROUGHT INDICATORS

In the second phase the more suitable indicators are defined for the different types of drought: meteorological drought, agricultural drought, hydrological drought and, finally, operative drought. The correspondent indicators are described in the following paragraphs.

5.1 Meteorological Drought Indicators

The most internationally used meteorological drought indicator is the SPI index (Standardized Precipitation Index), which is based on the historical rainfall series statistical analysis by means of the adjustment of an incomplete Gamma function with two parameters. SPI's negative values indicate lower rainfall than usual, whereas positive values indicate heavier rainfalls than usual.

SPI index is used as a short term indicator, to compare the rainfall dropped during the last week, last month or last months, with the distribution function corresponding to the same week, month or months, respectively, whereas it is considered a medium or long term indicator, when the rainfall of the last 12, 24 or 36 months is compared with the corresponding distribution functions to the same period of time.

This index is applied individually in each meteorological station; nevertheless, for the analysis of big river basins, a distributed index must be calculated along

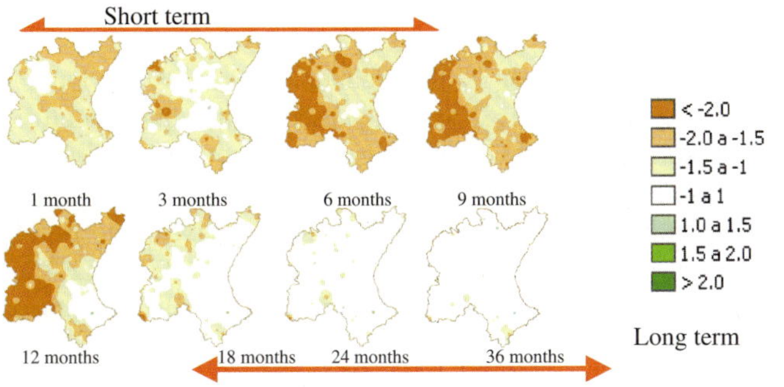

Figure 13. Short term (1, 3, 6 months) and long term (18, 24, 36 months) SPI distributed index calculated for September 2005

the whole extension of the basin, through different periods of time, from the short term, 1, 3 or 6 months, up to the long term, 24 or 36 months.

Figure 13 shows the SPI for short and long term distributed in the whole area of the CHJ calculated for September 2005. The 2004/05 hydrological year is characterized by extreme low rainfalls (12 months SPI), and corresponds to the year when the drought began, since the SPI of 24 or 36 marks normal values. Finally the short term SPI values (1, 3, 6 months) display how drought persists during the last months.

Regional SPI index is more suitable for the droughts historical characterization, since includes the whole basin surface. Figure 14 shows different drought periods occurred in the last 30 years, when long-term indicators (24 and 36 months) emphasize the 1980-82 and 1992-94 periods, and short-term indicators the beginning of the current drought.

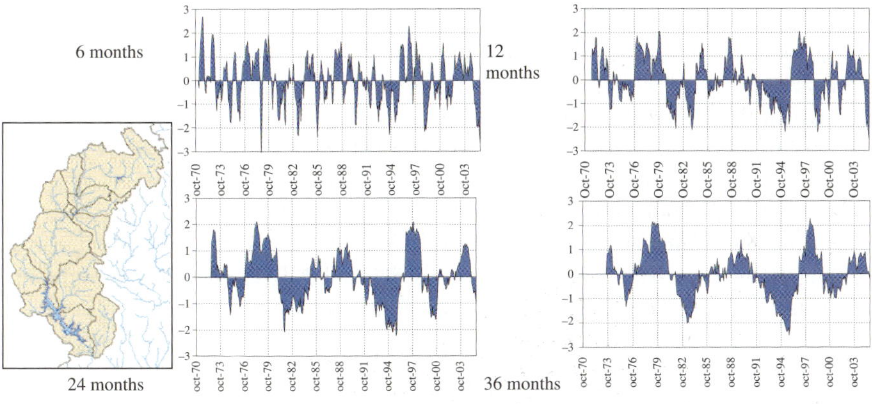

Figure 14. SPI regional index across the historical series

5.2 Agricultural Drought

The classic agricultural drought indicator is the Palmer index, (PDSI) Palmer Drought Severity Index (Palmer, 1965). PDSI index is based on the analysis of the soil moisture balance by reproducing the hydrologic cycle using the information of rain and temperature, and a set of parameters that must be calibrated.

The agricultural drought indicator in use is based on the results of the runoff "*Patrical*" model (Pérez, 2005), calibrated and validated for the whole area of the CHJ, which reproduces the hydrologic cycle and provides the soil moisture content at every moment in each point of the basin.

By comparing the soil moisture value in a certain month with the soil moisture distribution function in the same month, Figure 15, it is possible to determinate if the soil moisture content is abnormally dry, abnormally humid or if there is a common situation.

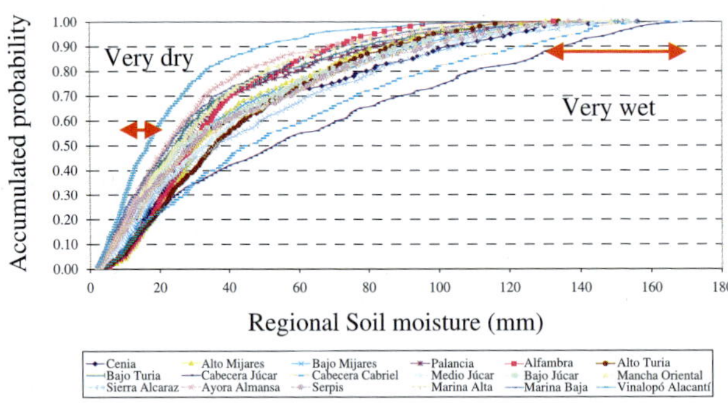

Figure 15. Accumulated soil moisture distribution function in the different climatic regions

5.3 Hydrological Drought Indicators

As far as drought advances it is transmitted to the hydrology, reducing the natural inflows of the water resource system. The hydrological drought indicator in use is the statistical analysis of the natural inflow series, more specifically, the position that the current inflow occupies in the whole historical natural inflow record. Nowadays monthly inflow series of the last 65 years are available.

Last year, besides being a slightly rainy year in the whole basin, it has been the year with less rainfall of the whole historical period in the Júcar River headwaters, concretely the Júcar and Cabriel headwaters registered a 312 and 261 mm rainfalls, opposite to the expected values of 595 and 536 mm, respectively.

Due to that situation, Júcar natural water inflows up to Tous reservoir have been the third worst of the whole historical series, Table 1.

In the current hydrological year, 2005/06, the situation has been getting worse, as it appears in the Table 2. From December 2005 up September 2006 the total inflow to the reservoirs registered in the current hydrological year, from October 2005,

Table 1. Natural inflows registered in the year 2004/05 in different points of the Júcar system and with the historical series

Last natural inflows (hm^3/year)	2004/05	Ranking in the historical series (1= minimum)
Alarcón	161,32	5
Contreras	134,90	6
Alarcón-Molinar *	213,44	4
Contreras-Molinar-Tous	174,35	26
Tous-Sueca	254,99	48
Total up to Tous	**684,00**	**3**
Total whole River	**939,00**	**6**

represents the historical minimum of the last 65 years (see *Natural accumulated inflows in Tous reservoir*).

5.4 Operative Drought Indicators

Operative drought evidence is given by the institutional indicators of the CHJ (see Figure 16), which at June 1, 2006 displayed an emergency situation in the Júcar system (more information is available on the web site: www.chj.es). Operative

Table 2. Natural inflows registered in different points of the Júcar system and their relation with the historical series

Inflow hm^3	oct-	nov-	dic-05	jan-06	feb-06	mar-	apr-	may-
Alarcón	6, 29	13, 82	9, 60	11, 88	8, 10	25, 18	16, 62	11, 64
Contreras	7, 47	8, 01	8, 18	8, 38	7, 56	10, 63	9, 28	9, 65
Alarcón-Molinar	17, 94	16, 67	16, 02	18, 23	17, 37	17, 29	18, 75	21, 88
Contreras-Molinar-Tous	16, 55	7, 13	14, 29	14, 19	14, 23	12, 30	7, 60	13, 01
Tous-Sueca	21, 89	21, 03	19, 79	19, 73	14, 37	14, 12	6, 89	15, 33
Total hasta Tous	**48, 25**	**45, 62**	**48, 09**	**52, 69**	**47, 26**	**65, 41**	**52, 25**	**56, 18**
Total	**70, 14**	**66, 66**	**67, 89**	**72, 42**	**61, 63**	**79, 53**	**59, 14**	**71, 51**
Natural accumulated inflow up to Tous since October	**48, 25**	**93, 88**	**141, 97**	**194, 66**	**241, 92**	**307, 33**	**359, 58**	**415, 75**
Ranking in the historical series	oct-	nov-	dic-05	jan-06	feb-06	mar-	apr-	may-
Alarcón	9	17	6	5	1	21	11	4
Contreras	3	2	1	2	1	5	2	3
Alarcón-Molinar	3	4	2	7	6	10	24	29
Contreras-Molinar-Tous	35	7	21	20	29	21	14	24
Tous-Sueca	42	47	51	48	21	13	4	36
Total hasta Tous	4	1	1	2	1	9	8	5
Total	**11**	**6**	**4**	**7**	**2**	**7**	**3**	**5**
Natural accumulated inflow up to Tous since October	**4**	**2**	**1**	**1**	**1**	**1**	**1**	**1**

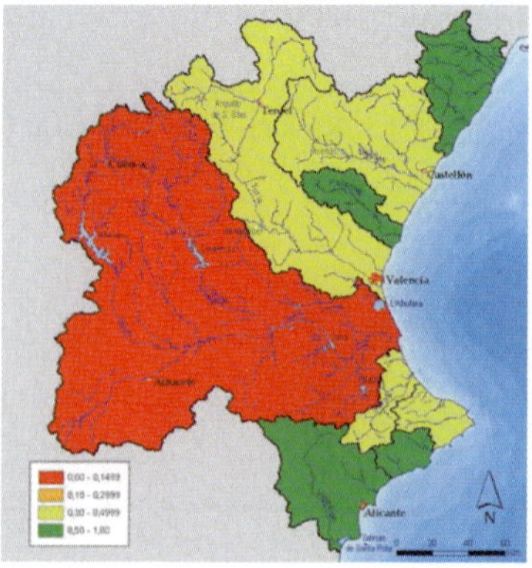

Figure 16. Operative drought indicator in the CHJ water systems

drought indicators include the possibility that human activities can be affected directly or indirectly by the drought; in this case the variable of the analysis is the volume of stored water in reservoirs and aquifers.

6. STAGE 3. WATER SYSTEM MANAGEMENT SIMULATION. RISK ANALYSIS

The Decision Support System Aquatool (Andreu et al., 1996), more specifically the Management Simulation Module Simges (Andreu et al., 1992) is used in the third phase as operative drought indicator of the Júcar system (Figure 17) (OPH, 2002) to foresee with certain anticipation the behaviour of the water system and the possible failures under different hypotheses.

The forecast for the present year (2005/06) is made considering the hypothesis of giving to the different users the same amount of water of the last year (2004/05) subject to the condition of having the same inflows as the last years, the third worst natural inflows of the last 65 years. This hypothesis is defined as the repetition of the last campaign, and provides the decision makers, who come from different professional backgrounds, with an easy to understand information.

Under these hypotheses the water storages in the system's reservoirs would become exhausted by the middle of July 2006, Figure 18, reaching the technical reservoir exploitation limit of 55 hm^3. Clearly the application of the management simulation models constitutes a system operative drought indicator.

Statistically the probability that current year inflows are identical to those happened last year is poor, and that is the reason why the previous analysis has only a didactic purpose but it is very improbable that takes place.

These technologies, known as "decision support methodologies by means of risk analysis", use the statistical properties of the natural hydrologic cycle in order to define the most probable inflow scenarios considering the current situation of the natural water system.

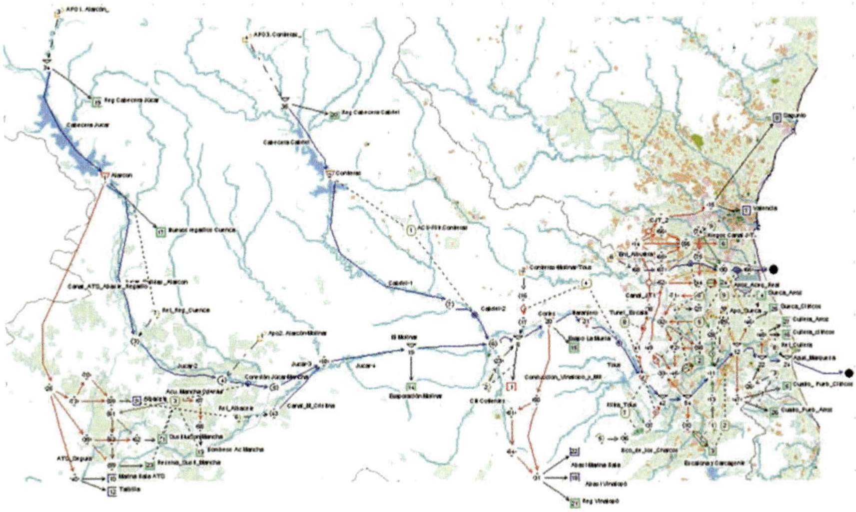

Figure 17. Jucar water system simulation model

6.1 Modelling the Júcar System for the Risk Analysis

The evaluation of the probability related to the system condition at the end of the hydrological year and the forecast of the future water supplies are calculated every month considering the hydrological information available at the moment.

The forecasts calculated in February 2006, using the water storage data registered at the end of January and the last two months natural inflows (December 2005 and

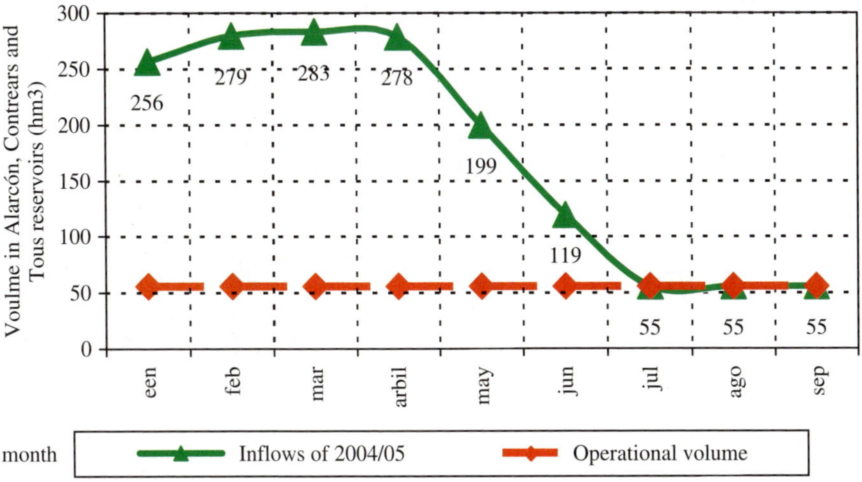

Figure 18. Water reserves forecast in the Júcar River reservoirs through 2006

January 2006), and considering the hypothesis of keeping the same water supplies to the users as the last hydrological year (Figure 19), show a 50% probability that water storages fall at the end of September under 64 hm^3. Taking into account that the technical storage limit is 55 hm^3 (minimum volume), there is a high probability that the system will be technically unable to supply water by the end of the hydrological year.

The technical emptying is expected in August 2006, and it is associated with a high risk of failure on water demand supply (Figure 20), reaching a 60% failure probability in case of agricultural demands.

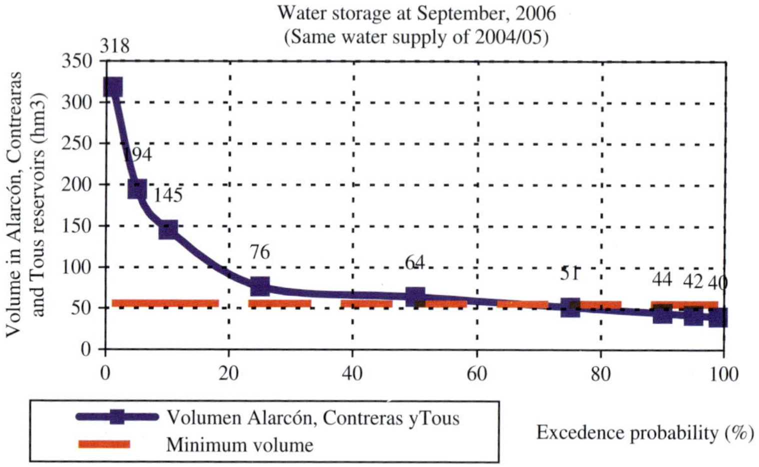

Figure 19. Water storage probabilities related to the Alarcón Contreras and Tous reservoirs at the beginning of October 2006

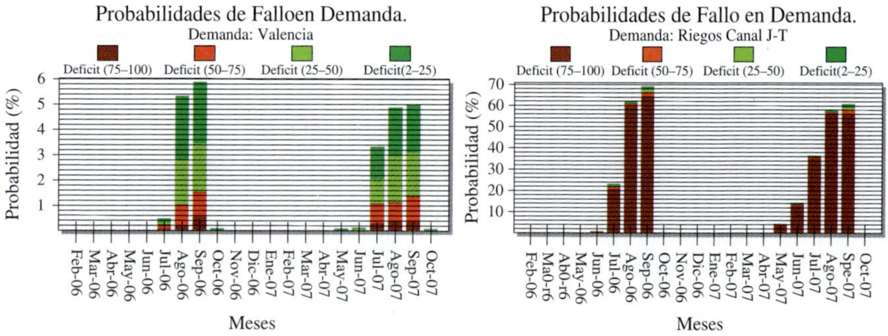

Figure 20. Failure probability through the 2006 year for the city of Valencia and the Canal Jucat-Turia agricultural area water demands, considering the repetition of the 2005 year water supplies

7. STAGE 4: MITIGATION MEASURES DEFINITION AND EFFECTIVENESS ESTIMATION

Due to the extreme drought in the Júcar area a Permanent Drought Commission has been constituted by the *Royal Decree 1265/2005, of October 21, 2005*, to simplify de decision process and the definition of the measures to apply during the current drought. As a support for the Commission a Drought Management Office has been created, whose purposes are: technical support for the definition of a measure plan, technical support for drought monitoring, technical support to evaluate the effectiveness of the measures and others.

The drought evolution monitoring and the measure effectiveness evaluation are performed by the Júcar River management simulation model (Figure 17) (OPH, 2002). The model calculates the probabilities related to the system conditions at the end of the campaign, and the failure risk related to each water demand, considering different management options and different possible measures.

The forecast results are displayed, every month, to the Permanent Drought Commission, in order to help the decision process.

The concerns about the forecast, which show a high probability of reaching the minimum technical storage limit by the August 2006, led to the approbation by the Commission of the following mitigation measures:

- the use of drought wells in the Ribera del Júcar area (Figure 21);
- the reuse of purified waste water for agricultural uses;
- the application of measures to save water;
- others.

The execution of the approved measures will make possible to save water during the current campaign and to store a sufficient volume of water in the Alarcón,

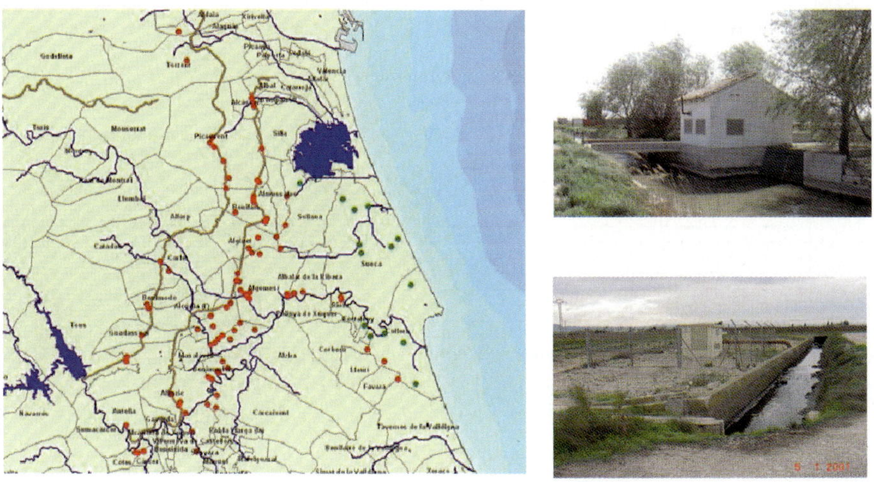

Figure 21. Drought wells and pumping stations for the reuse of water in the Ribera del Júcar area

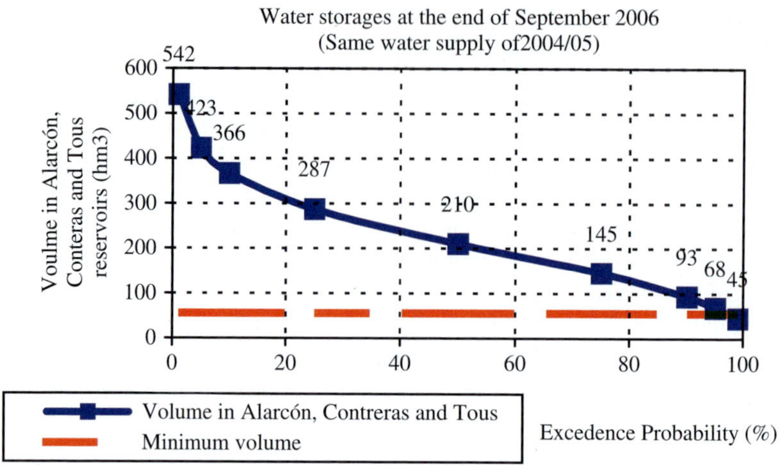

Figure 22. Probabilities of exceeding the volume of stored water in the Alarcón Contreras y Tous reservoirs, at the end of September 2006, calculated considering the execution of the approved measures

Contreras y Tous reservoirs (Figure 22) for the beginning of the next hydrological year in case the drought persists another year.

The approved measures reduce the risk of shortage for the local demands (Figure 23).

Finally, as it happened in the previous step while the simulation models were used as operation drought indicators, the storage forecast expected in case of complete application of the approved measures and in case of no intervention is calculated and compared considering the last year inflows. The results, Figure 24, are unlikely but easy to understand by experts and users.

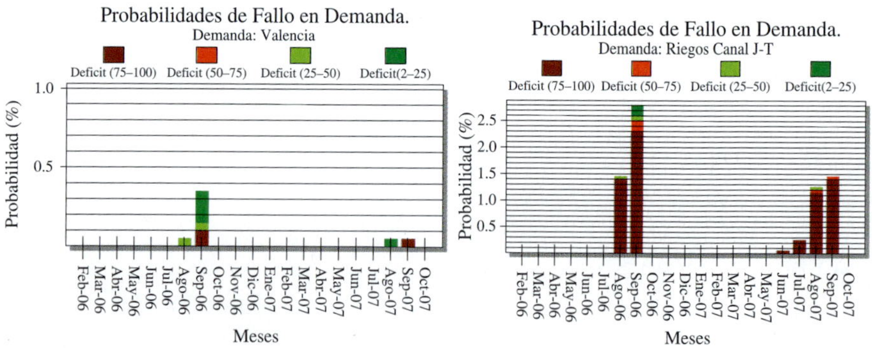

Figure 23. Failure probabilities through the 2006 campaign in the Valencia urban area and in the Canal Jucar-Turia agricultural area, considering the approved measures

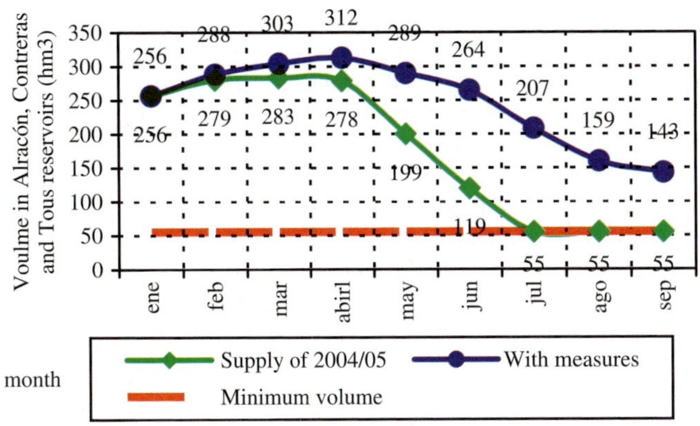

Figure 24. Water storage forecast in the Júcar system through the 2006 campaign, with and without the drought mitigation plan

8. CONCLUSION

A methodology developed for the water management institutions to foresee and to mitigate the effects of droughts has been described in the present chapter. The methodology includes a set of tools - from the indexes included in the drought early alert system up to the water system management simulation models– whose purpose is to define as well as evaluate the efficiency of the drought mitigation measures.

This methodology has been applied to the Júcar River basin in the hydrological year 2004-05, during one of the worst hydrological droughts of the modern history. The main advantage of the methodology is the ability to calculate, in the short and medium term, the probabilities related to the future water storages and supplies depending on the mitigation measures that are going to be adopted. The forecasts provided using this methodology increase stakeholders' sensitivity to the water saving and improve the efficiency of the water management policies.

REFERENCES

Andreu, J. Capilla, J. and Ferrer, J. (1992) *Modelo Simges de simulación de la gestión de recursos hídricos, incluyendo utilización conjunta. Manual del usuario.* SPUPV-92.1097, ISBN 84-7721-169-8.

Andreu, J. Capilla, J. and Sanchis, E. (1996) *AQUATOOL,* A Generalized Decision-Support System for Water-Resources Planning and Operational Management. *J. Hydrology.* 177: 269–291.

CHJ (2004) *Jucar Pilot River Basin, Provisional Article 5 Report Pursuant to the Water Framework Directive.* Confederación Hidrográfica del Júcar, Ministerio de Medio Ambiente.

McKee, T. B., N. J. Doesken, and J. Kleist, (1993) The relationship of drought frequency and duration to time scales. Preprints, *8th Conference on Applied Climatology*, 17–22 January, Anaheim, California, American Meteorological Society, 179–184

OPH (2002) *Consultaría y asistencia para el estudio de utilización conjunta de los recursos hídricos superficiales y subterráneos de las cuencas media y baja de los ríos Júcar y Turia*. Oficina de Planificación Hidrológica, Confederación Hidrográfica del Júcar.

Palmer, W. C. (1965) *Meteorological Drought*. Research Paper 45, U.S. Department of Commerce Weather Bureau, Washington, D.C.

Pérez, M. A. (2005) *Modelo distribuido de simulación del ciclo hidrológico y calidad del agua, integrado en sistemas de información geográfica, para grandes cuencas. Aportación al análisis de presiones e impactos de la Directiva Marco del Agua*.

Sánchez, S. T., Andreu, J. and Solera, A. (2001) *Gestión de Recursos Hídricos con Decisiones Basadas en Estimación del Riesgo*. Ed.: Universidad Politécnica de Valencia. I.S.B.N.84-9705-037-1.

CHAPTER 11

MIXED SIMULATION-OPTIMIZATION TECHNIQUE FOR COMPLEX WATER RESOURCE SYSTEM ANALYSIS UNDER DROUGHT CONDITIONS

G. M. SECHI AND A. SULIS

CRIFOR – UNICA, University of Cagliari, Italy

Abstract: The chapter illustrates a Decision Support System (DSS) for multi-reservoir water resource systems, named WARGI (Water Resources system optimization aided by Graphical Interface). WARGI was upgraded in the SEDEMED II Project to implement a mixed approach that uses optimization and simulation techniques which identifies short and medium-term actions to be implemented in system management according to a proactive approach. Under drought conditions, the DSS aims to predict in advance the consequences of management assumptions in a predefined system configuration. The optimization module considers future hydrological and demand scenarios to modify the set of decision variables to be adopted by the simulation module. Lastly, WARGI's application to the water resource system of Southern Sardinia (Italy) is illustrated

Keywords: Simulation and optimization of water resource systems, Drought management

1. INTRODUCTION

Appropriate management of complex water supply systems, when operating under water scarcity conditions, requires adoption of modelling tools helping decision-makers to identify sets of actions able to mitigate the impacts of drought on users. Usually, mitigation actions translate into variations in management rules and structural actions on system configuration, following either a reactive or a proactive approach to the drought phenomenon (see for instance Yevjevich et al., 1983; Rossi, 2000).

The reactive approach consists in measures adopted both during and after the drought period, once its consequences are perceived. While this approach is still today the most common response to drought emergencies, obviously, the actions undertaken are, as a rule, of brief duration and non-structural in nature, entail high economic and environmental costs for the community and often do not reduce the system's vulnerability to similar future events.

The proactive approach consists in a series of measures which may also include structural actions on the system, coherently developed in a planning strategy; they

G. Rossi et al. (eds.), Methods and Tools for Drought Analysis and Management, 217–237.

are designed well in advance of the start of the drought period, and implemented also during and after. The aim of such measures is to reduce the system's vulnerability and increase its reliability under drought conditions.

The identification of the best set of measures requires availability of appropriate modelling tools organized in Decision Support Systems (DSS). Within a proactive approach, an appropriate combination of long term and short term measures can be identified by considering alternative options whose effectiveness is evaluated with comparative analysis procedures. Usually, assessment of system performance under a predefined infrastructural setup and with assigned management techniques is, as a rule, achieved by means of simulation models, while optimization models are employed to assess "theoretical" system efficiency when management is hypothetically implemented by an ideal manager.

In previous works (Sechi and Zuddas, 2000; Sechi et al., 2004; Sechi and Sulis, 2005; Sulis, 2006) we illustrated how, in developing the WARGI DSS we had attempted to analyze complex water systems under different scenarios in a user-friendly manner through the implementation of a specialised graphical interface. WARGI combines possibilities of evaluating system efficiency by adopting a representation adhering to reality, as allowed by simulation models, with the exploratory potential afforded by optimization models. The structure of WARGI consists of various independent macro-modules (some of which were already described in Sechi et al., 2004) linked together through transfer of encoded variables (Figure 1). The toolkit is implemented in Linux environment and encoded in C++ and Tcl/Tk.

As was illustrated in Manca et al. (2004) and Sechi and Sulis (2005), the system initialisation and input module (Initialization and Hydrologic Data Input) handles value definition of the main parameters and the creation and possible modification of system elements. This module processes data coming from the Graphical User Interface module, transfers data required for formulating the optimization module (WARGI-OPT) and implements the simulation algorithm (WARGI-SIM). Software construction by means of independent modules also makes it possible to use the toolkit either for system optimization alone (WARGI-OPT) or for simulation alone (WARGI-SIM). If, moreover, analysis with scenario optimization is required, the Scenario Generation module passes to the WARGI-OPT module (Pallottino et al., 2005) parameters for model construction.

Water supply system analysis with the use of non-conventional resources is implemented in the WARGI-QUAL module (Salis et al., 2005a, Salis et al., 2005b) which links use of resources available to compliance with quality requirements. The procedure associated with solution of the optimization model determines creation of a standard .mps (mathematical programming standard) file, which is used as interface with the solving code. The WARGI-OPT module communicates model structure to the Solver module, responsible for connection with the user-selected software for resolving the optimization problem and enables management of information

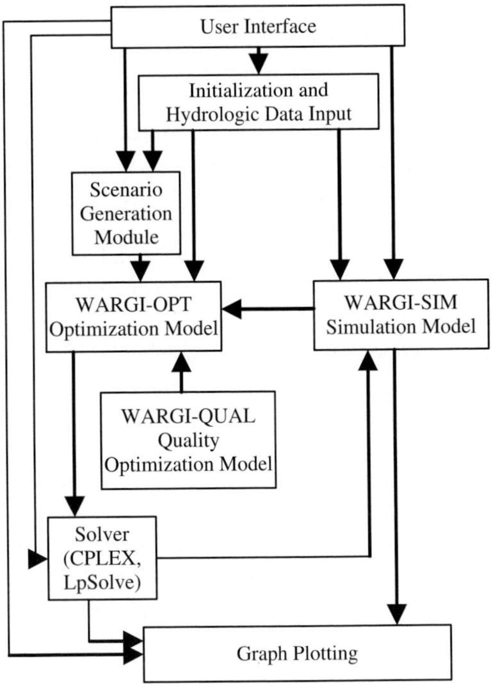

Figure 1. WARGI toolkit components

pertaining to results obtained. The user is able to view results in graphic format through the information visualised in the Graph Plotting module.

The application potential of WARGI has been tested on some real systems in the framework of previous European Projects WAMME (WAMME, 2003) and SEDEMED (SEDEMED, 2003). The WARGI-SIM simulation module was also used in the Water-Basin Plan (*Piano Stralcio di Bacino*) of the Region of Sardinia (RAS, 2005) to identify supply possibilities of multi-sector systems and the deficits associated with various use types in the drought-prone hydrological scenarios taken as reference for the Plan.

In the rest of this chapter, we will limit description of the WARGI toolkit to those aspects regarding interaction between the simulation module (within which infrastructural setup and management preferences are defined) and the optimization module which, with regards to different hydrological and demand scenarios, identifies short and medium-term actions to be implemented in the system management. We will briefly recall the methodological context and the scientific literature, thereafter, we illustrate interactions between the two modules and analysis of results obtained on the Flumendosa-Campidano system, used as a test-case in the SEDEMED II Project.

2. INTERACTION BETWEEN SIMULATION AND OPTIMIZATION IN WATER RESOURCE SYSTEM MODELLING

In recent decades simulation models have been extensively applied to real case water resource systems and are indeed still the mainstay for these analyses (Hufschmidt and Fiering 1966; Yeh 1985; Loucks and Van Beek 2005). In contrast, the use of optimization models for real-case water resource system management is rare. Nevertheless, optimization modelling is well documented in academic and research literature (Loucks et al., 1981; Yeh 1985; Labadie 2004). In general terms simulation models provide the response of a system with an assigned set of resource characteristics, resource availabilities and demands and predefined operating rules. Therefore, simulation models are mainly used to assess the economic benefits which may be generated by alternatives that are examined in a sequence of simulation steps. Again in general terms, simulation models do not define optimal operating rules to be used in system management; for their identification it is necessary to use iterative procedures of the trial-and-error type.

The search for optimal operating rules may thus prove rather time consuming, in view of the large number of alternative scenarios to be explored. It is desirable to use a modelling approach combining the representational flexibility of simulation models with the efficient state exploration of mathematical optimization models. In research-oriented applications, various strategies are employed to combine simulation and optimization (Wurbs 1993). Dorfman (1965) argued that using optimization to screen the full range of alternatives offers significant economies in the analysis of large river basin systems. Jacoby and Loucks (1972) analyzed the difficulties and advantages of the joint use of these two types of models in both static and dynamic planning problems. Stedinger et al., (1983)presented a brief state-of-the-art review of screening models developed over the last two decades, analyzing their ability to identify cost-efficient designs. Simonovic (1992) made a review of state-of-the-art analysis techniques for reservoir management and operation in large scale, complex problems highlighting the gap existing between research studies and their practical application to real systems.

In the search for effective operating policies for multi-purpose multi-reservoir systems, Karamouz et al., (1992) generalized a dynamic programming model already described by Karamouz and Houck (1982) forapplication to single reservoirs. Lund and Ferreira (1997) presented a network flow based optimization model (HEC-PRM by USACE) coupled with a simplified simulation model for refining and testing operating rules. Oliveira and Loucks 1997), and subsequently Ahmed and Sarma (2005), used genetic algorithm. Randall et al., (1997) described a linear programming model embedded in a month-by-month simulation model to be used for water supply planning. Nalbantis and Koutsoyiannis 1997) proposed and tested a parametric rule for a system of reservoirs. The parameters are estimated by optimization, using simulation to evaluate the objective function value for each trial set of parameter values. Sinha et al., (1999), and Neelakantan and Pundarikanthan (2000) developed non-linear programming models for the solution of the linked optimization-simulation model useful for solving reservoir operation and design problems.

The process employing optimization to reduce the range of designs and policies requiring simulation and more in-depth evaluation is often called "preliminary screening" (Loucks and Van Beek, 2005). Simulation models overcome the limits inherent in optimization models, but they work well only when a relatively small number of alternative combinations of physical facilities and operating procedures needs to be evaluated. Unfortunately, in a complex water resource system the number is quite large and the 'trial and error' process of simulation becomes very time consuming. The use of preliminary screening techniques, which couple analysis with an optimization model to shrink the range of alternative scenarios, highlights the limits of the optimization model (which is simplified so as to be mathematically tractable and computationally efficient in large-size systems) in effectively representing a real-scenario system and clearly indicate the best options. Indeed, we cannot be certain that the two models (simulation and optimization), when applied to the same system, will produce comparable performance index values. Coming down to details, there is no correspondence between the operating rules obtainable from optimum flows given by an optimization model and those we must introduce in a simulation model.

On the other hand, although one cannot univocally formulate interactions between optimization and simulation, clearly, simple reliance on simulation based on the "common sense" and experience of the manager in the choice of operating rules is fraught with danger, and may impair objective assessment of potential system efficiency.

3. THE COUPLED SIMULATION-OPTIMIZATION APPROACH IN WARGI

As an alternative to the simulation-only approach, which requires 'a priori' definition of operating rules, or to the *off-line* use of optimization for preliminary screening of operating rules to be adopted in simulation, in developing WARGI we have adopted a modelling format based on full integration between simulation and optimization modules.

As regards the **simulation module**, this approach does not require the input of precise operating rules, but simply definition of *preferences* and *priorities* by the manager. The preliminary requirement for this approach is to achieve in the input phase an adequately detailed description of all components of the water system. This enables the model to identify all implicitly present functional constraints. Thanks to the graphical interface of WARGI-GUI, the user can show the physical system in the form of an oriented graph consisting of nodes and arcs, using mask-windows to enter data on the characteristics of each component.

Definition of *preferences* by the manager implies for each demand node the possibility of providing an ordered list of resources from which a supply flow may be activated to meet the demand. These preferences may be based on physical, economic, legal, contractual or quality considerations, and, at all events, are subject to assessment of other conditions that make transfer admissible and cost-effective. These conditions can be easily exemplified with reference to reservoir resources for which the manager can define a "target storage", which may also be differentiated according to periods, and constraints on volumes to be set aside for specific uses. Constraints on volumes may be activated when reservoir status drops below critical levels which the user may associate with the reservoir, thus defining "reserved volumes" for specific use types. Optimal "target storage" and "reserved volume" values are defined through a specific WARGI procedure. Preference criteria may also be assigned to the resource: for instance, we can ensure that supply from flowing water is given priority in use over water that can be collected in reservoirs.

Definition of *priorities* by the manager enables subdivision of the volume required by each use type into priority bands by means of percent rates. Priority bands are identified for each demand node. The supply procedure identifying the volumes transferred in the network from resources to demands thus takes place following established priority levels. When water available is scarce, this makes it possible to avoid imbalances in meeting demands. Once transfer from resources to demands in the simulation time-step has been completed, the definition of resource targets enables WARGI-SIM to activate supplementary flows which make it possible to minimise deviation from the target configuration. Thus, resource-demand flows are activated in the simulation module through implementation of instruction sequences using a combination of procedures identified, first and foremost, on the basis of the preference and priority criteria established by the manager. Secondly, transfer amounts are defined through interaction between simulator and optimization modules.

Use of the **optimization module** interacting with simulation basically refers to two aspects.

The first aspect is linked to identification of the *optimal path* (essentially the minimum cost path on the oriented system graph) to transfer water to demands having equal priority from resources having equal preference. For resolution of this problem in WARGI we used a modified version of Dijkstra's algorithm (Dijkstra, 1959).The user can associate with each arc of the graph representing the system the costs, benefits and constraints on transfer possibilities. The constraints are the capacity thresholds of transfer pipelines and plants present in the system. Transfer costs or benefits may be dependant on transfer methods (e.g. gravity flow pipelines or pumping systems) or on the manager's preferences. Economic values and constraints associated with arcs determine the optimal path for flow transfers. Once the lowest cost path has been saturated, an alternative path is searched out on the graph (of gradually increasing cost) between the same resource and demand nodes. If resource-demand transfer is not possible, the lower-preference resource is sought. As already mentioned, flow transfers are always activated considering decreasing demand priorities. The aim being, in the event deficits arise, to prevent them from affecting higher priority bands in demand nodes.

The second interaction aspect with the optimizer regards definition of *programmed reductions* in supply to demands. For this purpose, flow configurations provided by the WARGI-OPT optimization module can be seen as a reference target for the simulation phase. In a predefined hydrological and demand scenario, flow configuration obtained in optimization can in fact be seen as being that obtained by an ideal manager with perfect knowledge of future events; thus the simulator can operate so as to minimize overall system management costs on the entire time horizon of the scenario (Sechi et al., 2004). This aspect is essential for implementation of mixed optimization-simulation modelling and makes it possible to determine the best combination drought mitigation measures within a proactive approach.

In view of the specific aims of the SEDEMED II project, in the section below we will provide further details on interaction between optimization and simulation with reference to the latter aspect, felt to be especially relevant in the management of water resource systems under drought conditions.

4. IDENTIFICATION OF OPERATING RULES UNDER RESOURCE SHORTAGE CONDITIONS

The aim of DSS WARGI is to implement a proactive approach in system management operating rules to mitigate the impacts of droughts on users. These changes, which anticipate the occurrence of resource shortages, are defined by processing in the simulation module the forecasts provided by the optimization module, based on the assumption of reference hydrological scenarios. Figure 2 plots the interaction between the WARGI-OPT optimization module and the WARGI-SIM simulation module in the analysis of a system on a time horizon covering T periods. During the simulation process (at synchronization instants τ_i between the

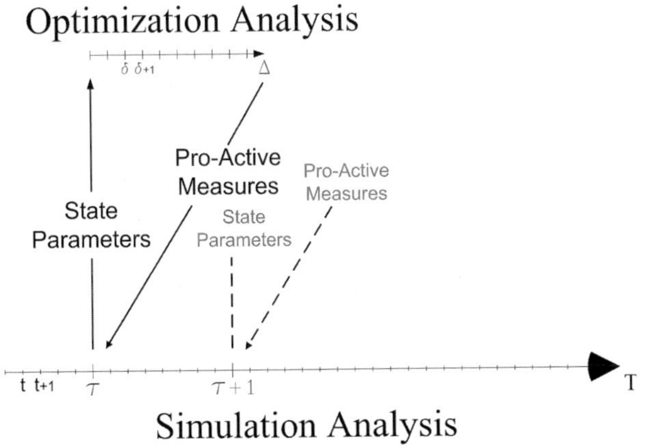

Figure 2. Interaction between the optimization model and simulation in WARGI

simulator and the optimizer) WARGI-OPT defines a system evolution hypothesis on a time horizon Δ, reduced with respect to the overall simulation time horizon T. For this reduced time horizon, on which the optimizer operates, we assume a hydrological scenario (resource input) and a demand scenario (resource output). With regards to these scenarios, the flow configuration provided by the optimizer may be considered as that set by an ideal manager of the system. Based on the information provided by the optimizer, starting from synchronization instant τ_i the simulator is able to define a set of measures which alter the value of decision variables in the subsequent period. At all events, these measures do not extend beyond the subsequent synchronization step. Thus, simulation explores the system's response with a set of operating rules modified so as to handle resource shortage situations identified in the optimization phase.

The time parameters (defined by the users by means of the graphical interface) in the simulator-optimizer interaction, are as follows:

- T time horizon of simulation for overall system analysis;
- Δ time horizon of optimization;
- $t = 1, .., T$ unit simulation step;
- $\partial = 1, .., \Delta$ unit optimization step;
- $\tau = \tau_1, \tau_n$ instants of simulation-optimization synchronization.

The optimization time horizon Δ must be representative of hydrological variability and demand trends. In particular, in the multi-reservoir systems examined in the SEDEMED II project, its extension is linked to the regulation capacity of the surface reservoirs present in the system. In WARGI, the user defines the system's configuration and operating procedures by means of graphical interface, specifying his preference and priority criteria as illustrated in the previous paragraph. The most immediate response from interaction of simulation with the optimizer consists in the adoption of limitation procedures for demands with lower priority levels. However, procedures modifying the values of reserved volumes or of target storages in reservoirs may also be adopted.

We should point out that in the optimization module, simulation preference and priority criteria are transformed into attribution of transfer-costs and operating constraints along the graph of the system, guiding the solver towards an optimal flow configuration in line with priority and preference criteria provided by the manager. Moreover, due to the above, the system configuration inputted to the optimizer must have a level of adherence to reality similar to that used for simulation.

Other aspects requiring special attention regard definition of the forecast hydrological scenarios on which the optimizer works and assignment of costs associated with possible deficits, differentiated by priority bands. These two aspects – hydrological criticality and deficit costs – will be expanded on below and in the following sections, illustrating analyses performed on a real case.

What emerges from the above discussion is that changes to the operating rules in WARGI-SIM mainly consist in redefinition of the demand configuration. Other actions may however be introduced by the user acting on preference and priority criteria. At each synchronization instant τ_i, WARGI-OPT defines a set of possible demand reductions in the various centres of the system, in accordance with the preference criteria and critical level of the adopted hydrological and demand scenario. This demand reduction action allows definition of *programmed deficit* volumes and is a *proactive* procedure in water crisis management. This consumption measure is based on assessment of the magnitude of water shortage and aims to minimise the vulnerability of higher priority demands in the water resource system during drought periods.

On closing the simulation procedure, the overall benefit in resource use can be graphically represented as the area enclosed below the marginal benefit curve identified by the volume supplied. In a predefined situation of admissible elasticity of demand, essentially linked to the programmed reduction period, the loss in benefits consequent on reduction in demand can be expressed by the following integral formula:

$$D = \int_{Q_r}^{Q_s} P_s \cdot \left(\frac{Q_s}{Q}\right)^{1/E} dQ \qquad (1)$$

where Q_s indicates request for complete demand satisfaction, P_s the resource marginal cost under conditions of completely fulfilled demand, Q_r reduced supply during the drought period.

To calculate D we need one point of the demand-curve and elasticity $E$of demand in relation to the programming period. Several static or dynamic models for evaluation of the water demand law (see for instance Agthe and Billings, 1980) and for estimating elasticity (e.g. Moncur, 1987) in the various use types are found in literature. Elasticity is a function of demand type and of proactive or reactive approach. Clearly, E will be very low for demands in high priority bands (e.g. urban uses), where demand varies only slightly in relation to price; also, as a rule the value of elasticity in a short term demand curve is lower than in a long-term curve. Thanks to use of a post-processor integrated in DSS WARGI, we can define for each priority level the losses in unit benefit with respect to long-term demand (identified by WARGI-OPT) and to short-term demand (identified by WARGI-SIM) which are defined, respectively, as "*programmed deficit cost*" and "*non-programmed deficitcost*".

As already stated, a second level of interaction between optimization and simulation may involve redefinition of reserved volumes in surface reservoirs as well as redefinition of operating rules. This further level of interaction is subsequent to the above mentioned cost considerations. The assignment of reserved volumes is generally effective on time horizons covering up to a number of years. In real-case applications, which we will address in the following section, we operate with WARGI-OPT in a forecasting context associated with hydrological scenarios of different levels of criticality. Moreover, we work on an optimization time horizon Δ that must be significant with respect to the possibility of system regulation. The correct assumption on time horizon Δ is a critical aspect in defining configuration of the reserved volumes for demands with higher priority.

5. APPLICATION OF THE MIXED APPROACH TO THE WATER RESOURCE SYSTEM OF SOUTHERN SARDINIA

As explained in the introduction, the application potential of mixed simulation-optimization techniques for analysis of complex water resource systems was assessed in the framework of the SEDEMED II Project considering the water resource system of Southern Sardinia, as defined in the Water-Basin Plan (Piano Stralcio di Bacino) of the Region of Sardinia (RAS, 2005). This system has a storage capacity from surface reservoirs totalling $723 \cdot 10^6$ m^3; aggregate residential demand of $100 \cdot 10^6$ m^3/year; aggregate irrigation demand of $234 \cdot 10^6$ m^3/year; aggregate industrial demand of $19 \cdot 10^6$ m^3/year. Time horizon T in the system analysis is set at 54 years, in line with hypotheses made by the Regional Authorities (RAS, 2005). As regards the system, hydrological inflow data on 19 selected sections are available on a monthly basis.

Based on the observed hydrological series, we can define hydrological scenarios with different extension and criticality (in terms of the sum of inflows to reservoirs and to diversion sections) to be considered in the optimization module. Critical hydrological series covering between one and three years, are graded from the driest to the wettest (Table 1).

Before applying the mixed simulation-optimization technique, we carried out a simulation-only analysis using the WARGI-SIM module alone. System simulation was implemented considering a time horizon $T = 54$ years divided into unit time intervals t of one month each. Results obtained provide an assessment of the system's capability to meet water shortage situations when no programmed supply reduction measures are implemented in the system to mitigate drought impacts. This phase thus provides an element of comparison allowing assessment of the benefits of the measures adopted in the following mixed optimization-simulation approach.

Numerous criteria may be used to quantify levels of system performance obtained by the system using predefined operating rules (Hashimoto et al., 1982a; Hashimoto et al., 1982b). These criteria may be obtained from individual indices or from

Table 1. Hydrological scenarios assessed on the basis of critical observed annual flow (10^6 m^3/year) in the system

Optimization Scenario		1st Critical Case	3rd Critical Case	5th Critical Case	7th Critical Case	9th Critical Case	11th Critical Case	13th Critical Case
1 Year	1st Year	134.8	176.3	208.0	211.6	214.9	221.8	232.1
2 Years	1st Year	223.8	211.6	134.8	176.3	214.3	312.8	300.7
	2nd Year	134.8	208.0	300.7	298.4	290.8	215.9	267.7
	Average	179.3	209.8	217.8	237.3	252.6	264.3	284.2
3 Years	1st Year	214.9	134.8	309.8	290.8	498.2	298.4	479.2
	2nd Year	265.9	300.7	211.6	239.4	223.8	364.5	176.3
	3rd Year	164.1	267.7	208.0	221.8	134.8	232.0	298.4
	Average	215.0	234.4	243.1	250.7	285.6	298.3	318.0

combinations of different indices. Using simulation alone, estimation of the best combination of measures was made starting from the following performance indices for the various uses (residential, industrial and irrigation):

1. vulnerability, expressed as percent value of maximum annual deficit on the total value of annual demand;
2. time reliability, expressed as percent value of the months in which deficit is less than predefined thresholds (15% and 25%);
3. volumetric reliability, expressed as deficit rate on the demand value for the entire period of analysis.

To simplify analysis, but also in agreement with hypotheses contained in the RAS (2005), all residential demand was placed in the higher priority band; the second priority band grouped all industrial demand, while irrigation uses were placed in the third band. Results obtained with WARGI-SIM, using simulation alone, are summarised in Table 2.

As a consequence of breakdown into priority bands, most of the system's resource deficit will concern irrigation use. Moreover, the prolonged periods of drought present in the hydrological series, associated with lack of adequate measures for protecting higher priority resources will engender intolerable stress conditions for residential uses (maximum annual deficit of 34.36% and 34 months in which mean monthly deficit exceeds the 25% threshold).

In the mixed simulation-optimization technique, the optimization module is recalled by the simulation at yearly intervals (synchronization interval of one year) and with synchronization instant τ placed at the start of the irrigation period (1st of April). The optimization time interval ∂ is also established as one month. As shown in Table 1, the optimization time horizon Δ has been varied from 1 to 3 years. Time horizon and criticality of hydrological scenario are essential parameters for defining the type and degree of measures for facing up to system failures in a proactive approach. Indeed, by considering different optimization horizons and scenarios we can determine the effectiveness of mitigation measures during the drought period seen in the light of immediacy of intervention.

Table 2. Values of performance indices relative to simulation alone

	Residential	Industrial	Irrigation
Max annual deficit (% Demand)	34.36	40.69	70.94
Time reliability (% Months) Deficit $\leq$ 15%	94.50	94.65	93.40
Time reliability (% Months) Deficit $\leq$ 25%	94.65	94.65	93.40
Volumetric reliability (% Demand)	96.12	94.82	91.85

System performances indexes were expanded also by considering the reduced volumetric reliability, obtained as percent value of non-programmed deficit on demand reduced according to the consumption containment implemented in the optimization module. Tables III show performance index values generated by the mixed approach, when the simulation adopts demand reduction rules deriving from optimization associated with different hydrological scenarios. More critical scenarios establish greater supply reductions in the lower priority bands (essentially irrigation demand) as a measure to lessen the vulnerability of residential uses to future drought periods. Any assumption of highly pessimistic scenarios engenders excessively drastic irrigation consumption containment actions, and conversely, over-optimistic scenarios provide no information useful for the adoption of prevention measures.

Optimal combination of prevention measures cannot be based only on previous quantitative performance indices, but must also be supported by analysis of the economic impact of implementation of these actions on different demand categories. However, the consequences of failure to fulfil demand can prove difficult to measure in purely economic terms (e.g.: ecologic, environmental, social). In our application of WARGI to the Southern-Sardinia system, we carried out prior sensitivity analysis of overall system performance value associated with varying configurations of programmed and non-programmed unit deficit costs.

A summary of the results obtained is contained in Figure 3 which shows the trend in the Objective Function (OF), with varying criticality of the optimization scenario and the ratio between programmed and non-programmed deficit costs. Based on results obtained with sensitivity analysis and available demand elasticity values, we then defined short and long-term demand curves for the various types of water uses. The system's economic performance was evaluated including in the OF, besides deficit costs, management and operating costs linked to demand satisfaction of different water use types. As was to be expected, as the ratio between non-programmed and programmed deficit cost increases, the system's optimal configuration is obtained by means of a combination of short and medium-term measures associated with hydrological scenarios of growing criticality. Moreover, an increase in deficit costs for residential water demand, if considering a fixed ratio between programmed and non-programmed costs, worsens the system's economic performance level when lower criticality scenarios are adopted.

Results illustrated in Figure 4 show the trend of economic performance, expressed as OF value, with varying criticality levels of the hydrological scenario used in the WARGI-OPT optimization model. In these analyses too, we based our calculations on a forecast horizon of one year, and we expressed scenario criticality as the ratio between aggregated hydrological inflows in the scenario and mean inflow in the overall simulation horizon. In particular, in Figure 4 we can see how implementation of deficit prevention measures based on overly pessimistic hypotheses on future hydrology formulated in WARGI-OPT determines, in the simulation phase, the adoption of excessively cautious system management rules with respect to actual water availability in the system. On the other hand, excessively optimistic

Table 3a. Performance index values using hydrological scenarios with different criticality and annual optimization horizon

	1st Critical case			3rd Critical case			5th Critical case			7th Critical case		
	Res	Ind	Irr	Res	Ind	Irr	Res	Ind	Irr	Res	Ind	Irr
Max annual deficit (% Demand)	1.91	55.78	90.81	6.94	22.82	77.41	15.94	23.59	68.66	15.99	23.61	69.92
Time reliability (% Months) Deficit ≤ 15%	99.84	96.23	66.04	99.37	95.75	67.92	98.74	98.74	71.38	98.74	98.43	71.38
Time reliability (% Months) Deficit ≤ 25%	100.0	96.23	66.04	99.37	99.37	71.54	98.74	98.90	73.11	98.74	98.58	73.11
Volumetric reliability (% Demand)	99.96	98.45	80.84	99.58	98.97	83.21	98.78	97.97	85.04	98.58	97.73	85.29
Reduced volumetric reliability (% Programmed demand)	99.96	100.00	100.00	99.63	99.43	99.72	98.78	98.90	99.78	99.13	98.66	99.06

	9th Critical case			11th Critical case			13th Critical case			Average year		
	Res	Ind	Irr	Res	Ind	Irr	Res	Ind	Irr	Res	Ind	Irr
Max annual deficit (% Demand)	15.99	30.81	69.93	21.66	30.56	72.27	21.72	24.85	73.50	34.35	40.69	70.94
Time reliability (% Months) Deficit ≤ 15%	98.58	89.62	71.38	98.43	83.81	71.38	97.96	97.96	71.38	94.34	94.65	93.40
Time reliability (% Months) Deficit ≤ 25%	98.74	98.58	73.11	98.43	98.11	73.11	98.11	98.11	75.00	94.65	94.81	94.30
Volumetric reliability (% Demand)	99.00	96.98	85.18	97.41	95.82	86.36	97.96	96.94	86.40	96.22	94.84	91.86
Reduced volumetric reliability (% Programmed demand)	99.11	98.68	99.06	98.99	98.30	98.80	98.76	98.18	98.58	96.22	94.84	91.86

hydrological scenarios engender insufficient long-term prevention measures and hence require implementation of short term measures, leading to significant penalisations and a consequent increase in OF values. The at times irregular trend in the OF (in particular when flows-ratio value is close to 0.5) indicates that, in a complex multi-use and multi-resource system, prevention measures should be better identified by using scenarios characterization that consider hydrological criticality not only in terms of mean annual value of total inflow in the system, but also as regards its space and time variability.

Table 3b. Performance index values using hydrological scenarios with different criticality and biennial optimization horizon

	1st Critical case			3rd Critical case			5th Critical case			7th Critical case		
	Res	Ind	Irr	Res	Ind	Irr	Res	Ind	Irr	Res	Ind	Irr
Max annual deficit (% Demand)	0.00	0.00	71.76	6.88	12.15	62.79	6.91	47.49	60.16	15.71	18.13	62.78
Time reliability (% Months) Deficit $\leq$ 15%	100.00	100.00	52.83	99.69	99.69	60.38	99.69	97.96	62.26	99.21	99.21	66.04
Time reliability (% Months) Deficit $\leq$ 25%	100.00	100.00	58.49	99.69	99.69	66.04	99.69	97.96	66.04	99.21	99.21	71.70
Volumetric reliability (% Demand)	100.00	100.00	77.83	99.74	99.61	82.63	99.78	98.72	82.79	99.45	98.90	84.65
Reduced volumetric reliability (% Programmed demand)	100.00	100.00	100.00	99.79	99.69	99.78	99.79	99.75	99.72	99.48	99.22	99.46
	9th Critical case			**11th Critical case**			**13th Critical case**			**Average year**		
	Res	Ind	Irr	Res	Ind	Irr	Res	Ind	Irr	Res	Ind	Irr
Max annual deficit (% Demand)	17.82	27.67	67.53	19.00	29.83	68.31	26.22	28.13	71.77	34.36	40.22	70.95
Time reliability (% Months) Deficit $\leq$ 15%	98.74	98.43	66.04	98.58	98.43	66.04	97.64	97.64	69.95	94.50	94.65	93.40
Time reliability (% Months) Deficit $\leq$ 25%	98.90	98.43	73.58	98.90	98.43	73.58	97.80	97.64	75.16	94.65	94.65	93.40
Volumetric reliability (% Demand)	98.94	98.15	85.66	98.98	97.87	85.81	98.23	97.17	87.65	96.24	94.85	91.86
Reduced volumetric reliability (% Programmed demand)	99.23	98.64	99.07	99.13	98.55	98.99	98.36	97.71	97.71	94.24	94.85	91.86

Table 3c. Performance index values using hydrological scenarios with different criticality and triennial optimization horizon

	1st Critical case			3rd Critical case			5th Critical case			7th Critical case		
	Res	Ind	Irr	Res	Ind	Irr	Res	Ind	Irr	Res	Ind	Irr
Max annual deficit (% Demand)	0.00	0.00	60.06	3.89	34.44	56.76	6.93	10.88	51.91	7.06	15.21	53.09
Time reliability (% Months) Deficit $\leq$ 15%	100.00	100.00	35.85	99.84	98.11	49.06	99.69	99.53	54.72	99.53	99.37	56.60
Time reliability (% Months) Deficit $\leq$ 25%	100.00	100.00	58.49	99.84	98.11	64.15	99.69	99.53	66.04	99.53	99.37	66.04
Volumetric reliability (% Demand)	100.00	100.00	75.75	99.93	99.35	79.76	99.72	99.44	82.16	99.64	99.30	82.75
Reduced volumetric reliability (% Programmed demand)	100.00	100.00	100.00	99.93	99.89	99.92	99.78	99.54	99.63	99.67	99.40	99.57

	9th Critical case			11th Critical case			13th Critical case			Average year		
	Res	Ind	Irr	Res	Ind	Irr	Res	Ind	Irr	Res	Ind	Irr
Max annual deficit (% Demand)	23.85	27.09	65.09	32.37	38.89	70.46	34.21	37.48	71.10	34.36	40.22	70.95
Time reliability (% Months) Deficit $\leq$ 15%	98.43	98.43	66.04	96.70	97.17	69.65	95.91	96.23	74.84	94.50	94.65	93.40
Time reliability (% Months) Deficit $\leq$ 25%	98.43	98.43	73.58	96.86	97.17	76.57	96.23	96.38	94.65	94.65	94.65	93.40
Volumetric reliability (% Demand)	98.76	98.17	86.55	97.04	95.81	88.83	97.08	95.39	89.85	96.24	94.86	91.86
Reduced volumetric reliability (% Programmed demand)	98.84	98.43	98.27	98.07	97.50	97.09	97.40	96.58	95.62	96.24	94.86	91.86

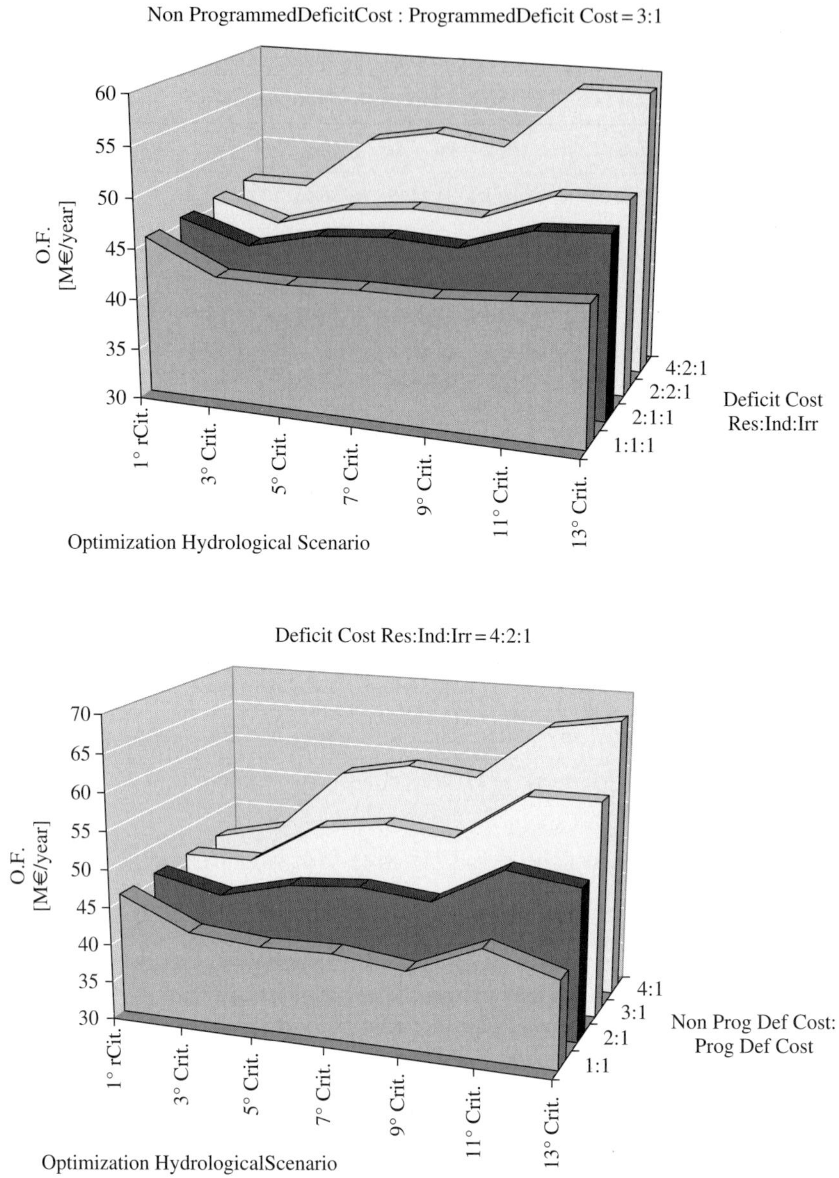

Figure 3. Sensitivity analysis of the system's economic performance with varying deficit costs

The mixed technique applied to the water resource system of Southern Sardinia shows that the best combination of short and long-term drought mitigation measures can be obtained with reference to a hydrological forecasting scenario with annual inflow close to 40% of historical series mean annual inflow.

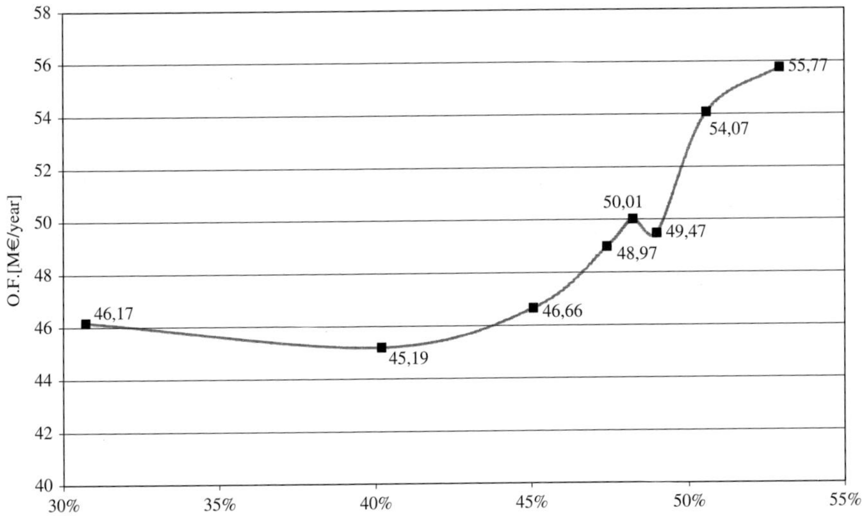

Figure 4. Economic performance trend relative to varying criticality of annual forecasting in hydrological scenario in WARGI-OPT

There follows, from the above discussion, that another key parameter in simulation-optimization interaction is extension Δ of the forecasting time horizon adopted in the optimization module. Again for the Southern-Sardinia system, in Figures 5 and 6 we show economic performance trends when the optimization time horizon Δ is extended to cover two and three-year periods. Clearly, as the time horizon is extended, we can consider more optimistic hydrological scenario in order to define appropriate measures.

Optimal management in simulation is obtained by considering in the system a mean hydrologic inflow in the two-year and three-year periods amounting to, respectively, 46% and 49% of mean annual value in the reference historical series. However, it should be noted that extension of the forecasting period also generates increased slope in the cost function when minimum value is exceeded. This factor should alert us to the need to avoid making excessively optimistic hydrologic scenario assumptions.

6. CONCLUSIONS

Results obtained indicate that the mixed simulation-optimization technique implemented in WARGI DSS provides effective support in the definition of optimal operating rules for the management of complex water resource systems. In particular, the mixed approach makes it easy to add to the model sets of actions (programmed deficits) aimed at mitigating drought impact on higher priority demands. The technique developed in WARGI DSS requires the manager to define his preference criteria to resources and priority criteria to demands. Updating of the

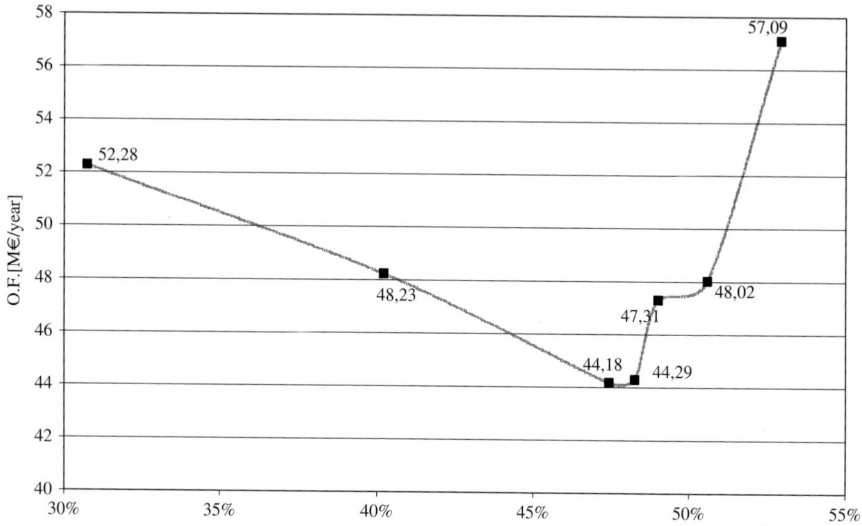

Figure 5. System economic performance trends relative to varying criticality in the two-year forecasting scenario

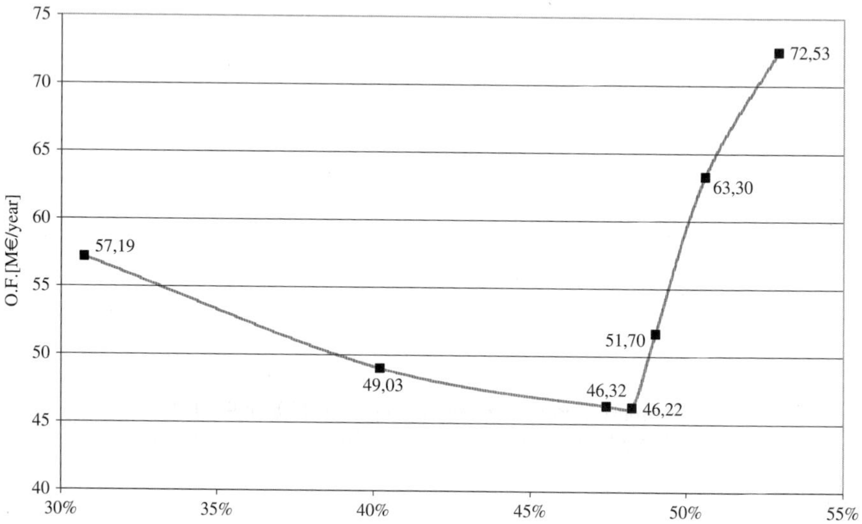

Figure 6. System economic performance trends relative to varying criticality in the three-year forecasting scenario

operating rules is carried out by the optimization module which makes forecasts on synthetic hydrologic inputs scenario sets. Effectiveness of drought mitigation actions is conditional on the criticality and extension of scenarios adopted in this phase. The report highlights the fact that application of this technique to real cases requires careful sensitivity analysis in assuming related values.

ACKNOWLEDGMENTS

This research has been developed with financial support from European Project SEDEMED II. The authors thank all partners involved for providing their collaboration in developing this research. The authors also express deep thanks to Prof. Rossi for his insightful comments and suggestions that greatly improved this chapter.

REFERENCES

Ahmed, J.A. and Sarma, A.K. (2005) Genetic Algorithm for Optimal Operating Policy of a Multipurpose Reservoir. *Water Resources Management* 19(2): 145–161.

Agthe, D.E. and Billings, R.B. (1980) Dynamic models of residential water demand. *Water Resources Research* 16(3): 476–480.

Dijkstra, E. (1959) A note of two problem in connexion with graph. *Numeriche Mathematik* 1: 269–271.

Dorfman, R. (1965) Formal models in the design of water resource systems. *Water Resources Research* 1(3): 329–336.

Hashimoto, T., Loucks, D.P. and Stedinger, J.R. (1982a) Robustness of Water Resources Systems. *Water Resources Research* 18(1): 21–26.

Hashimoto, T., Stedinger, J.R. and Loucks, D.P. (1982b) Reliability, Resiliency, and Vulnerability Criteria for Water Resource System Performance Evaluation. *Water Resources Research* **18**(1): 14–20.

Hufschmidt, M.M. and Fiering, M.B. (1966) *Simulation Techniques for the Design of Water Resource Systems*, Harvard University Press, Cambridge, Mass.

Jacoby, H.D. and Loucks, D.P. (1972) Combined Use of Optimization and Simulation Models in River Basin Planning. *Water Resources Research* 8(6): 1401–1414.

Karamouz, M. and Houck, M.H. (1982) Annual and Monthly Reservoir Operating Rules Generated by Deterministic Optimization. *Water Resources Research* 18(5): 1337–1344.

Karamouz, M., Houck, M.H. and Delleur, J.W. (1992) Optimization and Simulation of Multiple Reservoir Systems. *Journal of Water Resources Planning and Management* 118(1): 71–81.

Labadie, J.W. (2004) Optimal Operation of Multireservoir Systems: State-of-the-Art Review. *Journal of Water Resources Planning and Management* 130(2): 93–111.

Loucks, D.P., Stedinger, J.R. and Haith, D.A. (1981) *Water Resource Systems Planning and Analysis*, Prentice-Hall, Englewood Cliffs, NJ.

Loucks, D.P. and Van Beek, E. (2005) *Water Resources systems Planning and Management: an Introduction to Methods, Models and Applications*, UNESCO Press, Paris.

Lund, J.R. and Ferreira, I. (1997) Operating Rule Optimization for Missouri River Reservoir System. *Journal of Water Resources Planning and Management* 122(4): 287–295.

Manca, A., Sechi, G.M., Sulis, A. and Zuddas, P. (2004) Complex water resources system optimization tool aided by graphical interface.In: P. Liong and Babovic (Eds), *6th International Conference of HydroInformatics*, World Scientific Publishing Company, Singapore.

Moncur, J.E. (1987) Urban water pricing and drought management. *Water Resources Research* 23(3): 393–398.

Nalbantis, I. and Koutsoyiannis, D. (1997) A parametric rule for planning and management of multiple-reservoir systems. *Water Resources Research* 33(9): 2165–2177.

Neelakantan, T.R. and Pundarikanthan, N.V. (2000) Neural Network-Based Simulation-Optimization Model for Reservoir Operation. *Journal of Water Resources Planning and Management* 126(2): 57–64.

Oliveira, R. and Loucks, D.P. (1997) Operating rules for multireservoir systems. *Water Resources Research* 33(4): 839–852.

Pallottino, S., Sechi, G.M., and Zuddas, P. (2005) A DSS for Water Resources Management under Uncertainty by Scenario Analysis. *Environmental Modelling&Software* 20: 1031–1042.

Randall, D., Cleland, L., Kuehne, C.S., Link, G.W. and Sheer, D.P. (1997) Water Supply Planning Simulation Model Using Mixed-Integer Linear Programming "Engine". *Journal of Water Resources Planning and Management* 123(2): 116–124.

RAS (2005) *Piano Stralcio di Bacino della Regione Sardegna per l'Utilizzo delle Risorse Idriche. Autonomous Region of Sardinia*, Convention RAS-UNICA-EAF.

Rossi, G. (2000) Drought mitigation measures: a comprensive framework. In J. V. Vogt and F. Somma (Eds) *Drought and drought mitigation inEurope,* Kluwer, Dordrecht, The Netherlands, pp. 233–246.

Salis, F., Sechi, G.M., Sulis, A. and Zuddas, P. (2005a) Un modello di ottimizazione per la gestione di sistemi idrici complessi con l'uso di risorse convenzionali e marginali. *L'Acqua* 3:33–52.

Salis, F., Sechi, G.M., and Zuddas, P., (2005b) Optimization model for the conjunctive use of conventional and marginal waters, *Drought Management and Planning for Water Resources*, J. Andreu, G. Rossi, F. Vagliasindi and A. Vela (Eds), Taylor&Francis, Rome, 73–117.

Sechi, G.M. and Sulis, A. (2005) A mixed optimization-simulation technique for complex water resource systems analysis. *8th International Conference on Computing and Control for the Water Industry*, Exeter, UK.

Sechi, G.M., Sulis, A. and Zuddas, P. (2004) Una tecnica mista di ottimizzazione-simulazione per l'analisi di sistemi idrici complessi. *XXIX Convegno di Idraulica e Costruzioni Idrauliche*, Trento, Italy.

Sechi, G.M. and Zuddas, P. (2000) WARGI: Water Resources System Optimization Aided by Graphical Interface. In: W.R. Blain and C.A. Brebbia(eds), *Hydraulic Engineering Software*, WIT-PRESS, Southampton, UK, pp. 109–120.

SEDEMED (2003) *Drought and desertification in Mediterranean basins*, INTERREG IIIB MEDOCC Asse 4 - Ref. 2003-03-4.4-I-010.

Simonovic, S.P. (1992) Reservoir Systems Analysis: Closing Gap between Theory and Practice. *Journal of Water Resources Planning and Management* 118(3): 262–280.

Sinha, A.K., Rao, B.V. and Lall, U. (1999) Yield Model for Screening Multipurpose Reservoir System.*Journal of Water Resources Planning and Management* 125(6): 325–332.

Stedinger, J.R., Sule, B.F., and Pei, D. (1983) Multiple Reservoir System Screening Models. *Water Resources Research* 19(6): 1383–1393.

Sulis, A. (2006) *Un approccio combinato di ottimizzazione e simulazione per l'analisi di sistemi complessi di risorse idriche*, PhD Thesis, University of Cagliari.

WAMME (2003) *Water Resources Management Under Drought Conditions*, INCO-MED (DGXIII) Program, Contract N. ICA3-CT-1999-00014.

Wurbs, R.A. (1993) Reservoir-System Simulation and Optimization Models. *Journal of Water Resources Planning and Management* 119(4): 455–472.

Yeh, W.W.-G. (1985) Reservoir Management and Operation Models: A State-of-the-Art-Review. *Water Resources Research* 21(12): 1797–1818.

Yevjevich, V., Hall, W. and Salas, J. (1983) *Coping with Drought*, Water Resources Publication, Littleton, Colorado.

CHAPTER 12

RESERVOIRS WATER-QUALITY CHARACTERIZATION FOR OPTIMIZATION MODELLING UNDER DROUGHT CONDITIONS PART I – RESERVOIRS TROPHIC STATE CHARACTERIZATION

B. BEGLIUTTI[(2)], P. BUSCARINU[(2)], G. MARRAS[(2)], G.M. SECHI[(1)] AND A. SULIS[(1)]

[(1)]*CRIFOR – UNICA, University of Cagliari, Italy*

[(2)]*Agency of Flumendosa Basin, Cagliari, Italy*

Abstract: Definition of a synthetic index for classifying the quality of water bodies is a key aspect in integrated planning and management of water resources systems. In previous works (Salis et al., 2005; Sechi and Sulis, 2005), based on EU and Italian guidelines for water quality in surface reservoirs, a water system optimization modelling that considered a quality index based on the adoption of a small number of parameters has been proposed. This chapter analyses the feasibility of an index that considers quality parameters estimated both on the basis of the overall trophic state of the water body as on concentration density values of the most toxic species of algae. This index enables to describe the conditions limiting water use due to excessive nutrient enrichment in the water body and to the health hazard linked to toxic algae blooms. Application to the Flumendosa-Campidano water system highlighted the usefulness of this approach for obtaining a synthetic quality index in complex water resource systems

Keywords: Sustainable management, eutrophication, trophic state index (TSI), cyanobacteria, WARGI-QUAL software

1. INTRODUCTION

Water management requires analysis of quantitative as well as qualitative aspects in a common strategy. Quality management is a critical component of overall integrated water resources management. Over the past decades, a number of simulation and optimization models have been developed to manage multi-reservoir systems. Most of these models took into account only the quantity of water. Models defining water quality policy can be developed at different levels of complexity. At times, models are simple qualitative conceptual representations of the relationships between important variables and indicators of these variables. More quantitatively based models can be used to make predictions about the assimilative

G. Rossi et al. (eds.), Methods and Tools for Drought Analysis and Management, 239–261.

capacity of a water body or the effectiveness of certain best management practices. These complex models often suffer from limited computational feasibility and great computational complexity.

However, in complex water systems in Mediterranean countries, reservoirs can be considered as the most important element, and a critical factor in their management is related to the definition and monitoring of water quality. To identify and classify quality indexes it is also necessary to identify which uses are constrained by water quality and what the minimum acceptable quality levels for each type of use are. When these levels are not met, water users must be prepared either to pay an additional cost for water treatment or, at the least, run increasing risks inherent in the use of poorer quality water.

In the Mediterranean region, eutrophication is one of the most serious problems affecting the quality of water stored in reservoirs. The increase in nutrients leads to greater productivity of the water system that may result in an excessive increase in algal biomass or other primary producers such as macrophytes. Excessive algal biomass can seriously affect water quality, especially if it creates anaerobic conditions. Therefore, a simplified approach to the need of including water quality aspects in any given model may focus on the trophic state of reservoirs. Trophic state is a multidimensional concept, involving different aspects; nevertheless single criterion alternatives to multi-parameter indices have also been developed.

This approach can be chosen because the commonly adopted trophic criteria are mutually correlated. Several researchers (i.e.: Sakamoto 1966; Carlson, 1977; Schindler, 1978; Smith, 1982) have proposed relationships between algal biomass, nutrient concentration and nutrient load. Carlson (1977) suggested that algal biomass is the key indicator for trophic state classification and Chlorophyll-a is the best for estimating algal biomass in most reservoirs. Classification of the quality of water held in reservoirs based on trophic state alone enables us to make initial, important assessment of quality related to overall algal biomass concentrations in the ecosystem; this classification can be inserted in qualitative-quantitative system modelling.

Through calculation of the Trophic State Index (TSI) (Carlson, 1977) it is possible to formulate some general hypotheses about algal development in the period of maximum stratification. TSI can be easily used to insert management constraints on water use possibilities, to predict short-term trophic evolution in different climate scenarios and to check relationships between parameters (Sechi and Sulis, 2005). This approach can be adopted in simplified models for completely mixed lakes without stratification. Use of TSI alone does not provide data about algal composition nor does it allow us to establish whether the algal blooms will be made up of a specific type of algae. Some families of microscopic phytoplanktonic algae produce algal toxins during water body eutrophication phenomena. In particular, in lakes toxin-producing phytoplankton belongs almost exclusively to Cyanophytes. Toxic blooms of Cyanophyte algae is a growing problem throughout the world. Monitoring of the toxicity of these blooms showed that almost half are indeed toxic (Sivonen et al., 1990). The most frequently found toxins, microcystins, are new, oncogenic-risk substances which cannot be eliminated by standard water purification processes. Special filters or costly treatments are needed to prevent them from entering the supply network. In such cases, limitations on resource use based solely on TSI values may not be sufficient. A combined TSI and algal classification system could be used to provide a single approach for the purpose of classifying reservoirs in a complex water system.

2. RESERVOIR TROPHIC STATE CHARACTERIZATION

The modelling of trophic conditions of water bodies must take into account complex phenomena that are most notably related to human activity in the basins. Trophic status is a multidimensional concept involving a wide range of data types: morphological, physical, chemical and biological (Brylinsky and Mann, 1973). To assess the eutrophication rate, primary productivity is frequently measured. To compensate the lack of this information for immediate application or to estimate future trends under changing conditions, several models and statistical analyses have been developed. Experimental studies in lakes have shown that nutrients (nitrogen and phosphorus) play an important role as factors influencing phytoplankton production. An indirect measure of phytoplankton production is represented by Chlorophyll-a concentration. Chlorophyll-a (Chl-a) is a simpler and more useful estimator than cell number or cell volume.

Several researchers (Sakamoto, 1966; Carlson, 1977; Schindler, 1978; Smith, 1982) have found a log-linear relationship between Chl-a and total phosphorus (TP) concentrations. The input-output models developed by Dillon and Rigler (1974) and by Vollenweider (1975) can be used to predict phytoplankton biomass (expressed as Chl-a). Walker (1977) quantified the uncertainty associated with phosphorus-loading models. Smith (1982) developed a theoretical framework that reduces the error of chlorophyll prediction in the Dillon-Rigler model by taking into account the influence of total phosphorus/total nitrogen weight ratio. Smith showed that total nitrogen concentration influences chlorophyll concentration, even in lakes where phosphorus alone had previously been presumed to be a limiting factor. In these simple input-output models, other factors that can also influence trophic status, such as biological interaction, internal nutrient loading and physical conditions are not assessed. Although these models are only applicable in discrete periods of the year under the very restrictive assumption that a lake is a mixed system, these approaches are the most common in literature.

When analysing complex phenomena, understanding how these factors are related, and inserting them into a model using analytical relations require great effort. Brylinsky and Mann (1973) analyzed the interactions between a large number of variables using a combination of hierarchical models and statistical analyses. Hierarchical models provided a framework for identification of the relationships between these variables that were tested using correlation and factor analysis. The variables were grouped into those influencing primary production through nutrient availability (depending on nutrient input, characteristics and area of drainage basin, nutrient dilution, precipitation, evaporation, surface area and mean depth, and nutrient distribution, wind, surface area and temperature range of water) and through energy availability (depending on total incident radiation, reflection and turbidity). They used simple or multiple regression analysis for comprehension of the type and degree of the relationship between morphological, chemical, physical and biological variables. The results of regression analysis showed that phytoplankton biomass and chlorophyll-*a* are closely correlated with each other and with phytoplankton production. Multiple regression using biological indices to estimate phytoplankton production also showed that chlorophyll-*a* and total phosphorus explain 45% of phytoplankton production variance.

Recent literature contains various attempts to analyse these aspects in a simplified manner enabling description of the trophic condition of lakes in a mathematical optimization modelling approach for complex water system management.

Following Sakamoto's study (1966), Dillon and Rigler (1974) used chlorophyll-*a* as a simple and useful estimator of phytoplankton biomass and found, for 19 lakes in southern Ontario, a relationship between the average concentration of chlorophyll in the summer by using measures of total phosphorus during the spring:

$$\log_{10}[Chl] = 1.449\log_{10}[TP] - 1.136 \tag{1}$$

Although the correlation coefficient is really high ($R = 0.95$), this is valid only if the N/P ratio exceeds 12. Schindler (1978) examined the relationship between mean annual chlorophyll concentration and mean annual concentration of total phosphorus ($R = 0.89$), as well as mean annual concentration of total phosphorus with phosphorus input corrected for water renewal ($R = 0.88$). He found that chlorophyll was quite accurately related to phosphorus input ($R = 0.76$). This is a significant factor in water quality management, because it makes it possible to predict the trophic state of lakes once phosphorus input and water renewal time are known. Smith (1982) extended this phosphorus-chlorophyll regression analysis taking into account the influence of the TN/TP ratio on annual mean chlorophyll concentration in Florida lakes even with TN/TP > 12. This model involves a simple log-linear combination of total P and total N ($R = 0.91$):

$$\log Chl = 0.374\log TP + 0.935\log TN - 2.488 \tag{2}$$

However nitrogen was not very significant for lakes with TN/TP > 35 alone.

Smeltzer et al., (1989) developed a regression of summer chlorophyll-*a* versus spring total phosphorus for Vermont lakes with Chl a values less than or equal to $0.15\,\text{mg/m}^3$ ($R = 0.81$):

$$\log Chl = -0.49 + 1.08\log TP \tag{3}$$

A number of studies have attempted to assess the hydrologic contributions of phosphorus to water bodies.

Walker (1991) presented a series of regression models relating concentration of measured water quality parameters (here indicated by C) to previous rainfall (HI_i = total precipitation in i-antecedent days) and mean daily water storage level (E_j):

$$\log_{10}[C] = A_0 + A_1 \log_{10}(HI_i + 0.01) + A_2 E_j \tag{4}$$

The testing of alternative expressions indicated that logarithmic transformation is preferable to linear expression of antecedent rainfall. Multiple regression function ($R = 0.66$) for total phosphorus was:

$$\log_{10}[P] = 1.857 - 0.213E_{30} - 1.091 \log(HI_{365} + 0.01) \tag{5}$$

As stated in the introduction, starting from reservoir trophic state characterization, we need to define a synthetic quality evaluation index for modelling optimization of multi-reservoir and multi-use systems.

3. A FIRST QUALITY INDEX EVALUATION BASED ON TROPHIC STATE

Recent national legislation in European countries has provided a set of guidelines for reservoir water quality classification based on a limited number of parameters, called "macro-descriptors", which define the trophic state of reservoirs and lakes. Italian legislation (Decree Law 152, 1999; Decree Law 258, 2000) indicates the following macro descriptors for water bodies:

- Chlorophyll-*a*
- Transparency
- Total Phosphorous
- Hypolimnion Oxygen

In accordance with the above, environmental state is summarized in a synthetic index based on five possible numerical values corresponding to the following quality evaluation (QE) index:

$$\text{QE} = 1 \rightarrow \text{excellent}$$

$$\text{QE} = 2 \rightarrow \text{good}$$

$$\text{QE} = 3 \rightarrow \text{acceptable}$$

$$\text{QE} = 4 \rightarrow \text{poor}$$

$$\text{QE} = 5 \rightarrow \text{bad}$$

To define the quality evaluation index in modelling multi-reservoir and multi-use systems, as mentioned in the introduction, we can use the Carlson (1977) Trophic State Index (TSI), which in recent years seems to have gained general

Table 1. Relations between TSI values, lake attributes and QE values (Table adapted from Carlson and Simpson, 1996)

QE	TSI (Chl)	Chl (μg/l)	Attributes
1	<30–40	<0.95–2.6	Oligotrophy
2	40–50	2.6–7.3	Mesotrophy
3	50–70	7.3–56	Eutrophy
4	70–80	56–155	Hypereutrophy
5	>80	>155	Over-Hypereutrophy

acceptance from the limnological community for obtaining a qualitative index for measuring stored water quality. TSI can be used for characterizing a lake's trophic state and can be evaluated using Chlorophyll-*a*, Total Phosphorus and Secchi disk transparency measurements. TSI values can be calculated from the following simplified equations:

$$\text{Total Phosphorus TSI (TP)} = 14.42\ \ln(\text{TP}) + 4.15 \tag{6}$$

$$\text{Chlorophyll-a TSI (Chl)} = 9.81\ \ln(\text{Chl}) + 30.6 \tag{7}$$

$$\text{Secchi disk TSI (SD)} = 60 - 14.41\ \ln(\text{SD}) \tag{8}$$

Total phosphorus and Chlorophyll-*a* are measured in micrograms per litre (μg/l) and Secchi disk transparency is measured in meters. The TSI scale shown in Table 1 ranges from 0 (ultra-oligotrophic) to 100 (hyper-eutrophic). High and/or increasing trophic status values indicate an increase in eutrophic conditions (higher productivity).

According to Carlson (1977), averaging TSI values makes no sense as, when available, TSI(Chl) is a better predictor than TP and SD and it is not logical to combine a good predictor with less effective predictors, which should only be used when Chl is not available (Carlson, 1983). SD should only be used if there are no better methods available to evaluate the TSI index.

Table 1, drawn from Carlson and Simpson (1996), reports TSI values along with water body attributes, Chl and the QE values. In a previous work (Sechi and Sulis, 2005) this simplified quality index evaluation, only based on trophic state, have been used within an optimization model to associate the quality index with possible water use in a multi-reservoirs system.

4. PROBLEMS ASSOCIATED WITH THE PRESENCE OF PHYTOPLANKTON IN WATER STORED IN RESERVOIRS

In internal low hydro-dynamism water bodies (such as regulation reservoirs with significant ratio between storage capacity and mean annual inflow), the microscopic organisms making up phytoplankton are the plant group playing the leading role in the primary production of living organic substance or biomass. Extensive algal blankets cover the seabed, while phytoplankton develops in the euphotic zone.

While phytoplankton plays a key role in lake ecosystems, its excessive proliferation becomes a serious problem in surface waters used as reservoirs for potable and/or recreational water supply. Immediate consequences of this excessive growth are numerous, ranging from simple abundance of suspended particles (phytoplankton itself, zooplankton, bacteria, fungi and detritus) to increased concentrations of ammonia, nitrites, hydrogen sulphide, methane, ethane and humic acids, and to bad flavour and smell in fish and water due to the presence of particular algae, and the possible development of toxic algae.

At times the phenomenon is so obvious that the naked eye can see the mass of microscopic algae which produce *blooms* giving particular colouring to the water body. The term '*algal blooms*' indicates a situation in which 80–90% of the mass of microscopic algae consists of one or two species. In particular, cyanobacteria may be considered as blooming when their cell number exceeds one million per litre. This coating of microscopic algae covers the surface of the water and decreases its transparency. This in turn prevents penetration of sunlight, something that, coupled with the thermal stratification typical of lakes in the Mediterranean area, inevitably engenders conditions of anoxia and hence the above mentioned consequences.

A further alarming aspect associated with eutrophication is the fact that algal species responsible for blooms belong to the taxonomic group of Cyanobacteria, present in surface waters all round the world. Cyanobacteria produce a wide range of toxins which, according to their effects, are classified as hepatotoxins, neurotoxins, skin irritants and other toxins (WHO, 1998).

Hepatotoxins and neurotoxins are produced by cyanobacteria (see Table 2) commonly found in surface water bodies and are released into the water only in the cell senescence or death stage. Thus, a water body populated by toxic algae will inevitably become enriched with their toxins which can affect the human beings or

Table 2. Name and producer organism for the cyanobacteria (Table adapted from Carmichael, 2000)

Name	Produced by
Neurotoxins	
Anatoxin-a	*Anabaena, Aphanizomenon*
Homo-Anatoxin-a	*Oscillatoria (Planktothrix)*
Anatoxin-a(s)	*Anabaena, Oscillatoria (Planktothrix)*
Paralytic Shellfish Poisons (Saxitoxins)	*Anabaena, Aphanizomenon*
Liver Toxins	
Cylindrospermopsin	*Aphanizomenon, Cylindrospermopsis*
Microcystins	*Anabaena, Aphanocapsa, Microcystis, Oscillatoria (Planktothrix)*
Nodularins	*Nodularia*
Contact irritant-dermal toxins	
Aplysiatoxin	*Schizothrix*

animals using the water. Literature reports frequent cases of deadly poisoning of many animal species, both wild and domestic, following consumption of water with cyanobacteria blooms, the poisoning being due to neurotoxins or hepatotoxins.

Neurotoxins are produced by algae such as *Anabaena flos-aquae, Anabaena spiroides, Anabaena circinalis, Aphanizomenon flos-aquae, Oscillatoria, Gomphosphaeria, Trychodesmium.* These algae can at the same time produce hepatotoxins but the latter act more slowly, therefore the clinical picture is dominated by neurotoxins.

Hepatotoxins are more frequent than neurotoxins and are produced by cyanobacteria belonging to a number of genera, the most common being *Microcystis, Anabaena, Oscillatoria, Nostoc, Nodularia* and*Aphanizomenon* (Harada et al., 1999). Among hepatotoxins, the most abundant are microcystins, 50 types of which are known. Their effects on human beings and animals may be summarised as:

- Acute hepato-toxicosis due to direct ingestion;
- carcinogenic effect, if ingested in sub-acute doses over a long period of time (hepatic cancer);
- Allergic pneumonias, if breathed in.

UNESCO recently adopted the threshold of 1μg/l (acute intoxication risk) for consumption of drinking water from eutrophicated reservoirs contaminated with cyanobacteria. International literature recommends the threshold of 1μg/l for chronic hazard from consumption over long periods of waters from contaminated reservoirs (Ueno et al., 1996).

In Italy, at the present time, microcystins affect each year the reservoirs of seven regions out of twenty. While the World Health Organization has addressed the problem of cyanobacteria blooms, neither EU nor Italian legislation have established concentration limits for cyanotoxins in water for human consumption and/or recreational use. With regards to WHO guidelines on drinking water quality, considering drinking water consumption of 2 litres water/day for a person weighing 60kg, we can obtain a reference value of 1μg/l. A density of 100×10^6 cell l^{-1} of cyanobacteria could produce microcystin concentration from 10μg/l to 20μg/l, equivalent to 10 to 20 times the cautionary threshold value recommended by the WHO, depending on the toxic potential of the algal species characterising the bloom.

On the assumption that a 3-stage water purification system lowers phytoplankton density by 90%, in this study we adopted a threshold value for potable use (and for cautionary purposes, also for irrigation use) of 100×10^6 cell l^{-1}. These values are of course influenced by lack of data and absence of clear legislative standards as to the risk engendered by the danger of these organisms for human health.

5. EVALUATION OF WATER QUALITY IN THE FLUMENDOSA-CAMPIDANO RESERVOIRS

The Flumendosa-Campidano water system extends over south-eastern Sardinia, reaching to the centre of the Island. Hydrology is typically Mediterranean, with alternation of several droughty years with years marked by intense rainfall. Besides being the most extensive on the island, this water system is also the most complex, since it interconnects with other systems and it is a multi- reservoir and multi-use system.

Its pivot is a series of reservoirs, linked in a cascading sequence, whence depart pressure pipelines for residential use and open channels supplying irrigation water to the Campidano plain. An extended description of main characteristics of the system and the connections scheme made using the graphical interface of WARGI-DSS (Sechi and Sulis, 2005) are reported in the second part of this chapter.

Since the early 1990s, the Regional Water System Management Board (*Ente Autonomo del Flumendosa, EAF*) has conducted an intensive monitoring program to identify water quality status in the most important lakes in the system, particularly the following four lakes, whose main characteristics are shown in Table 3: 1) *Flumendosa at Nuraghe Arrubiu*; 2) *Mulargia at Monte su Rei*; 3) *Is Barroccus*; 4) *Cixerri at Genna Is Abis*.

Measurements of the main chemical and biological parameters have been performed by the EAF since the appearance of algae blooms in reservoirs in the early 1990s. These data seem to suggest that lake eutrophication and high mineral nutrient load in inflowing water are indeed the major water quality problems of the system.

Of the four main artificial reservoirs in the system, only the two larger basins, Flumendosa and Mulargia, have an intermediate trophic level between oligotrophy and mesotrophy. The other two lakes, Cixerri and Is Barrocus, are in a state of more or less advanced eutrophication, at times hovering between eutrophy and hypertrophy. EAF measured sampling parameters selected to represent trophic conditions of the lakes, particularly Chlorophyll-a (Chl), Total Phosphorus (TP) and Secchi Disk transparency (SD). Measurements were taken at least once a month from January 1994 to December 2003 at different depths in all lakes. Time evolution of these parameters is reported in Figure 1.

Table 3. Main characteristics of reservoirs in the Flumendosa-Campidano system

	Cixerri	Mulargia	Flumendosa	Is Barroccus
Total catchment basin [km^2]	426	1183.16	1004.51	93
Reservoir surface at maximum level [km^2]	4.90	12.40	9.00	6.30
Elevation at maximum level [m asl]	40.50	259.00	269.00	414.55
Elevation at maximum regulation level [m asl]	39.00	258.00	267.00	413.00
Volume at maximum level [$10^6 m^3$]	32.20	347.70	316.40	14.04
Volume at maximum regulation level [$10^6 m^3$]	23.90	320.70	292.90	11.96

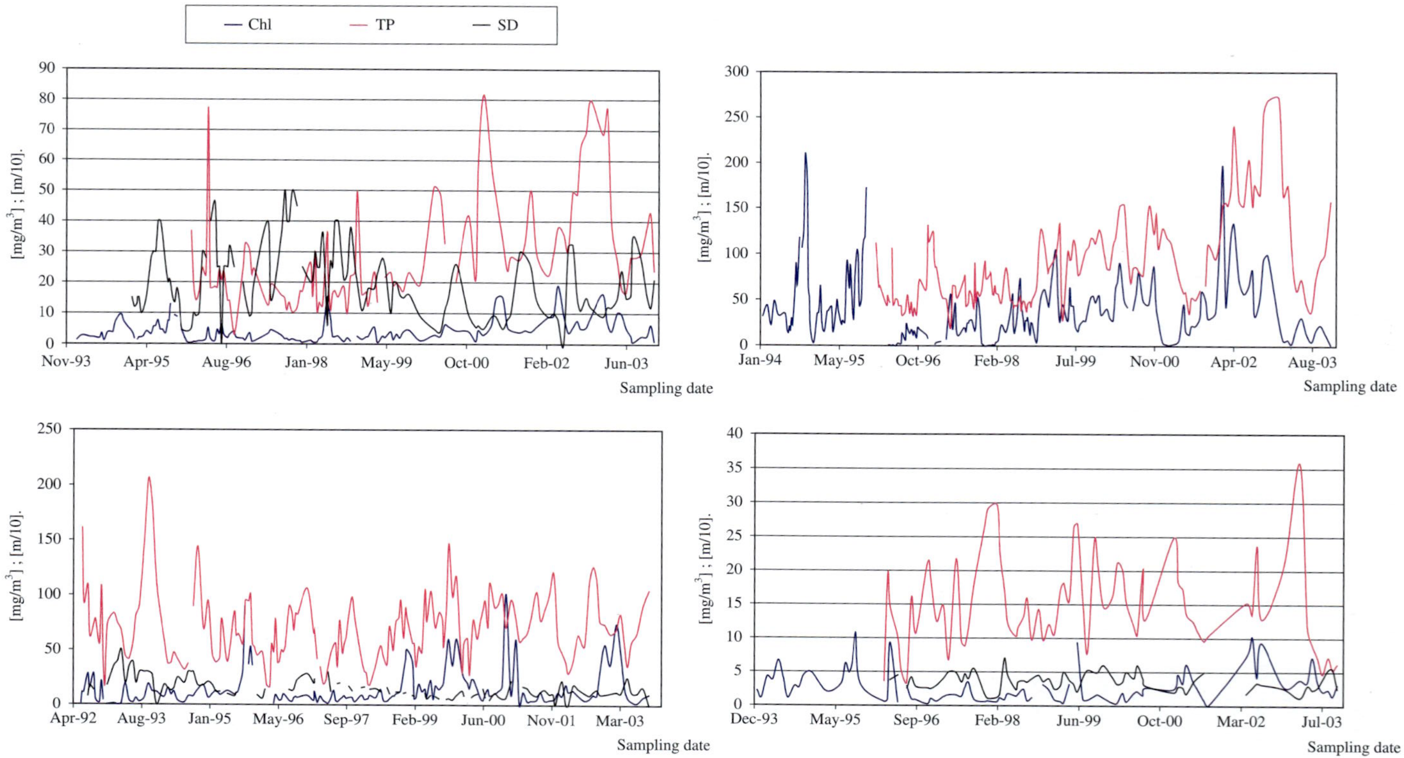

Chl
TP
SD
[mg/m3] ; [m/10].
Sampling date
Nov-93
Apr-95
Aug-96
Jan-98
May-99
Oct-00
Feb-02
Jun-03
Jan-94
May-95
Oct-96
Feb-98
Jul-99
Nov-00
Apr-02
Aug-03
Apr-92
Aug-93
Jan-95
May-96
Sep-97
Feb-99
Jun-00
Nov-01
Mar-03
Dec-93
Sep-96
Jun-99
Mar-02
Jul-03

Table 4. Statistical summaries of TSI data evaluated from samples

Lake	Water Quality Parameter	Number of Samples	Average Annual Sampling Frequency	Max	Min	Mean	Standard Deviation
Flumendosa	TSI(Chl)	87	8.7	53.9	13.7	37.4	8.5
	TSI(TP)	72	7.2	55.7	22.4	41.7	7.0
	TSI(SD)	69	6.9	58.6	32.0	42.5	5.6
Cixerri	TSI(Chl)	217	21.7	83.1	18.2	61.5	12.4
	TSI(TP)	164	16.4	85.0	46.7	66.7	7.3
	TSI(SD)	145	14.5	87.3	38.3	66.5	8.4
Is Barroccus	TSI(Chl)	165	13.8	75.9	18.8	52.7	9.3
	TSI(TP)	171	14.3	81.0	44.9	63.9	6.7
	TSI(SD)	126	10.5	70.0	36.8	55.0	7.0
Mulargia	TSI(Chl)	139	13.9	59.6	23.8	41.9	7.9
	TSI(TP)	101	12.6	67.7	21.4	50.0	8.2
	TSI(SD)	122	13.6	73.2	36.8	52.1	8.6

As previously explained, a trophic state index (TSI) based on Chl, TP and SD can be obtained fo r the classification of lakes. Using the previous equations (6–8), TSIs have been calculated from observed data. Statistical TSI values reported in Table 4 are based on the average of these evaluations.

As regards phytoplankton composition, algal species in the four lakes mostly belong to five classes: Cyanophyceae, Chlorophyceae, Bacillariophyceae (Diatoms), Cryptophyceae and Conjugatophyceae.

To assess the contribution of each class to total density, for each lake algal density values collected between 1996 and 2005 were processed as follows:

- mean annual density values were calculated for each algal class (absolute and percent value);
- minimum and maximum values reached by each class of algae in the period under study (1996–2005) were recorded;
- for each class of algae a different value scale was then established. To this scale, in Figure 2, a distinguishing shade of colour has been assigned, whereby the lower the density, the lighter the colour, and vice versa;
- the most representative species in each group of algae were identified.

Figures from 3 to 6 show the data processing output. Graphic depiction enables us to gain an idea of how the density of each class of algae changed over time in each lake.

Figure 1. Temporal distribution of mean Chrolophyll-*a* concentration (**Chl**). Total phosphorous concentration (TP), **Secchi** Disk transparency (SD) for the main lakes in the Flumendosa-Campidano system (**Mulargia** (*a*), **Cixerri** (*b*), Is **Barroccus** (*c*), **Flumendosa** (*d*))

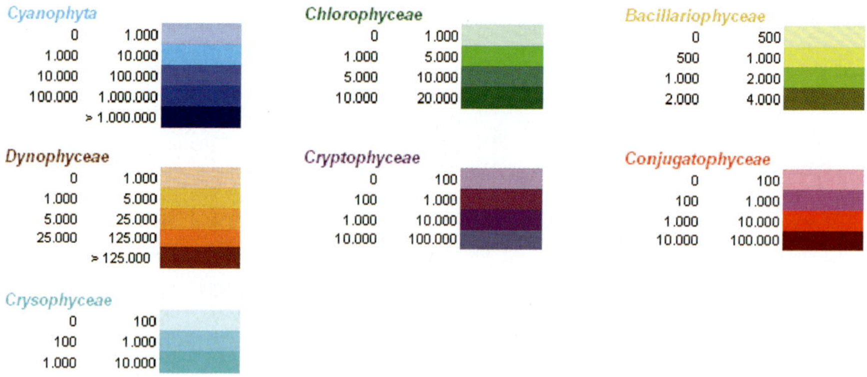

Figure 2. Chromatic scale showing density of the different species of algae

Phytoplanktonic composition in the Cixerri lake between 1996 and 2005 (Figure 3) consisted almost entirely of cyanobacteria. Mean annual values of total density and cyanobacteria density show the same decreasing trend after 2000, when they peaked with more than 700×10^6 cell l^{-1}, reaching their lowest levels in 2005 with 27×10^6 cell l^{-1} and 15×10^6 cell l^{-1} respectively. Only in the last two years, 2004 and 2005, do we see a growing presence of cryptophyceae and chlorophyceae.

As regards phytoplankton in the Flumendosa (see Figure 4), we see marked predominance of cyanobacteria in 1999 and from 2001 to 2005. In 1996, there was prevalence of diatoms, in 2000 of chlorophyceae, while in 1997 and 1998 all three classes of algae were well represented. The general trend shows an increase in total density starting from 1997 and in the density of cyanobacteria starting from 1996; both values then decreased between 2002 and 2005. In this case too, annual mean values of total density and cyanobacteria density both peaked in 2002 with

Cixerri Reservoir

	Cyano	Chloro	Diatom	Crypto	Dyno	Conjug	Cryso	Total Density
1996	98,8	1,2	0,1	0,0	0,0	0,0	0,0	$368*10^6$
1997	92,2	5,3	0,7	1,5	0,0	0,3	0,0	$40*10^6$
1998	97,0	0,8	0,3	0,3	0,1	1,6	0,0	$45*10^6$
1999	97,4	1,7	0,8	0,1	0,0	0,0	0,0	$385*10^6$
2000	99,6	0,1	0,1	0,0	0,0	0,2	0,0	$742*10^6$
2001	99,5	0,4	0,0	0,1	0,0	0,0	0,0	$362*10^6$
2002	82,0	13,3	0,4	4,2	0,0	0,1	0,0	$101*10^6$
2003	89,5	4,7	0,9	4,7	0,1	0,1	0,0	$39*10^6$
2004	42,9	4,4	0,1	52,3	0,0	0,0	0,0	$149*10^6$
2005	56,4	20,8	14,6	7,6	0,2	0,2	0,1	$27*10^6$
Annual Mean (%)	**85,5**	**5,3**	**1,8**	**7,1**	**0,3**	**0,0**	**0,0**	

Figure 3. Algal density in lake Cixerri

Flumendosa Reservoir

	Cyano	Chloro	Bacill	Crypto	Dyno	Conjug	Cryso	Total Density
1996	2,8	12,2	84,6	0,0	0,1	0,4	0,0	1,6*10^6
1997	24,9	30,8	34,8	7,4	0,1	2,0	0,0	1*10^6
1998	30,8	46,5	15,8	3,9	0,0	2,8	0,0	2*10^6
1999	72,2	19,6	5,8	2,2	0,0	0,2	0,0	5*10^6
2000	6,0	89,6	2,7	1,5	0,0	0,2	0,0	7*10^6
2001	92,5	6,2	0,9	0,3	0,0	0,0	0,0	24*10^6
2002	93,6	0,3	5,3	0,2	0,0	0,2	0,4	30*10^6
2003	95,2	1,3	2,6	0,7	0,0	0,1	0,1	21*10^6
2004	96,2	0,4	1,8	1,5	0,0	0,1	0,0	20*10^6
2005	83,4	2,6	12,7	1,1	0,0	0,1	0,1	10*10^6
Annual Mean (%)	**59,7**	**20,9**	**16,7**	**1,9**	**0,6**	**0,0**	**0,1**	

Figure 4. Algal density in lake Flumendosa

about 30×10^6 cell l^{-1}. Total density and density of cyanobacteria dropped to their minimum values in 1996 and 1997 respectively.

Phytoplankton in lake Is Barrocus (see Figure 5) displays more variable composition. Only in five out of the 10 years under scrutiny were cyanobacteria the dominant class, with percent values never dropping below 64% of total density. In the remaining five years, their presence was counterbalanced by chlorophyceae and diatoms. As in lake Cixerri, maximum and minimum values of total density and cyanobacteria density were recorded in 2005 and 2000 respectively.

The phytoplanktonic composition of lake Mulargia (see Figure 6) was dominated by cyanobacteria throughout almost the whole period under study, apart from the years 1998 and 1999, which also recorded the lowest total density values.

In 1998 and 1999 cyanobacteria were well balanced by chlorophyceae and diatoms were also present, albeit to a far lesser extent. In lake Mulargia, as in

Is Barrocus Reservoir

	Cyano	Chloro	Bacill	Crypto	Dyno	Conjug	Cryso	Total Density
1996	86,5	9,3	3,4	0,7	0,0	0,1	0,0	35*10^6
1997	92,2	3,6	2,4	1,7	0,0	0,1	0,0	29*10^6
1998	32,6	45,3	13,8	5,8	0,0	2,2	0,0	7*10^6
1999	47,4	32,3	10,3	8,3	0,3	0,8	0,0	17*10^6
2000	99,7	0,2	0,0	0,0	0,0	0,0	0,0	2*10^6
2001	37,7	42,2	8,4	11,0	0,0	0,6	0,0	7*10^6
2002	64,4	8,6	19,8	6,5	0,0	0,5	0,2	14*10^6
2003	3,9	28,8	29,5	35,6	0,0	0,9	0,9	8*10^6
2004	6,4	30,4	44,7	17,7	0,1	0,4	0,2	5*10^6
2005	93,6	1,7	2,1	2,5	0,0	0,0	0,0	46*10^6
Annual Mean (%)	**56,4**	**20,2**	**13,4**	**9,0**	**0,1**	**0,6**	**0,1**	

Figure 5. Algal density in Lake Is Barrocus

Mulargia Reservoir

	Cyano	Chloro	Bacill	Crypto	Dyno	Conjug	Cryso	Total Density
1996	85,7	11,1	1,9	1,2	0,0	0,1	0,0	$19*10^6$
1997	95,9	3,4	0,5	0,1	0,0	0,1	0,0	$50*10^6$
1998	38,6	50,4	6,2	1,9	2,3	0,6	0,0	$4*10^6$
1999	46,1	44,7	7,3	1,6	0,0	0,4	0,0	$6*10^6$
2000	88,6	8,7	1,9	0,5	0,1	0,1	0,0	$138*10^6$
2001	99,2	0,2	0,6	0,0	0,0	0,0	0,0	$160*10^6$
2002	98,3	0,0	0,2	0,0	0,0	1,4	0,0	$172*10^6$
2003	98,1	0,4	0,9	0,4	0,0	0,2	0,1	$38*10^6$
2004	94,5	3,5	0,8	1,0	0,0	0,1	0,0	$36*10^6$
2005	81,3	13,0	3,7	1,6	0,1	0,1	0,1	$11*10^6$
Annual Mean (%)	**82,6**	**13.5**	**2,4**	**0,8**	**0,3**	**0,3**	**0,0**	

Figure 6. Algal density in Lake Mulargia

Is Barrocus and Cixerri maximum and minimum total density and cyanobacteria density were recorded in the same years, that is 1998 and 2002 respectively.

A further level of investigation provided information on the most important species within each class of algae in each of the four lakes studied.

In lake Cixerri *Planktothrix sp.* (see Table 5) was the dominant cyanobacterium species in terms of both sheer quantity and constant presence. *Anabaena flos-aquae* and *Aphanizomenon flos-aquae* were found almost every year, but with markedly lower density. *Microcistis aeruginosa* was more sporadic, although, when present, it reached considerable density values. Among chlorophyceae, *Pediastrum simplex* and *Coelastrum pseudomicroporum* were the most important species, followed by those belonging to genus *Scenedesmus* spp. Diatoms were almost exclusively of one species, since only *Cyclotella* spp. was constantly present and had significant density values. Two cryptophyceae, *Rhodomonas minuta* and *Cryptomonas* sp. together with one conjugatophycea, *Cosmarium* sp. contributed to characterising the phytoplankton of the Cixerri lake (Sechi et al., 1998).

In lake Flumendosa (see Table 6) *Planktothrix* sp., absent in 1996, was always present over the period 1997 to 2005 and was moreover the most abundant species. Among chlorophyceae the two most common species were *Coelastrum reticulatum* and *Oocystis* sp. followed by *Pediastrum simplex.* The highest mean density value was recorded by *Volvox aureus* which however was only found in 2000. Only one diatom recorded sizeable density values: *Cyclotella* spp. The picture is completed by two cryptophyceae: *Rhodomonas minuta*, always present, and *Cryptomonas* sp., found starting from 2001, and by two conjugatophyceae: *Closterium aciculare* and *Closterium gracile,* present in all years albeit with relatively low density values.

Lake Is Barrocus (see Table 7) is the only one of the four lakes where no species belonging to genus *Planktothrix* sp were ever found. In fact, the Cyanobacteria class is represented by *Aphanocapsa* sp., *Aphanotece* sp., two species of the *Anabaena* genus (*A.aphanizomenoides, A. flos-aquae*), *Aphanizomenon flos-aquae* and *Microcystis* sp. None of these was however constantly present, being

Table 5. Annual density values of different species of algae in lake Cixerri

Cixerri Reservoir

	CYAHOPHYCEAE					CHLOROPHYCEAE		
	Plank spp.	Mic aer	Ana f-a	Ana f-a	Ana pla	Ped sim	Coel pse.	Scen spp.
1996	339.219	18.742	0	4.901	172	2.527	208	16
1997	13.486	57	21.996	771	0	817	479	116
1998	0	0	43.556	116	0	0	291	0
1999	366.119	0	0	0	0	5.311	375	284
2000	736.051	0	0	0	0	445	60	322
2001	7.231	247.995	51.958	907	52.084	171	349	101
2002	44.865	2.290	1.571	18.636	13	141	10.403	1353
2003	3.144	0	1.552	1.129	32	90	104	1427
2004	0	0	3.852	55.953	0	1.856	115	2644
2005	227	0	983	10.265	0	1.659	218	1454
Annual Mean	151.034	26.908	12.547	9.268	5.230	1.302	1.260	772

	DIATOMEAE		CRYPTOPHYCEAE		COHJUGATOPH.	
	Cyc sp.	Cyc atom	Rho min	Cryp sp.	Cosm sp.	Clos sp.
1996	135	0	86	0	0	0
1997	174	0	544	54	84	1
1998	125	0	0	126	667	1
1999	2843	0	88	439	29	66
2000	70	0	39	109	868	229
2001	135	0	28	192	8	0
2002	225	0	1558	2692	71	0
2003	111	0	1175	678	19	0
2004	113	0	77415	374	24	0
2005	957	2794	1547	398	22	18
Annual Mean	489	279	8.248	506	179	32

found together or alternatively in the different annual cycles. Among chlorophyceae *Oocystis* sp. was the most abundant, followed by *Chlorella* sp., *Pediastrum simplex* and *Dyctiosphaerium* sp. As to diatoms, besides *Cyclotella ocellata*, always present, we found in 2001 *Fragilaria crotonensis*. Cryptophyceae are here too almost exclusively represented by *Rhodomonas minuta* and *Cryptomonas* sp. Finally, *Closterium aciculare* and *Closterium gracile* were the most representative conjugatophyceae.

In lake Mulargia (see Table 8) the most important cyanophyte was *Planktothrix* sp.. High density values were also recorded for *Microcystis* sp. The most abundant chlorophycea was *Pediastrum simplex* followed by *Coelastrum reticulatum* and *Oocystis* sp. The most representative diatom was *Cyclotella* sp. Among cryptophyceae, the most abundant was once again *Rhodomonas minuta,* always accompanied by *Cryptomonas* sp. The only conjugatophyceae always present were *Closterium aciculare*, *Staurastrum gracile* and *Closterium gracile,* all of which however contributed but little to total density.

Table 6. Annual density values of different species of algae in lake Flumendosa

Flumendosa Reservoir

	CYAHOPHYCEAE			CHLOROPHYCEAE			
	Plank sp	Apha f-a	Aph tec	Vol aur	Coe ret	Ped sim	Ooc sp
1996	0	0	1	0	33	0	0
1997	110	95	18	0	0	296	33
1998	617	0	0	0	71	68	617
1999	1.495	316	1.975	0	22	532	37
2000	331	0	0	5115	293	12	40
2001	22.389	0	0	0	1501	0	9
2002	27.788	0	0	0	44	0	12
2003	19.590	0	0	0	0	0	17
2004	15.376	1.935	0	0	0	0	4
2005	8.123	0	0	0	95	0	26
Annual Mean	9.582	235	199	512	206	91	80
	DIATOMEAE		**CRYPTOPHYCEAE**		**CONJUGATOPH.**		
	Cyc sp	**Cyc oc**	**Rho min**	**Cryp sp**	**Clo aci**	**Clo gra**	
1996	1274	0	0	0	3	3	
1997	224	0	80	0	20	2	
1998	74	0	81	0	53	5	
1999	263	0	116	0	9	1	
2000	177	0	95	0	14	2	
2001	220	0	61	14	7	2	
2002	725	0	38	34	26	9	
2003	435	0	126	26	1	2	
2004	307	7	199	26	5	4	
2005	0	1254	65	21	3	4	
Annual Mean	370	126	86	12	14	3	

Data processed up to this point indicate that in these four lakes, by far the most dominant class in the period under consideration was that of cyanobacteria, responsible for extensive and regular blooms in the system's lakes.

Within this class, dominant genera were mainly: *Planktothrix, Microcystis,* and *Anabaena* (see Table 9). During algal bloom episodes, their concentration can be as high as several million cells per litre, thus creating a significant restraint on use of the water resource. If we add to this the fact that all three genera are known as being toxin-producing, we can easily see how important it is to monitor the quality of our surface water bodies, also from the viewpoint of their phytoplankton composition (see Table 10).

As illustrated in Table 1, attribution of QE values in a simplified approach can be made on the basis of TSI values by considering Chl-a and, when this is not available, by using TP and SD measures. Nevertheless, data analysis on cyanophyte concentration in the lakes in the period 1996–2005 highlights the necessity to integrate the attribution of QE values also considering dominant species concentration.

Table 7. Annual density values of different species of algae in lake Is Barrocus

Is Barrocus Reservoir	CYAHOPHYCEAE						CHLOROPHYCEAE			
	Aph cap	Aph tec.	Ana aph	Ana f-a	Apha f-a	Micr sp	Ooc sp.	Chl sp.	Ped sim	Dyet sp
1996	23.806	0	0	3.267	0	3.353	836	1.079	1	186
1997	0	23.558	0	3.157	343	0	592	36	81	38
1998	28	1.556	0	476	0	0	642	587	265	227
1999	1.195	0	832	0	5.370	0	813	402	2.003	427
2000	489	0	1.902	21	0	0	0	0	0	0
2001	588	502	829	536	0	0	483	538	0	69
2002	0	0	6.765	1.750	0	0	140	5	25	241
2003	0	6	0	0	0	215	471	30	12	0
2004	0	0	0	105	0	14	240	279	0	745
2005	40.977	488	0	0	0	1.801	97	18	17	0
Annual Mean	6.708	2.611	1.033	931	571	538	431	297	240	193

	DIATOMEAE			CRYPTOPHYCEAE		CONJUGATOPH.	
	Cyc oc	Cyc spp.	Fra cro	Rho min	Cry sp.	Clo aci	Clo gra
1996	0	1.200	0	236	22	3	20
1997	706	0	0	351	148	7	7
1998	520	0	0	271	112	100	27
1999	1.470	0	0	691	697	94	11
2000	0	0	0	0	0	0	0
2001	106	0	316	151	618	0	25
2002	228	0	2.131	383	511	2	29
2003	0	2.268	37	2.027	911	10	21
2004	776	842	153	528	203	0	6
2005	702	5	0	1.023	82	0	6
Annual Mean	451	432	264	566	330	22	15

Table 8. Annual density values of different species of algae in lake Mulargia

Mulargia reservoir

	CYAHOPHYCEAE		CHLOROPHYCEAE		
	Plank sp.	Mic sp	Ped sim	Coe ret	Ooc sp.
1996	10.182	116	735	582	396
1997	47.500	0	847	0	32
1998	104	0	1.241	154	16
1999	666	1.056	2.618	81	14
2000	94.374	16.076	7.469	1.099	654
2001	29.518	127.971	26	179	10
2002	167.960	423	0	9	3
2003	28.567	2.811	0	0	114
2004	33.219	344	6	61	65
2005	8.179	129	0	711	179
Annual Mean	42.027	14.893	1.294	288	148

	DIATOMEAE		CRYPTOPHYCEAE		CONJUG
	Cyc spp.	Ast for	Rho min	Cry sp.	Mou sp.
1996	350	0	217	10	0
1997	248	2	63	5	0
1998	214	0	77	2	0
1999	410	0	62	36	0
2000	2163	291	604	100	0
2001	838	0	25	18	0
2002	74	81	40	34	2344
2003	256	1	76	46	39
2004	101	2	94	41	0
2005	323	1	65	37	0
Annual Mean	498	38	133	33	238

With the use of coupled criteria, using the maximum density value of toxic cyanobacteria given by WHO, the quality index of the water held in a general reservoir $j \in res$, at time t can be evaluated as:

$$QE_j^t = F(TSI(Chl_j^t), D(cyano)_j^t) \tag{9}$$

or, alternatively:

$$QE_j^t = F(TSI(TP_j^t), D(cyano)_j^t) \tag{10}$$

$$QE_j^t = F(TSI(SD_j^t), D(cyano)_j^t) \tag{11}$$

Figure 7 shows the frequency distribution of QE in the 5 classes for the lakes monitored. In general terms, Flumendosa reservoir shows high or good environmental status throughout the study period. Sporadic episodes of extensive cyanobacteria blooms were recorded in lake Mulargia and lake Is Barrocus. Mean TSI values

Table 9. List of the most important species recorded in the different lakes in the period 1996–2005

	Cixerri	Flumendosa	Is Barrocus	Mulargia	Simbrizzi
Cyanohactesisa	*Planktothrix spp.* *Microcystis aeruginosa* *Anabaena flos-aquae* *Aphanizomenon flos-aquae* *Anabaena Planctonica*	*Planktothrix spp.* *Aphanizomenon flos-aquae* *Aphanotece sp.*	*Aphanacapsa sp.* *Aphanotece sp.* *Anabaena aplianizomenoides* *Anabaena flos-aquae* *Aphanizomenon flos-aquae* *Microcyslis sp.*	*Planktothrix spp.* *Microcystis sp.*	*Aphanizomenon flos-aquae* *Planktothrix spp.* *Anabaena aphanizomenoides* *Aphanizomenon sp.* *Planktoiyngbia sp.* *Aphanotece sp.*
Chlamphyceae	*Pediastrum simplex* *Coelastrum pselldomicroporum* *Scenedesmus spp.*	*Coelastrum reticulatum* *Pediastrum simplex* *Oocystis sp.*	*Oocystis sp.* *Chlorella sp.* *Pediastrum simplex* *Dyctiosphaerium sp.*	*Pediastrum simplex* *Coetastrum reticulaturn* *Oocystis sp.*	*Chlorella sp.* *Oocystis sp.* *Scenedesmus spp.*
Demomance	*Cyclotella spp.* *Cyclotella atomas*	*Cyclotella spp.* *Cyclotella ocellate*	*Cyclotella ocellata* *Cyclotella spp.* *Fragitaria crotonensis*	*Cyclotella spp.* *Asterionella formosa*	*Melosira ambigua* *Cyclotella spp.* *Synedra spp.* *Melosira granulate var. ang.*
Cryptophyceae	*Rhodomonas minuta* *Cryplomonas sp.*	*Rhodomonas minuta* *Cryptomonas sp.*	*Rhodomonas minuta* *Cryptomonas sp.*	*Rhodomonas minuta* *Cryptomonas sp.*	*Cryplomonas sp.* *Rhodomonas minuta*
Conjugatophyceae	*Cosmarium sp.* *Ciosterium sp.*	*Closterium aciculare* *Closterium gracile*	*Closterium aciculare* *Closterium gracile*	*Mougeclia sp.*	*Cosmarium sp.*

Table 10. List of dominant cyanobacteria, their seasonality and toxicity (Sp=Spring; S=Summer; F=Fail; W=Winter)

Reservoir	Dominant species	Density (cell * $10^3\,l^{-1}$)	Seasonality	Tossix
Flumendosa	*Planktothrix spp.*	7.343	Sp/S/F	+
Mulargia	*Plaktothrix spp.*	33.846	Sp/S/F	+
	Microcystis sp.	14.108	W/F	+
Is Barrocus	*Aphanocapsa sp.*	6.708	S	–
	Aphanotece sp.	2.611	S/F	–
Cixerri	*Planktothrix spp.*	151.034	Sp/S	+
	Microcystis aeruginosa	26.908	S	+
	Anabaena flos-aquae	12.547	S	+

at Is Barrocus were sensibly higher than at Mulargia without this fact leading to higher cyanophyte concentrations or a greater number of blooms. The quality of lake Is Barrocus was classified in 55% of months under study as acceptable and in 7% as poor and incompatible with direct residential use.

Lake Cixerri has acceptable quality levels in winter and spring thanks to water renewal and mixing. In the summer, anoxic degradation of the algal biomass and extensive cyanophyte blooms indicate the need for careful monitoring of the health risk linked to the use of this resource.

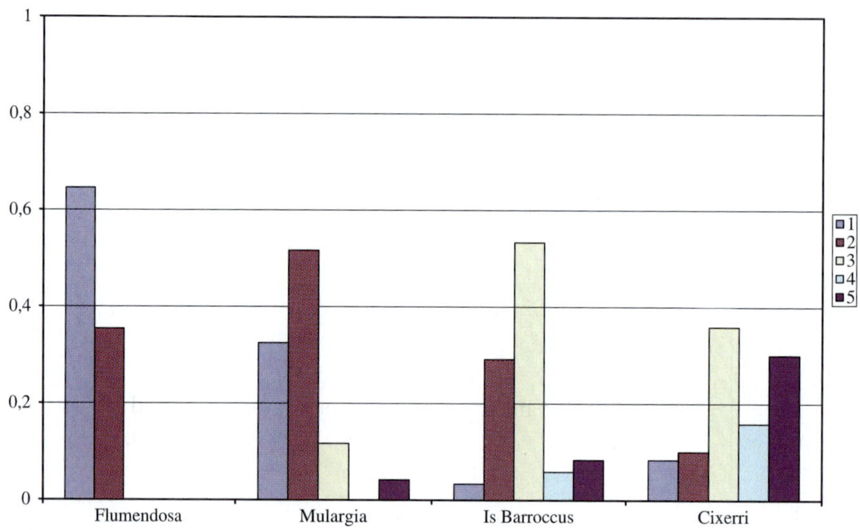

Figure 7. Frequency distribution of QE values on the basis of TSI estimation in the Flumendosa-Campidano lakes

Analysis of TSI values and cyanophyte concentration in the period 1996–2005 confirms the complementary nature of information provided by these indicators. TSI is a descriptor of total algal biomass, while cyanophyte concentration enables us to provide more detailed indications about the composition of algal species. Current legislation does not indicate toxin concentration thresholds in water for residential use. Absence of precise legislative guidelines in this regard is due to incomplete knowledge of this phenomenon. It is quite probable that known cyanotoxins are only part of existing toxins, while others are still to be identified. While most cyanobacteria blooms are toxic, this species may be associated with both toxic and non-toxic blooms. On the other hand, the impossibility of assessing adequately the toxicological aspect of the algal blooms phenomenon highlights the limitation inherent in using TSI alone when blooms are toxic. Some toxins with potentially lethal effects on human beings pass unharmed through standard water purification systems and reach our homes with concentrations several orders of magnitude higher than the guidelines laid down by WHO. In some months, an intermediate TSI value in Table 1 is compatible with high cyanophyte density values and vice versa. The use of one of these indicators alone would not provide sufficient information for imposing limitations on resource use.

6. CONCLUSIONS

Definition of minimum quality requirements with regards to possible uses is an essential element in the management of complex, multi-user and multi-resource water systems, since it limits actual availability of the resource for use. This should also be linked to the level of treatment required for the various uses. EU and Italian legislation identify a small number of parameters for classifying the ecological status of water bodies. Ecological status classification in reservoirs is performed based on trophic status. In the field of limnology, definition of a synthetic index representing the complexity of chemical, physical and biological factors concurring to establish trophic conditions of reservoirs is based on the use of the TSI (Trophic State Index) developed by Carlson (1977). This index assesses increase in total biomass (concentration of chlorophyll, primary production, nutrient concentration) but it does not assess the dynamics and composition of the algal population, let alone its possible toxicological implications. Algal blooms in lakes, albeit a well-known phenomenon, have not been adequately addressed in legislation regulating the use of water for various purposes. Since in Italy toxic algae blooms occur each year in seven regions out of twenty, and the toxins thus produced are carcinogenic risk substances, it seems obvious that a limiting factor for water use must be adequately established based on the density values of the various toxic algal species. Our study proposes a simplified criterion for classifying quality of water stored in reservoirs and lakes, based on the use of TSI, calculated from the values of macro-descriptor parameters, and density of the dominant and most toxic class of algae. The aim is to define a synthetic water quality index (QE) as a decision-making aid for resource planning and management. Modelling developments oriented to insert the QE index in the system optimization have been presented in (Sechi and Sulis, 2005) and in (Sechi and Sulis, 2007). The usefulness in including the index in the WARGI-QUAL model optimization has been confirmed in the application to Flumendosa-Campidano system.

ACKNOWLEDGMENTS

This research has been developed with financial support from European Project SEDEMED II. The authors thank all partners involved for providing their collaboration in developing this research. The authors also express deep thanks to Prof. Rossi for his insightful comments and suggestions that greatly improved this chapter.

REFERENCES

Brylinsky, M. and Mann, K.H. (1973) An analysis of factors governing productivity in lakes and reservoirs. *Limnology and Oceanography* 18(1): 1–14.

Carlson, R.E. (1977) A trophic state index for lakes. *Limnology and Oceanography* 22(2): 361–369.

Carlson, R.E. (1983) Discussion on "Using differences among Carlson's trophic state index values in regional water quality assessment", by Richard A. Osgood. *Water Resources Bulletin* 19: 307–309.

Carlson, R.E. and Simpson, J. (1996) *A coordinator's Guide to Volunteer Lake Monitoring Methods*, North American Lake Management Society. Madison, WI-USA.

Carmichael, W.W. (2000) Health effects of toxin producting cyanobacteria: "the cyanohabs". *Freshwater harmful algal blooms: health risk and control management*, Istituto Superiore di Sanità, Rome, pp. 70–80.

D.Lgs n. 152, Disposizioni sulla tutela delle acque dall'inquinamento e recepimento della direttiva 91/271/Cee concernente il trattamento delle acque reflue urbane e della direttiva 91/676/Cee relativa alla protezione delle acque dall'inquinamento provocato dai nitrati provenienti da fonti agricole, *Gazzetta Ufficiale*,124, 1999.

D.Lgs. n. 258, Disposizioni correttive ed integrative del decreto legislativo 11 maggio 1999, n. 152, in materia di tutela delle acque dall'inquinamento, a norma dell'articolo 1, comma 4, della legge 24 aprile 1998, n. 128, *Gazzetta Ufficiale*, 218, 2000.

Dillon, P.J. and Rigler, F.H. (1974) The phosphorus-chlorophyll relationship in lakes. *Limnology and Oceanography* 19(5): 767–773.

Harada, K., Kondo, F. and Lawton, L. (1999) Laboratory analysis of Cyanotoxins. In: I. Chorus and J. Bartram (eds), *Toxic Cyanobacteria in Water*, E & FN Spon Publishers, London, pp. 369–399.

Salis, F., Sechi, G.M., Sulis, A. and Zuddas, P. (2005) Un modello di ottimizazione per la gestione di sistemi idrici complessi con l'uso di risorse convenzionali e marginali. *L'Acqua* 3: 33–52.

Sakamoto, M. (1966) Primary production by phytoplankton community in some Japanese lakes and its dependence on lake depth. *Archives Hydrobiology* 62: 1–28.

Schindler, D.W. (1978) Factors regulating phytoplankton production and standing crop in the world's freshwater. *Limnology and Oceanography* 23(3): 478–486.

Sechi, N., Buscarinu, P., Pilo, E. and Luglio, A. (1998) Il fitoplancton del Lago Cixerri nei primi cinque anni di esistenza. *XII congresso A.I.O.L.*, pp. 41–51.

Sechi, G.M. and Sulis, A. (2005) Multi-reservoir system optimization using chlorophyll-a trophic indexes. *6th International Conference of European Water Resources Association*, Menton.

Sechi, G.M. and Sulis, A. (2007) Reservoirs water-quality characterization for optimization modelling under drought conditions: PART II , Water-quality optimization modelling. *In this book.*

Smith, V.H. (1982) The nitrogen and phosphorus dependence of algal biomass in lakes: an empirical and theoretical analysis. *Limnology and Oceanography* 27(6): 1101–1112.

Smeltzer, E., Walker, W.W. and Garrison, V. (1989) Eleven years of lakes eutrophication monitoring in Vermont: a critical evaluation. *Enhancing States' Lake Management Programs*: 53–62.

Sivonen, K., Carmichael, W.W., Namikoshi, M., Rinehart, K.L., Dahlem, A.M. and Niemela, S.I. (1990) Isolation and characterization of hepatoxic microcystin homologs from filamentous freshwater cyanobacterium Nostoc sp. Strain 152. *Applied Environmental Microbiology* 56: 2650–2657.

Ueno, Y., Nagata, S., Tsutsumi, T., Hasegawa, A., Watanabe, M.F., Park, H.D., Chen, G.C., Chen, G. and Yu, S.Z. (1996) Detection of microcystins, a blue-green algal hepatoxin, in drinking water sampled in Haimen and Fusui, endemic areas of primary liver cancer in China, by highly sensistive immunoassay. *Carcinogenesis* 17(6): 1317–1321.

Vollenweider, R.A. (1975) Input-output models with special reference to the phosphorus loading concept in limnology. *Schweizerische Zeitzsch-rift fur Hydrologie*, 37: 53–84.

Walker, W.W. (1977) *Some analytical models applied to lake water quality problems*. Ph.D. thesis, Harvard University.

Walker, W.W. (1991) Water quality trends at inflows to everglades national park. *Water Resources Bulletin* 27(1): 59–72.

World Health Organization (1998) *Guidelines for drinking water quality. Health criteria and other supporting information*, Vol. 2, Geneva.

CHAPTER 13

RESERVOIRS WATER-QUALITY CHARACTERIZATION FOR OPTIMIZATION MODELLING UNDER DROUGHT CONDITIONS PART II – WATER-QUALITY OPTIMIZATION MODELLING

G. M. SECHI AND A. SULIS
CRIFOR – UNICA, University of Cagliari, Italy

Abstract: In the southern regions of Mediterranean Europe, the greatest part of water resources for supply systems are derived from artificial reservoirs. A simplified approach to the need of inserting water quality aspects in the mathematical optimization model can be achieved considering the trophic state of reservoirs. In recent years, Trophic State Index (TSI) based on Carlson's (1977) classification appears to have obtained general acceptance as a reasonable manner of classifying stored water in reservoirs; it enables to insert quality constraint in the water management optimization model also considering large size multi-reservoir and multi-user systems. Using TSI index and concentration density values of the most toxic algae, water quality aspects can be easily considered in an optimization model specifically designed for large system management optimization. The usefulness of a mixed quantity-quality optimization approach has been confirmed by WARGI application to a real multi-reservoir water resources system in southern Sardinia

Keywords: Complex water systems, water quality indexes, trophic status, optimization, WARGI-QUAL software

1. INTRODUCTION

Mathematical optimization procedures available for large water resources systems are still not able to deal with all complexities of the real world even in cases when they can be easily incorporated in a simulation model. Nevertheless, since mathematical optimization procedures can be constructed to efficiently solve the optimization model and this can be considered as an adequate approximation of the real water systems management requirements, a subsequent simulation phase can greatly narrow down the search for the optimum (Loucks et al., 1981; Yeh, 1985; Simonovic, 2000; Labadie, 2004). Optimization results obtained by solving an adequately adherent model can be seen as the management reference-targets for simulation since these results can be considered as obtained by an ideal manager of

G. Rossi et al. (eds.), Methods and Tools for Drought Analysis and Management, 263–273.

the system. When considering the tools needed for adequately approaching mathematical optimization to real problems, water quality characterization requirements are critical components in constructing the optimization model.

The first essential element in the construction of this advanced optimization model is linked to the way for definition of water quality in the water bodies and to identification of constraints setting out correct system management criteria. The definition of minimum quality requirements should also be viewed in relation to possible use and the level of water treatment necessary. In particular, considering multi-resource and multi-use systems, waters coming from the different resources may be marked by constraints limiting use in respect of the different end-users. The transfer of this aspect of the problem to a management optimization system, in general terms, requires further simplification in the classification of the environmental state of the water body as single-element attribution of quality of the water to be considered, also in terms of possible interaction between the water bodies themselves and with demand centers.

Particularly when the greater part of water resources for supply systems are derived from artificial reservoirs, as frequently occurs in the regions of southern Mediterranean Europe, a simplified approach to the need of inserting water quality aspects in the mathematical optimization model can be achieved considering the trophic state of reservoirs, strictly related to their artificial nature. Modeling trophic conditions of water bodies needs to take into account complex phenomena that are closely related to human activities in the basins (Sechi and Sulis, 2005).

In Begliutti et al. (2007), coupled criteria for classifying reservoirs in complex water systems, has been defined using both Carlson's Trophic State Index (Carlson, 1977) and concentration density values of the most toxic species of algae. Trophic State Index (TSI), which in recent years appears to have attained general acceptance in the limnological community, can be evaluated using Chlorophyll-*a*, Total Phosphorus and Secchi disk transparency measurements. TSI doesn't give enough information about the consequences of algal blooms and more detailed analysis on toxic algal classification are necessary. Analysis of TSI values and cyanophyte concentrations in the period 1996-2005 (Begliutti et al., 2007) highlight the complementary nature of information provided by these indicators. TSI is a descriptor of total algal biomass while cyanophyte concentration enables us to provide more detailed indications about the composition of algal species. With the use of coupled criteria, a synthetic water quality index (QE) is inserted in the optimization model.

2. QUALITY OPTIMIZATION MODEL FOR COMPLEX WATER SYSTEMS

The increasingly important problems related to water scarcity have brought about the need to improve our knowledge of water quality phenomena, in particular when water resources are obtained from artificial reservoirs. The amount and characteristics of water flowing in affluent rivers to the reservoir typically produce significant changes in the trophic conditions of the reservoir, also related to its water retention time (OECD, 1982), which can be evaluated on the basis of inflowing and outflowing volumes and reservoir storage capacity. Modeling water quality on the basis of trophic conditions needs to take into consideration complex phenomena that include chemical, physical, morphological and biological aspects.

Water quality models assist decision makers in identifying and evaluating plans and management alternatives to achieve social, economic and environmental goals. As stated before, mathematical optimization models are required to screen many alternatives and need simplified forms of water quality classification. In general terms, variables introduced into the optimization model may be divided into flow (or operational), and project variables (Loucks et al., 1981; Labadie, 2004; Pallottino et al., 2005). Flow variables refer to the different types of resource transfer to the network. Project variables are on the other hand associated with works undergoing re-sizing and therefore refer to the main characteristics of these works. Referring to a "static" or single-period situation, we can represent the physical system by a direct network (basic graph) consisting of nodes and arcs. A dynamic multi-period network can be generated by replicating the basic graph for each period $t = 1, T$ in the time-horizon T and then connecting the corresponding reservoir nodes for different consecutive periods by additional arcs carrying water stored at the end of each period; we call these *inter-period arcs*.

Objective function (OF) in optimization model should incorporate operation, maintenance and repair (OMR) costs and construction costs, as well as costs for different measures addressed to assure sustainability and efficiency to the system and costs depending on water quality upgrade by purification processes (Loucks, 2000).

Optimization model constraints are introduced in order to show the relationships between system variables and the limits of their attributions. In modeling we usually identify three types of constraint: a) continuity constraints at the nodes; b) functional constraints on the arcs; c) bounds on the flow and project variables. In all three types of constraint, both flow and project variables may appear. The model may be represented in the following compact structure:

$$\min \gamma Y + c_i x_i + c_j (QE)_j x_j \tag{1.i}$$

s.t.:

$$Ax = b \tag{1.ii}$$

$$F(Y, x) \geq 0 \tag{1.iii}$$

$$l \leq x \leq u \tag{1.iv}$$

$$L \leq Y \leq U \tag{1.v}$$

$$QE \leq QE^{\min} \tag{1.vi}$$

where Y indicates the set of project variables and γ indicates associated costs; $[x_i, x_j]$ are the subsets of flow variables x related to costs c_i, representing OMR and user-defined costs to assure efficiency of the system, and to costs c_j representing purification costs. In particular, the purification costs can be fixed or dependent on water requirements as to use (costs for quality protection and management).

Constraints (1.ii) represent the continuity equations at the nodes, in which the vector b represents input and output at the nodes. Constraints (1.iii) express relations between the flow and project variables; the subsequent two relations (1.iv and 1.v) are the bounds attributed to the two sets of variables. Constraint (1.vi) refers to water quality; in particular it indicates that the reservoir quality index must meet the constraints associated with the water body to enable use of its water.

As illustrated in Begliutti et al. (2007), in this work the problem of water quality attribution has been addressed starting from the trophic state of water bodies using basically Chlorophyll-*a* measurements by means of which we can evaluate the TSI(Chl) trophic index and algal classification in particular regarding density of toxic species.

The quality constraint for reservoir j must ensure that water issued by or transferred from the reservoir in time t complies with the required quality index. Implicitly we consider linear approximation in the use of quality indices. For general reservoir j the quality constraint requires that reservoir waters seen as a whole guarantee higher quality (or at the least the same quality) than that requested for waters issued to end users:

$$
\begin{aligned}
QE_j^t y_j^t + \sum_{i=1,np} QE_i^t p_i^t + \sum_{i=1,nf} QE_i^t f_i^t \le QE_j^{t+1} y_j^{t+1} \\
+ \sum_{i=1,ng} QE_i^t g_i^t + \sum_{i=1,ne} QE_i^t e_i^t + \sum_{i=1,ns} QE_i^t s_i^t, \quad j \in res
\end{aligned} \tag{2}
$$

where:

y_j^t	flow variable representing volume held in reservoir j in time t;
p_i^t	water input to the reservoir during arc i of time t;
f_i^t	flow variables indicating water coming from node i;
g_i^t	flow variables indicating water outgoing towards node i;
e_i^t	flow variables representing the sum of losses from the reservoir;
s_i^t	flow variables representing downstream overflows and issues;
n_p, n_f, n_g, n_e, n_s	number of natural inputs to, flows going into, outgoing from, loosed by and overflowed from a reservoir.

Basically, this constraint means that overall incoming flows at the node must guarantee higher quality (or at least the same quality) as that required for outgoing waters. Evidently, linearity implies that in the product between quality index and flow rate only one of the two terms can be a variable, whereas the other will be constrained by pre-assigned values. In particular, for arcs outgoing from the reservoir we may impose quality values which must be guaranteed for the resource if exit flow rates are variable. No stricter water quality constraint has been assumed, as we consider simplification of non diffusive and conservative processes regarding the assumption of water quality.

To provide a user-friendly tool for water system modeling, WARGI (Sechi and Zuddas, 2000; Sulis, 2006) optimization software has been developed following formalization of the procedure explained above. WARGI makes it possible to characterize water bodies and quality of water for different uses, which are definitely pertinent to large-scale water system optimization problems.

The WARGI optimization tool has been developed in order to:

- guarantee simplicity of use in the input phase, in scenario setting and in processing output results;
- simplify modification of system configuration to perform sensitivity analysis and process uncertainty;
- prevent obsolescence of the optimizer exploiting the standard input format in optimization codes.

WARGI generates an MPS (Mathematical Programming System) file to feed the solver. MPS is now the standard format supported by most efficient commercial and non-commercial mathematical programming computer codes. Standard data format allows the use of any solver, possibly the most efficient or the most suitable or the most readily available in the work environment. Moreover, this prevents obsolescence of the used solvers and ensures portability of the DSS. The solver can be selected from a pre-established list (or configured if not in the list) and launched. In the present work, WARGI has been linked with CPLEX (1993) solver.

A graphical post-processor manages the great quantity of information provided by the DSS, solving the optimization problem. The user can select whatever is of interest to him in the environment under examination. It is possible to show flow configuration along arcs in any type of scenario: stored volumes in reservoirs, transfers to demand centres, deficits related to any demand centre, etc.

3. MODEL APPLICATION TO REAL SYSTEM IN SOUTH-SARDINIA

Application of the optimization model to a real multi-reservoir system considering quality constraints has been carried out in the *Flumendosa-Campidano* (Sardinia, Italy) water system. The schematization carried out using WARGI for the *Flumendosa-Campidano* system is shown in Figure 1. An explanation of symbols can be found in the palette-window shown in the same figure. The data of the system elements useful in this work are:

- ten reservoirs with a total capacity of $723 \cdot 10^6 m^3$;
- ten civil demand centers with a total request of $116 \cdot 10^6 m^3$/year;
- nine irrigation demands with a total request of $224 \cdot 10^6 m^3$/year;
- two industrial demands with a total request of $19 \cdot 10^6 m^3$/year;
- five pumping stations with a total potentiality of $441 \cdot 10^6 m^3$/year;

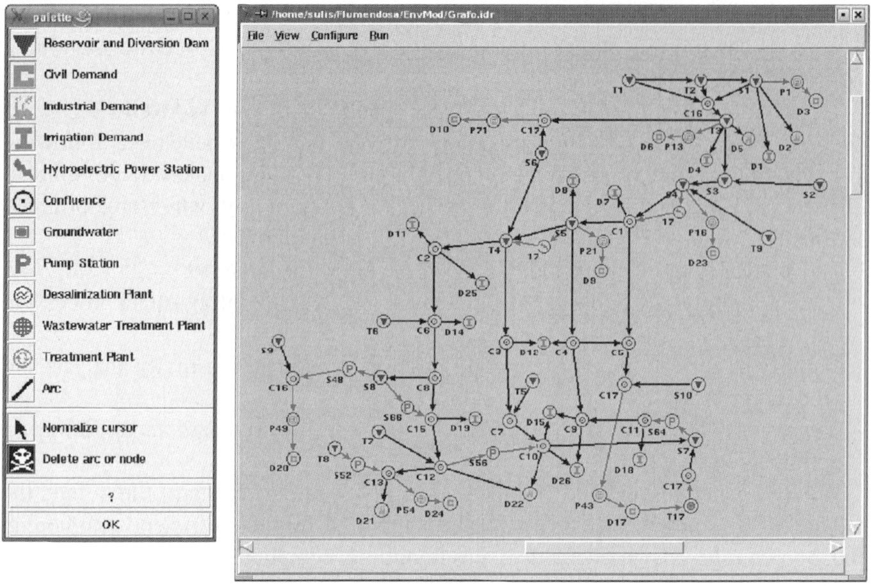

Figure 1. The Flumendosa-Campidano water system in WARGI schematization; main lake codes in the system are: Flumendosa (S3), Mulargia (S4), Is Barroccus (S6) and Cixerri (S8)

- nine water treatment plants with total treatment capacity of $116 \cdot 10^6 m^3$/year;
- one waste-water treatment plant with total treatment capacity of $23 \cdot 10^6$ m^3/year.

To properly characterize system performances, it is necessary to consider water system optimization in the framework of different synthetic quantity and quality data sets, characterizing reservoir water bodies in different hydrological, demand and management scenarios. The optimization approach has been performed using WARGI and considering a more extended time horizon than historical data-set presented in Begliutti et al. (2007). These synthetic data-set places the same time horizon of the Sardinia Region official hydrologic data-base ranging over 54 years in the period 1922–1975 (SISS, 1995). Historical data have been modified in accordance with the Regional Water Plan, rescaling the observed runoff series according to a mean estimation of runoff reduction in the last decade (RAS, 2005).

In Begliutti et al. (2007) it is defined the Quality Evaluation (QE) index in a reservoir $j \in res$, at time t, as a unique index evaluated by the following equation:

$$QE_j^t = F(TSI(Chl_j^t), D(cyano)_j^t) \tag{3}$$

It is to point out that when considering the synthetic data set extended 54 years, attribution of TSI and cyanophyte concentration values to the reservoirs in the absence of any kind of direct water-quality observation is, obviously, a difficult task and a high level of uncertainty is to be expected for these attributions needed to provide synthetically reconstructed quality data. Dealing with non observed periods, several types of relations based on regression analysis have been considered with the aim of modelling the dependency of TSI from other data characterizing reservoirs. Indeed, in order to determine the regression equation, a simple linear least-squares curve was determined between hydrological parameters and TSI values at each reservoir. More complicated approaches do not produce significantly better results, at least for the water bodies under consideration. The final equation used to model the relation between stored-water TSIs and inflows in reservoir is a multiple linear regression that incorporated month-explanatory TSI variation to allow the curve's slope to change depending upon the month of the year. As stated above, monthly TSI values in the period between 1996 and 2005, were used to calibrate the regression model in the following form:

$$TSI_t = a \cdot TSI_{(t-1)} + b \cdot T + c \cdot HI_t + d \cdot \sum_{i=1}^{N} HI_{t-i}^{1/i} + h \tag{4}$$

where:

- TSI_t : evaluated variable at current time t;
- $TSI_{(t-1)}$: evaluated variable at the antecedent time period;
- T: transformed time date;
- $HI_{t-i,i=0,\dots,N}$: hydrological inflow in the lake in the ith antecedent time period;
- a,b,c,d,h : regression coefficients.

Number of data pairs, correlation coefficient R and residual standard error are listed in Table 1.

A large number of chemical, physical, and biological factors and their interactions determine the cyanobacteria density in a reservoir (Pick and Lean, 1987; Dokulil and Teubner, 2000). It is widely accepted that TP and TN supplies are the most important factors enhancing the growth of cyanobacteria in situ. In general, nutrient concentrations influence the growth of cyanobacteria more than the TN:TP ratio. Several observations do not support the significance of the TN:TP ratio (Xie et al., 2003; Vaitomaa, 2006). The phosphorus-chlorophyll-a models of Dillon and Rigler (1974) and Vollenweider (1975), already used to implicate phosphorus as an important nutrient in phytoplankton biomass, have also been useful for predicting the concentration of cyanobacteria or microcystin in lakes. Using regression analysis, best fits for both chlorophyll-a and cyanobacteria suggest similar patterns of change in chlorophyll-a and cyanobacteria occur with increasing total phosphorus concentration (Haney and Ikawa, 2000).

Table 1. Regression results for TSI evaluation in Flumendosa, Mulargia, Is Barroccus and Cixerri lakes

	Flumendosa	Mulargia	Is Barroccus	Cixerri
Coefficient R	0.68	0.66	0.79	0.77
Standard Error	5.93	5.98	5.64	7.71
Number of samplings	67	98	117	111

Preliminary analyses have been made to verify interaction over time between chlorophyll-a and cyanobacteria in the main lakes of *Flumendosa-Campidano* system, and to evaluate the possibility of forecasting cyanobacteria density. In Begliutti et al. (2007), QE values attribution has been given in two different ways:

- adopting a simplified index attribution (given in Table 1 - in the first part of the chapter), considering the direct association between QE and TSI;
- adopting a coupled criteria, using TSI and maximum density value of toxic cyanobacteria given by WHO.

However, when considering extended time horizon and in the absence of measured values for water quality parameters, the simplified QE attribution could be adopted. As stated before, in this work we are focussing attention on synthetic data set for an extended time horizon optimization.

Considering demand centers quality requirements, the *Flumendosa-Campidano* is a multi-purpose water system (water supply for municipal, industrial, irrigation, hydropower and flood control) where different types of demand require different water quality standards. Minimum QE values that must be guaranteed for demands during the whole optimization period were assumed as follows: equal to 1 for urban use, 3 for industrial use, 4 for irrigation use and no restriction for hydropower.

In the system network in Figure 1 we have inserted water treatment plants between water sources and water demand centers. In the optimization model they make it possible to update the quality index attribution of water flowing along arcs with a treatment plant inserted.

To characterize system performances under synthetic data set extended 54 years, we constructed a synthetic TSI quality data scenario using equation (4) that makes it possible to attribute lake quality indexes on the basis of the considered hydrological series. As previously stated, this optimization approach has been performed using WARGI and utilizing a synthetic hydrologic series constructed, in accordance with the Regional Water Plan (SISS, 1995). Following results are referred to this synthetic quality data and hydrologic scenario.

The application of WARGI optimization using reconstructed quality data was also carried out considering two phases: in the first phase the system was analyzed using WARGI-OPT (WARGI quantity optimization module) without limiting resource use on the basis of its quality and in the second phase we introduced in WARGI-QUAL (WARGI quality optimization module) the estimated QE values. To emphasize the consequences of the introduction of quality constraints, an index of irrigation demand reduction was evaluated in order to annul the deficits in the first phase. This approach makes it possible to highlight modifications in the management criteria caused by the introduction of water quality requirements.

As was to be expected, the introduction of quality indices determined high deficit values in the system. Figure 2 shows the comparative evolution over time of total storage volumes, obtained by WARGI and WARGI-QUAL, summing storage in each of the 3 reservoirs, located in the northern area of the system, with similar water quality behavior. Comparison of storage volumes highlights the effect of quality constraints on reservoir management, particularly the significant

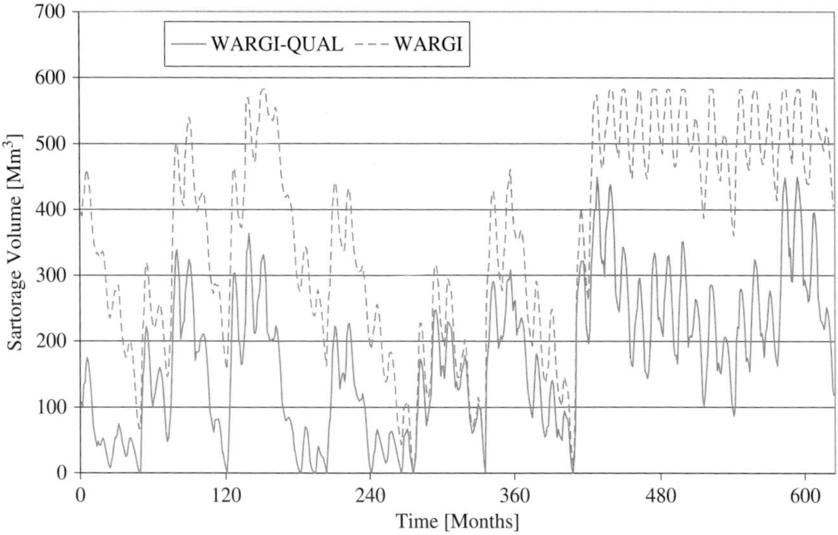

Figure 2. Evolution over time of total storage volumes in reservoirs S3, S4 and S6

reduction of resource availability due to the increased demand for water characterized by good status, mainly located in reservoirs in the northern area of the system.

The increase of deficit reported in Table 2, is the consequence of the adoption of reservoir operating rules and water supply constraints connected with water

Table 2. Performance indices of optimization using quality constraints

	Civil Demand	Industrial	Demand Hydroelectric	Demand Irrigation
Maximum annual deficit (% of demand)	0.08	0.00	33.33	71.66
Temporal reliability (% of months) Deficit = 0%	99.83	100.00	99.17	85.67
Temporal reliability (% of months) Deficit <5%	100.00	100.00	99.17	87.83
Temporal reliability (% of months) Deficit <15%	100.00	100.00	99.17	88.00
Temporal reliability (% of months) Deficit <25%	100.00	100.00	99.17	88.33
Volumetric reliability (%)	100.00	100.00	99.17	89.10

quality attribution for reservoirs. As can be seen in the table, lowest volumetric reliability is obtained mainly for irrigation demand (89.10%) since urban, industrial and hydroelectric demands were characterized by higher deficit penalization cost in the OF, in order to assign higher priority to water supply requirements. On the other hand, the urban and industrial centers in the Flumendosa-Campidano system are connected to efficient treatment plants that in some cases permit the use of resources characterized by high QE values.

4. CONCLUSION

The attribution of water-quality indexes modeling supply systems, constraining the possibility of water utilization, is an essential element in the assessment of system efficiency. These quality constraints may become particularly significant in conditions in which there is a shortage of resources, i.e. during droughts. In the approach described in this chapter, the quantification of the quality of water bodies and reservoirs has been obtained by examining the trophic state of the lakes which, particularly in the Mediterranean region, may be marked by eutrophication phenomena preventing their possible utilization. As described in Begliutti et al. (2007), both TSIs and algal classification could be used to provide a single index for the purpose of classifying reservoirs in a complex water system.

In the application, we illustrated the WARGI implementation of this modeling approach. The optimization tool is characterized by considering a user-friendly graphical interface. The WARGI package has been developed considering the possibility of using water quality characterization for solving large-scale water system optimization problems.

The usefulness of the water-quality optimization approach has been confirmed in the WARGI application. Results obtained by the application of the model have made it possible to evaluate the increase in deficit in the Flumendosa – Campidano system for irrigation purposes in the presence of constraints given by water quality. The benefit of resorting to mathematical programming optimization models for solving real cases of water resource planning and management problems, seems to be confirmed when factoring in quality constraints.

ACKNOWLEDGMENTS

This research has been developed with financial support from European Project SEDEMED II. The authors thank all partners involved for providing their collaboration in developing this research. The authors also express deep thanks to Prof. Rossi for his insightful comments and suggestions that greatly improved the chapter.

REFERENCES

Begliutti, B., Buscarinu, P., Marras, G., Sechi, G.M. and Sulis, A. (2007) Reservoirs water-quality characterization for optimization modelling under drought conditions: PART I, reservoirs Trophic state characterization. *In this book*.

Carlson, R.E. (1977) A trophic state index for lakes. *Limnology and Oceanography* 22(2): 361–369.

CPLEX Optimization, Inc. (1993) *Using the CPLEX Callable Library and CPLEX Mixed Integer Library*, Incline Village, Nevada.

Dillon, P.J. and Rigler, F.H. (1974) The phosphorus-chlorophyll relationship in lakes, *Limnology and Oceanography* 19(5): 767–773.

Dokulil, M.T. and Teubner, K. (2000) Cyanobacterial dominance in lakes, *Hydrobiologia* 438: 1–12.

Haney, J.F. and Ikawa, M. (2000) A survey of 50 NH lakes for microcystins (MCs), *Final Report*, University of New Hampshire, Durham.

Labadie, J.W. (2004) Optimal operation of multi-reservoir systems: state-of-the-art review. *Journal of Water Resources Planning and Management* 130(2): 93–111.

Loucks, D.P., Stedinger, J.R. and Haith, D.A. (1981) *Water Resource Systems Planning and Analysis*, Prentice Hall, Englewood Cliffs, NJ.

Loucks, D.P. (2000) Sustainable water resources management. *Water International* 25(1): 3–10.

OECD (1982) *Eutrophisations des eauxs. Méthodes de surveillance, d'évaluation et de lutte*, OECD, Paris.

Pallottino, S., Sechi, G.M. and Zuddas, P. (2005) A DSS for water resources management under uncertainty by scenario analysis. *Environmental Modelling & Software* 20(8): 1031–1042.

Pick, F.R. and Lean, D.R. (1987) The role of macronutrients (C, N, P) in controlling cyanobacterial dominance in temperate lakes, *New Zealand Journal of Marine and Freshwater Research* 21: 425–434.

RAS (2005) *Piano Stralcio di Bacino della Regione Sardegna per l'Utilizzo delle Risorse Idriche. Autonomous Region of Sardinia*, Convention RAS-UNICA-EAF.

Sechi, G.M. and Zuddas, P. (2000) WARGI: Water Resources System Optimization Aided by Graphical Interface. In: W.R. Blain and C.A. Brebbia (Ed.), *Hydraulic Engineering Software*, WIT-PRESS, pp. 109–120.

Sechi, G.M. and Sulis, A. (2005) Multi-reservoir system optimization using chlorophyll-a trophic indexes. *6th International Conference of EWRA*, Menton; also selected for a Special Issue of International Journal of Water Resources Management, in press.

Simonovic, S.P. (2000) Tools for Water Management - A View of the Future. *Water International* 25(1): 76–88.

SISS (1995) *Studio dell'Idrologia Superficiale della Sardegna,* Cassa per il Mezzogiorno – Regione Autonoma della Sardegna – Ente Autonomo del Flumendosa, Cagliari.

Sulis, A. (2006) Un approccio combinato di ottimizzazione e simulazione per l'analisi dei sistemi complessi di risorse idriche, *PhD Thesis*, University of Cagliari.

Vaitomaa, J. (2006) The effects of environmental factors on biomass and microcystin production by the freshwater cyanobacterial genera *Microcystis and Anabaena, Academic Dissertation*, University of Helsinki, Finland.

Vollenweider, R.A. (1975) Input-output models with special reference to the phosphorus loading concept in limnology. Schweizerische Zeitzschrift fur Hydrologie 37: 53–84.

Xie, L., Xie, P., Li, S., Tang, H. and Liu, H. (2003) The low TN:TP ratio, a cause or a result of Microcystis blooms? *Water Resources* 37: 2073–2080.

Yeh, W.-G. (1985) Reservoir management and operations models: a state-of-the-art review. *Water Resources Research* 25(12): 1797–181.

PART IV

MONITORING AND MANAGEMENT OF GROUNDWATER UNDER DROUGHT CONDITIONS

CHAPTER 14

METHODS AND TOOLS FOR GROUNDWATER MONITORING AIMED AT THE WATER RESOURCES MANAGEMENT UNDER DROUGHT CONDITIONS

V. FERRARA

Department of Geological Sciences, University of Catania, Italy

Abstract: Groundwater monitoring requires the realization of a network of observation points selected on the basis of hydrogeological and hydrodynamic characteristics of aquifers, by means of which it is possible to carry out a real time control of groundwater resources using indicators of the state and consequent forecasting instruments of drought phenomena. Such criteria has been utilized in the case of a hydrostructure chosen after a preliminary hydrogeological analysis of main aquifer systems of the eastern part of Sicily. The chosen hydrostructure belongs to the hydrogeological system of Mt. Etna and constitutes the north-eastern flank of the volcano in which a considerable amount of water resources are stored that feed important springs and aqueduct networks of large municipalities.

The hydrodynamic behaviour of the aquifer, analysed on the basis of piezometric data and discharge of the springs, has highlighted a different temporal response with respect to the input of rainfall. Based on the average storage capacity, the renewal rate of the dynamic resources indicates a poor regulating ability of the aquifer over the long term. Taking into account current exploitation of aquifer, a correct management of the available resources needs a careful and continuous observation of the conditions of the aquifer in relation to variations in pluviometric inputs

Keywords: Hydrogeology, aquifer, hydrodynamic behaviour, piezometric level, springs discharge

1. INTRODUCTION

Studies and researches on the main aquifers of eastern Sicily have been undertaken in order to define criteria and methods for realizing an efficient network of groundwater monitoring, aimed at forecasting and preventing the risk of drought by means of a correct management of the water resources. Such aquifers constitute the primary source of drinking water for a large population and water for the irrigation of large agricultural areas.

In relation to the variability of the yearly recharging of the aquifers, the availability of groundwater seems to be insufficient to satisfy the demand for the above mentioned water uses, with consequent shortage problems. The recurrent water

G. Rossi et al. (eds.), Methods and Tools for Drought Analysis and Management, 277–301.

crises that take place in Sicily over the years with less or variously distributed precipitation with respect to the long term average, highlight that there is an excessive and irrational withdrawal of groundwater without taking into account the availabilities of renewable resources contained in the aquifers.

One of the recent crises of particular gravity took place in the years 2001, 2002 and part of 2003, due to insufficient rainfall during the autumn and winter months, when most recharging of the aquifers normally occurs. The effects of the prolonged period of pluviometric anomaly ("drought") and the consequent failure to reconstitute the water reserves of the aquifers have determined severe lowering in piezometric levels with a drastic reduction in the production of wells and springs. This led to rationing the available water from the late spring on, with serious repercussions on water supplies for the civil and agricultural sectors, with a greater incidence on the densely populated areas in which the demand for water is particularly high and the extraction from the aquifers is carried out without suitable controls on the hydrodynamic balance of the groundwater. This highlights the lack of a rational management of the groundwater resources which leads in most of the aquifers to conditions of over exploitation, with associated processes of qualitative degradation of the groundwater resources. Such conditions tend to worsen in time, with expectable risk of availability reduction of these resources in the future.

The crises of water supplies characterizing some periods of the recent past are linked, besides climatic variations, to the absence of a suitable understanding of the hydrogeological and hydrodynamic characteristics of aquifers, on the volumes of exploitable resources and the actual extractions. In such conditions, it is not possible to carry out a correct management of the groundwater resources that would allow offsetting the effects of drought both with emergency interventions and water resource exploitation planning.

Groundwater monitoring, by means of a network of measuring points chosen on the basis of the hydrogeological, hydrodynamic and hydrochemical characteristics of the aquifers, is essential for achieving the above mentioned objectives. The data deriving from such activity, together with the quantification of annual recharging and the extractions from the aquifer systems, allow carrying out a real control of groundwater resources, using indicators of the state and consequent instruments to forecast drought phenomena.

2. DESIGN CRITERIA FOR A GROUNDWATER MONITORING NETWORK

2.1 Methodological Aspects

To attain useful information in order to plan a quality network for the monitoring of groundwater resources on a regional basis, the methodological scheme must consist of the following stages.

The first phase consists in the collection of existing information on the general geological and hydrogeological characteristics of the territory in order to individuate the aquifer systems. To this end, all the possible sources of information and data will be searched and analysed, such as scientific publications, geological and thematic maps and unpublished reports on hydrogeological surveys over wide sectors of the territory. This information will be organized in a database for a successive elaboration and integration with data deriving from the subsequent monitoring activity.

The second phase consists in the selection of areas with the aquifer systems of greatest interest for the importance of their water resources, according to the existing quantitative and qualitative data.

The third phase consists in the analysis of the hydrostructures that constitute the individuated aquifer systems, with definition of: boundaries, total volumetric dimensions and of the saturated part of the aquifer, permeability, ability to store, actual exploitations. The picture deriving from such activity will allow defining the hydrodynamic conditions and realizing conceptual models of the aquifers. It will also be useful to determine the characteristics of vulnerability to processes of qualitative degradation of groundwater eventually induced by human activity.

On the basis of the above information, the most suitable points for monitoring groundwater may be chosen, installing piezometers of adequate depth and/or using suitable existing wells for this objective, to be equipped for the continuous measuring of piezometric levels and possibly some chemical-physical characteristics of the water. The monitoring should also include the springs and infiltration galleries with continuous measures of the discharge and some qualitative characteristics of waters.

The data deriving from this research, together with the preliminary findings, will allow to estimate at each moment the state of the available resources, in conditions of both normal pluviometric input as well as lesser feeding by the aquifers, on the basis of the hydrogeological balance that considers all the input and output factors of the system. This will enable the risk assessment of water deficiency and the implementation of opportune actions, that can be adopted in real time, in order to avoid conditions of water crisis.

2.2 Project Elements

A ground waters monitoring network consists of a set of measuring points of water levels, constituted by wells and piezometers, arranged according to a variable sized mesh in relation to the hydrogeological characteristics of the territory and the goals that are to be achieved.

The primary aims are represented by the reconstruction of the piezometric surface and the measuring of its fluctuations, that allow:

- to individuate the directions of groundwater flow;
- to recognize the existence of geologic watersheds;
- to verify the relationships between surface and groundwater;
- to verify the effects of extractions from wells;

– to evaluate the variations in volume of water resources stored in the aquifer;
– to verify the existence of overexploitation conditions.

In the case of overlapping aquifers, the monitoring network will have to be constituted by specific measuring points for each aquifer.

The density of the measuring points must be established on the basis of the hydrogeological characteristics of the area and the size of the aquifer system.

The choice of the measuring points should take into account:
– easy accessibility;
– safety conditions for integrity of the equipment;
– possibility to repeat the measurements in the scheduled periods;
– knowledge of the hydrogeological characteristics of the site;
– suitable distance from wells in use, estimated on the basis of the radius of influence determined by pumping.

An optimal frequency of measuring cycles is quarterly, but should be no less than a six-month period of the hydrological year. It is opportune that a certain number of strategically located measuring points can be controlled with continuity, namely by means of sensors installed and connected to recording systems of the piezometric level, possibly with remote data transmission, with the aim of verifying variations of the water table in real time.

The monitoring network must also include stations measuring the discharge of the most important springs, equipped in such a way as to allow continuous measurement of the data or with a frequency established on the basis of the characteristics and the importance of the aquifer that feeds it. In particular, this survey should take place in the period of the hydrological year not affected by precipitation, so as to be able to verify the hydrological regime of the springs with opportune elaborations.

2.3 Hydrogeological Characterization

The necessary knowledge for the design of a groundwater resources monitoring network concerns the essential geological and hydrogeological characteristics of the aquifers, namely:
– lithostratigraphical and structural framework;
– hydrogeological characteristics;
– hydrodynamic conditions;
– estimate of the groundwater resources.

The activities necessary to acquire the above information are:
– geological and morphological survey, with the support of eventual geophysical surveys and stratigraphical data of wells and boreholes;
– analysis of the characteristics of permeability on outcrops and by means of pumping tests in wells and piezometers;
– census of the water points with measurements of level in wells and piezometers and measurements of spring discharge.

The elaboration of the acquired data, organized in a relational database, will allow drawing up maps and hydrogeological profiles and the calculation of the hydrogeological balance of the aquifers. To this end, the spatial domain of each aquifer through its geometric configuration (volume and boundaries), structural characteristics and feeding area will be defined analytically and characterized from the geologic point of view.

The hydrodynamic characterization of the aquifers will be first of all derived from the definition of the permeability or hydraulic conductivity of the lithostratigraphical succession and the structural framework, determined by tectonics and the morphology of the basement, which condition the modalities of the ground flow and therefore the hydraulic gradient of the water table. Necessary elements for this purpose are the levels of water measured in wells and piezometers, from whose correlation by means of geostatistical methods the motion field of the aquifers can be deduced.

The estimate of groundwater resources will be obtained through the calculation of the hydrogeological balance of the single aquifers by evaluating the inputs and outputs to the system. Elements of such calculation are, among the inputs, the average annual effective infiltration (hydrological year) of precipitation and the eventual contribution deriving from the interconnection with surface water and from the restitution of used waters to the aquifers. Among the output, the main elements to be quantified are represented by withdrawal from wells and the discharge from springs. A negative variation of the water reserves will indicate conditions of imbalance of the system determined by excessive and unregulated exploitations.

The results of this analysis will enable the choice of the monitoring network measuring points, whose data will allow controlling the conditions of the resources stored in the aquifer and carrying out forecasts of their availability in relation to the variability of the pluviometric contributions.

3. STUDY CASE

3.1 Choice of the Hydrostructure

The criteria and methodologies for designing a ground water monitoring network have been applied to the case of a hydrostructure chosen as significant for the presence of important water resources and large withdrawals from the aquifer for drinking water, as well as the availability of data deriving from previous measurements of the water levels and springs discharge.

The choice also derives from a preliminary hydrogeological analysis of the main aquifer systems of the eastern sector of Sicily, different for lithology, type and degree of permeability, water potential and quality of water (Ferrara, 1999a, Ferrara et al., 1999, Ferrara and Pappalardo, 2003). These are (Figure 1):

- Aquifer system of Peloritani Mts.
- Aquifer system of Etna
- Aquifer system of the Catania Plain
- Aquifer system of the Erei Mts.
- Aquifer system of the Hyblean Upland

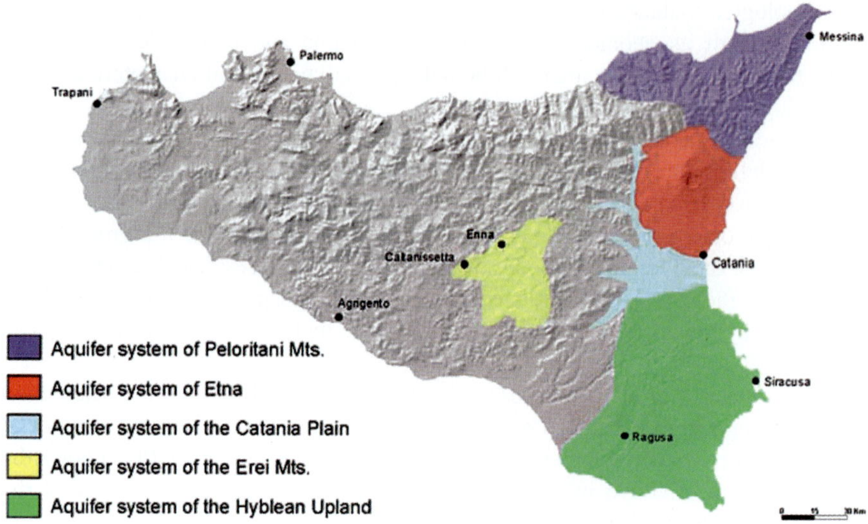

Figure 1. Main aquifer systems of eastern Sicily

The current conditions of such systems have been verified by comparing the piezometric levels of previous years and through an estimate of the water resources, considering the annual recharge and volumes of groundwater measured in the different periods for civil, agricultural and industrial purposes. It has been assessed that in all the analysed systems, at times, there has been underway a gradual lowering of the water table level, often accompanied by a worsening of the quality of the resource due to salinization or excess of some ionic components in the waters of deep circuits (Figure 2).

In the light of these results, it has been thought opportune to concentrate the attention on the aquifer system of Etna, being considered the importance of its groundwater resources and the existence of consistent withdrawals mainly for drinking purposes (Ferrara, 1975, 1991).

The hydrostructure chosen as the study case, among those making up the Etnean aquifer system, is constituted by the north-eastern flank of the volcano (Figure 3) in which notable water resources are contained that feed important springs near the Ionian coast. Large volumes of ground water are extracted from this structure to supply the aqueduct networks for great municipalities, which significantly affect the hydrodynamic equilibrium of the aquifer (Ferrara, 1999b, 2001).

Therefore, it was considered useful to analyse, by means of the data of previous measurements of level and discharge, the behaviour of the aquifer with appraisal of the input and output, in conditions both of normal recharging, as well as insufficient pluviometric input. This enables identifying the behaviour of the aquifer and obtaining useful information to define a model for managing the water resources that would allow avoiding water crises due to drought events.

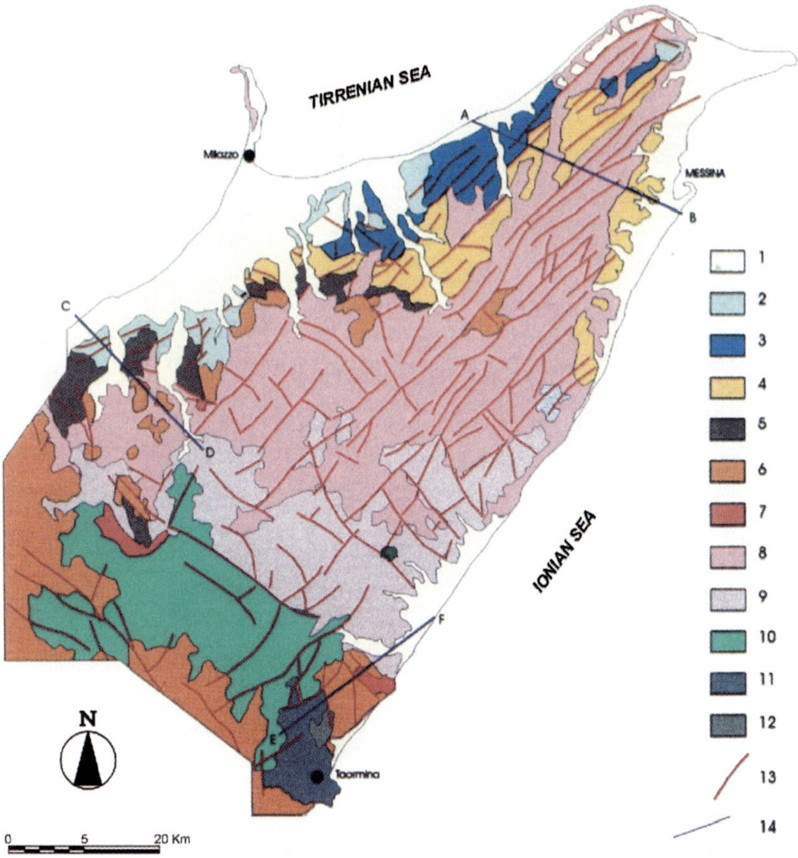

Figure 2a. Aquifer system of Peloritani Mts .1) Alluvium; 2) Clays; 3) Evaporites; 4) Clays, sandstones and conglomerates; 5) Varicoloured shales; C.d.O.flysch; Red sandstones and conglomerates; 8-12) Metamorphic rocks; 13) fault; 14) cross-section

3.2 Geological Lineaments

The rocks outcropping in the studied area are mainly represented by volcanic products resting on a sedimentary basement chiefly constituted by an alternation of fine grain sandstones and argillaceous marls (Piedimonte Formation) of Eo-Oligocene age, outcropping extensively at the north-eastern border of the area. To a more limited extent, Pleistocene grey blue marly clays are also present, whose outcrops are found in small strips between the volcanites and near the coast. Locally and along the shoreline, there are also alluvial and gravely-sandy beach deposits that cover both old lava flows, as well as argillaceous sediments of the basement.

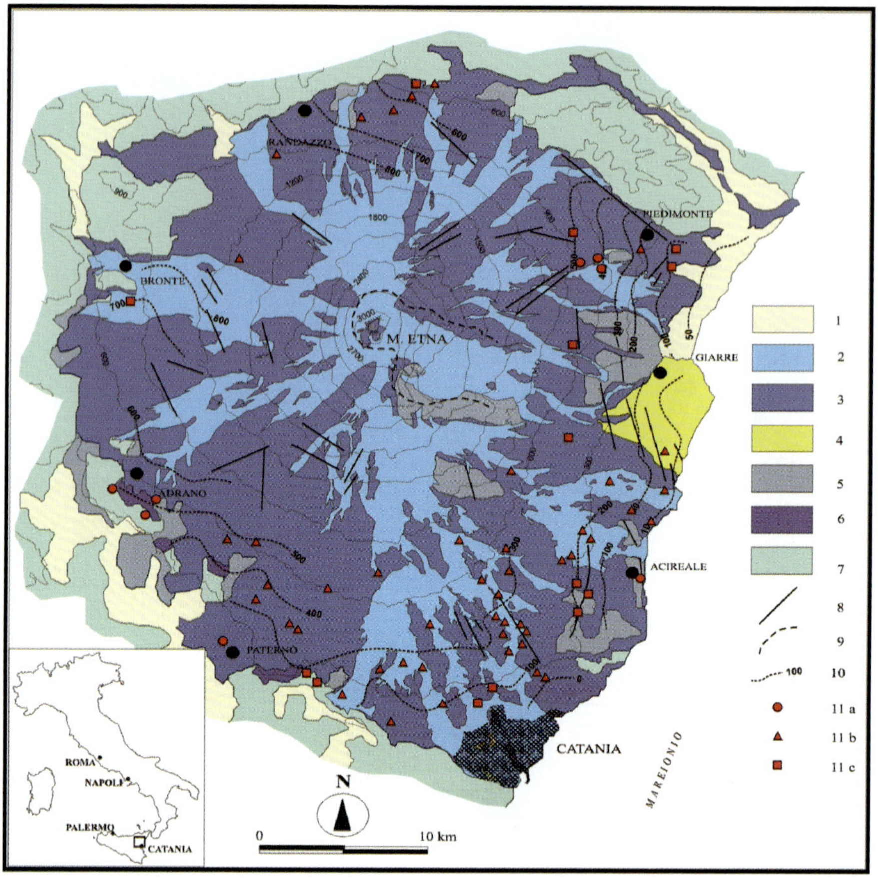

Figure 2b. Aquifer system of. Etna:.1) Alluvium; 2) Historic lavas; 3) Recent lavas; 4) Sandy conglomerates; 5) Older volcanic products; 6) Basal volcanites; 7) Sedimentary basement; 8) piezometric contour lines; 9) topography; 10) fault; 11) spring; 12) well; 13) infiltration gallery

The volcanic products are represented by lava flows of differing age and composition, in banks of varied thickness and generally of limited lateral extension. The oldest lavas constitute the prevailing lithological type, while the more recent products are represented by historical flows and pyroclastic deposits (lapilli, sands and ash) mixed with volcanic scoriae (Branca and Ferrara,2001).

From a structural profile, there are regional fault systems, oriented NE-SW and NNW-SSE, that have given rise to marked slopes, at times allowing the sedimentary base to outcrop. Particular significance, from a hydrogeological point of view, has the fault that raises the basement along the coastline constituting an obstruction for the ground outflows.

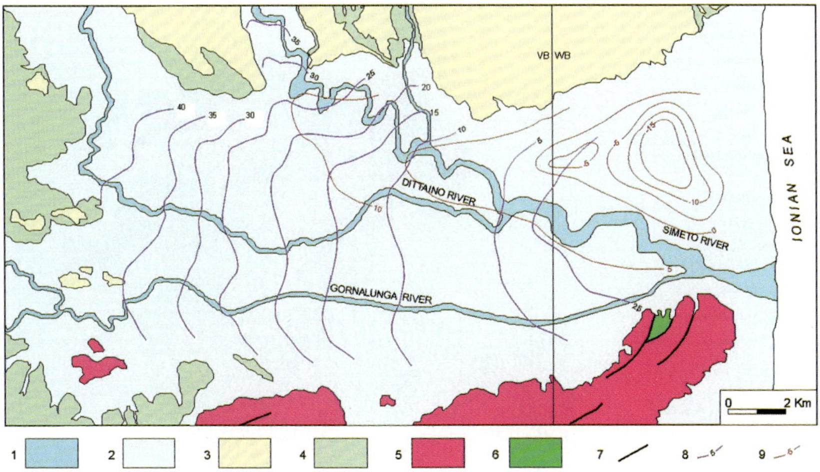

Figure 2c. Aquifer system of the Catania Plain. 1) Actual alluvial deposits; 2) Recent alluvial deposits; 3) Pleistocene sands and gravels; 4) Pleistocene clays; 5) Plio-Pleistocene volcanites and calcarenites; 6) Oligo-Miocene limestones; 7) fault; 8) water table contour lines; 9) piezometric surface contour lines

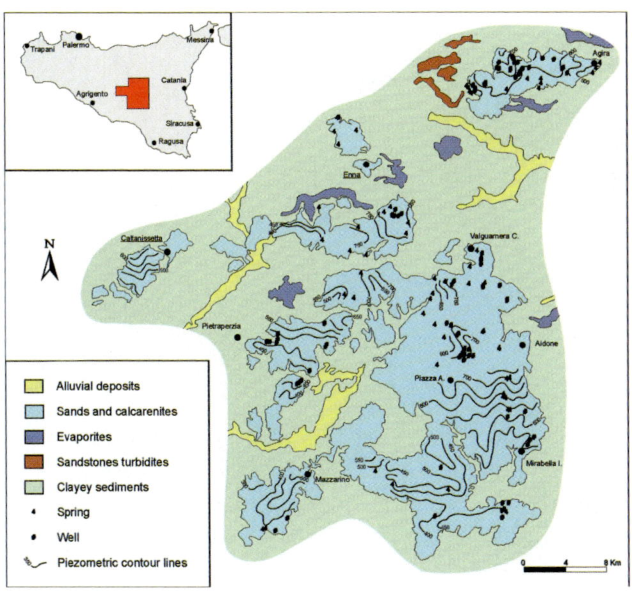

Figure 2d. Aquifer system of the Erei Mts

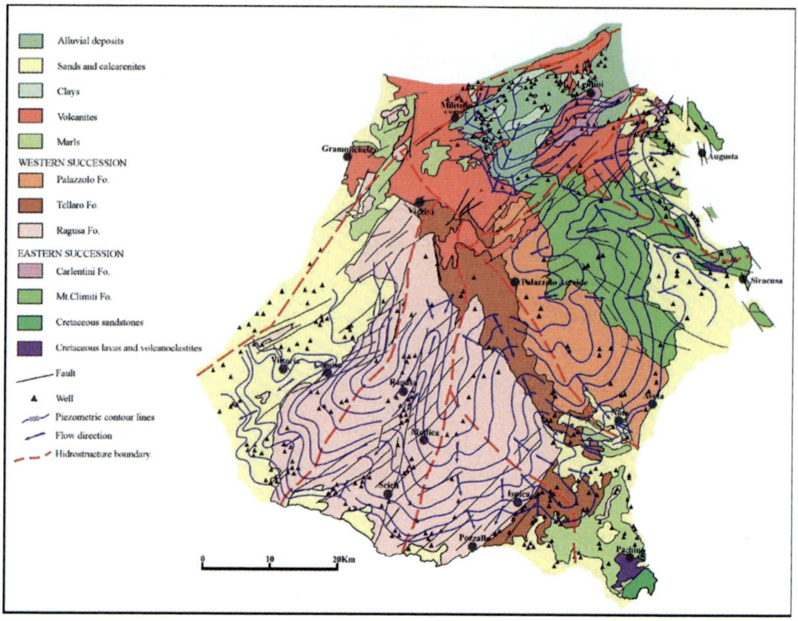

Figure 2e. Aquifer system of the Hyblean Upland

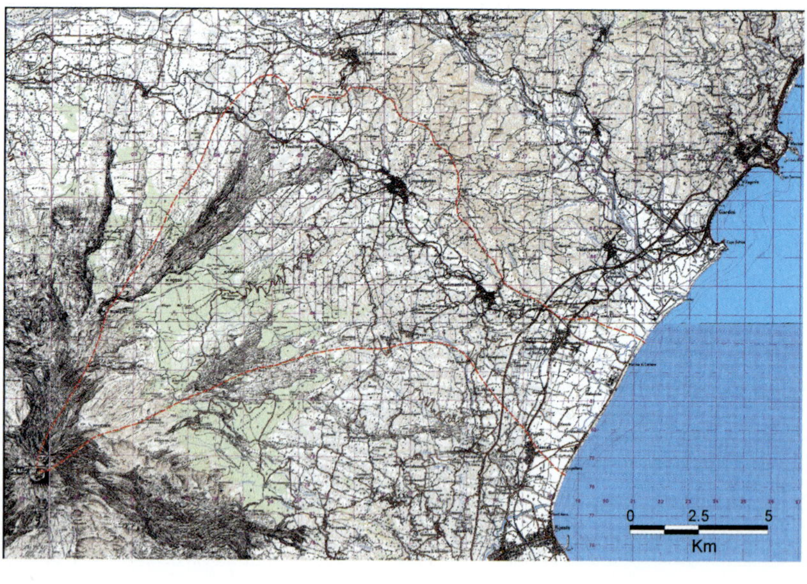

Figure 3. Hydrostructure of the north-eastern flank of Etna

3.3 Hydrogeological Characteristics

The terrains constituting the stratigraphical succession of the studied area can be distinguished as:

- Permeable terrains for their porosity, represented by alluvial deposits and loose pyroclastic products, with average degree of permeability, varying in relation to granulometric composition.
- Permeable terrains because of fissuring and porosity, represented by both old and recent lavas and by associated scoriae, with high to elevated degree of permeability in relation to the frequency and amplitude of the discontinuities and the voids present.
- Terrains of very low permeability or impermeable, mainly represented by prevailing argillaceous sediments making up the basement on which the volcanic rest.

The ensemble of the lava and scoriae constitute an aquifer of high permeability, fed by the infiltration of rainfall over a broad surface extending from the volcano summit to the Ionian coast, which represents the studied hydrostructure (Ferrara, 1991, 1999b).

The geological framework of the aquifer has been defined by means of field survey and ground data constituted by stratigraphic data of boreholes and by geophysical prospecting (Cassa per il Mezzogiorno, 1982) (Figure 4).

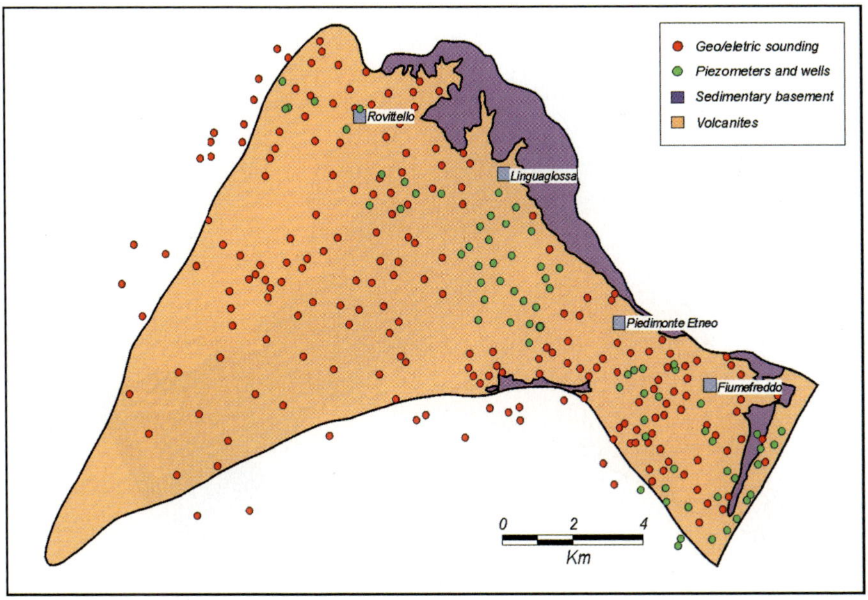

Figure 4. Subsoil explorations utilized for basement morphology and aquifer geometry reconstruction

The sedimentary basement morphology has been reconstructed and the geometry of the volcanic aquifer defined by using these data (Figure 5). To this end, the data available in numerous points of the hydrostructure have been reported on a georeferential basis, using a GIS (MapInfo Prof., 5) with relational database, and interpolated by means of various models (Geostatistical Kriging and Simple Natural Neighbour), obtaining a good coherence of the results from the semivariograms.

Using a CAD-DWG, a DEM has also been constructed of the hydrostructure that, compared to the morphology of the sedimentary basement, has allowed calculating the aquifer thickness (Figure 6) and drawing longitudinal and cross-sectional profiles in order to evidence its variability.

Inside the deep depression of the sedimentary basement (paleovalley), present in the piedmont zone and individuated by the abovementioned surveys, the aquifers are contained in a prevalently lava succession that rests on the sedimentary basement represented by the arenaceous-marly alternation of the Piedimonte formation. In this succession, there are local semi-confined conditions due to the variable distribution of discontinuities in the lava banks, as well as to the presence of clastic materials between several lava flows. These materials constitute levels of different thickness within the volcanic succession and their composition varies in relation to the origin of the materials and emplacement modalities. They comprise both sandy-pebbly deposits with mixed fine material as well as muddy-sandy deposits with rare pebbles and blocks. Their nature is partly volcanic and partly sedimentary,

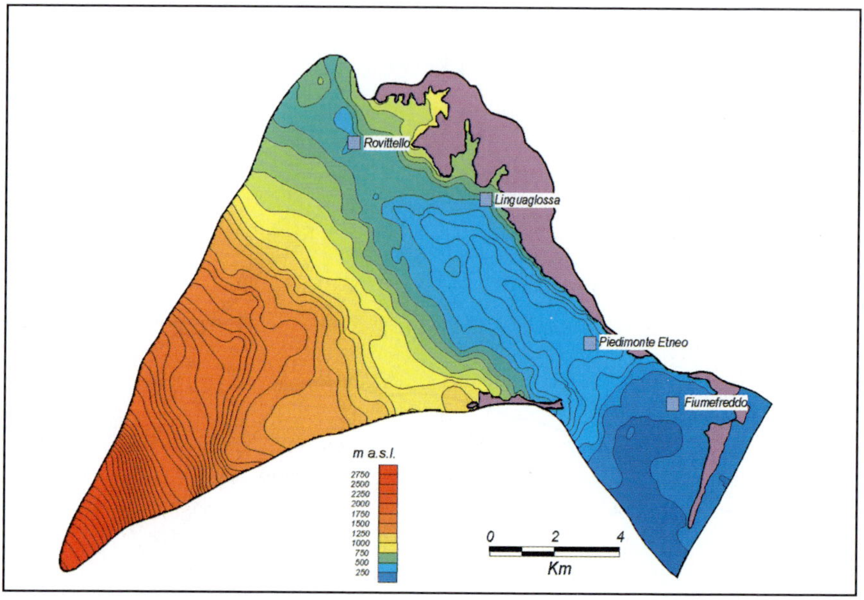

Figure 5. Morphology of sedimentary basement

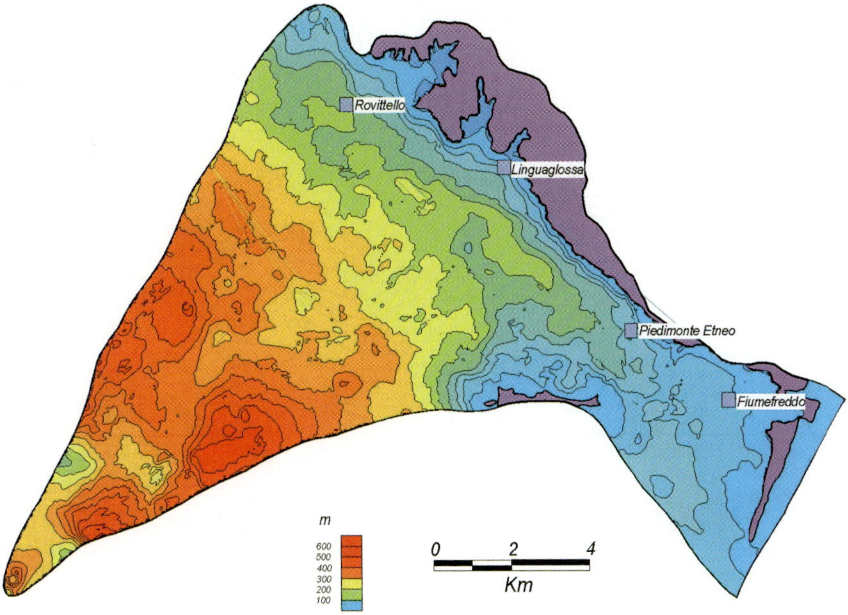

Figure 6. Distribution of aquifer thickness

comparable to paleosoils formed in the interval of time between the emplacement of successive lava flows. Other clastic deposits, with thickness up to several meters, are referable to phenomena of obstruction of torrents by lava flows (Branca and Ferrara, 2001).

As a consequence of the various factors highlighted above, the hydrodynamic parameters of the aquifer present a notable variability from zone to zone. For the definition of such parameters, the data of pumping tests performed in boreholes have been used together with measurements of the water level in nearby piezometers. On the basis of the obtained data, the main hydrodynamic parameters of the aquifer have been calculated, using the more commonly employed formulas for the transitory regime in fully penetrating wells in free and semi-confined aquifers. In particular, the calculation of the transmissivity has allowed obtaining a mean value equal to $T = 5,09 \times 10^{-2}\ m^2/s$ (Ferrara, 1999b) (Figure 7).

The modalities of the water flow are comparable, similarly to karstic systems, partly to a system of interdependent drains and partly to a dispersive system. In the first type, the presence of lava tubes and zones with open and interconnected fractures constitutes preferential paths for the ground flow. In the second type, the lava masses affected by a tight mesh of fissures and fractures and the associated scoriaceous materials constitute zones in which waters move with reduced speed. However, these latter have an important function in the ground circulation since they contain consistent water reserves that feed the drainage network with continuity, also in periods of less rainfall than average (Ferrara and Pennini, 1994).

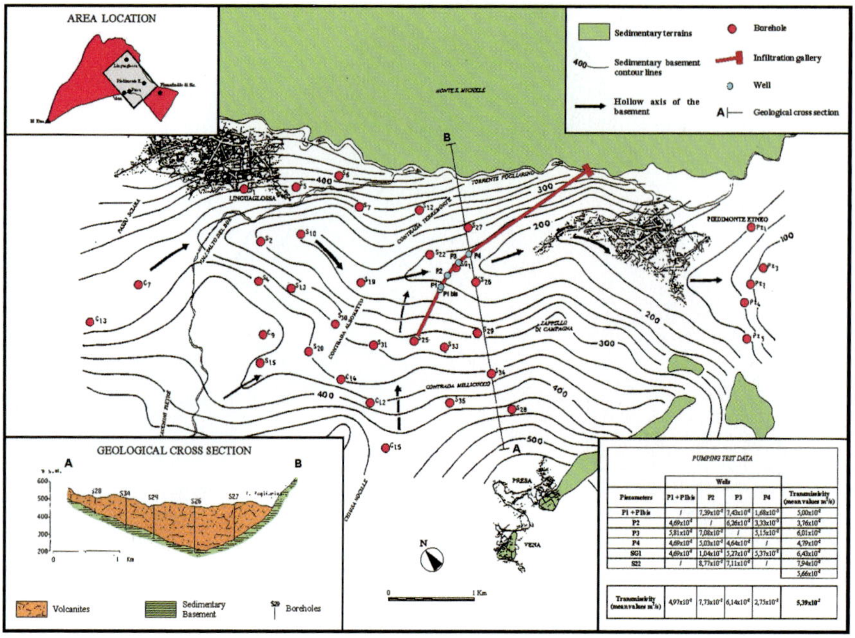

Figure 7. Aquifer transmissivity evaluation

The piezometric surface, reconstructed by correlation of the water levels measured in previous years in wells and piezometers of the piedmont and coastal zone, is located at an average height of 320 m a.s.l. in the hilly zone and approximately 50 m downslope from Piedimonte town, gradually diminishing until almost zero near the coast where the springs are. The mean hydraulic gradient may be evaluated at around 5% upslope and diminishes considerably downslope, until little more than 1% near the shoreline (Figure 8).

3.4 System Behaviour

The monitoring activity carried out in the period June 1993–August 1996 in 12 piezometers of the piedmont zone (Regione Siciliana-Ass.LL.PP., 1994) has highlighted different temporal response and amplitude of fluctuation of the water levels to the pluviometric input, indicative of the variability of the groundwater circulation determined by the lithostratigraphic and structural conditions of the aquifer. Some piezometers showed a greater sensibility, while others were less influenced by rainfall effects.

In all piezometers, however, positive variations of level took place during the hydrological year 1995–96 (Figure 9). In this year, the precipitation was in fact considerably higher than the long term average (1921–2000), with a concentration in the months from November ‘95 to March ’96 that favoured a greater volume of water

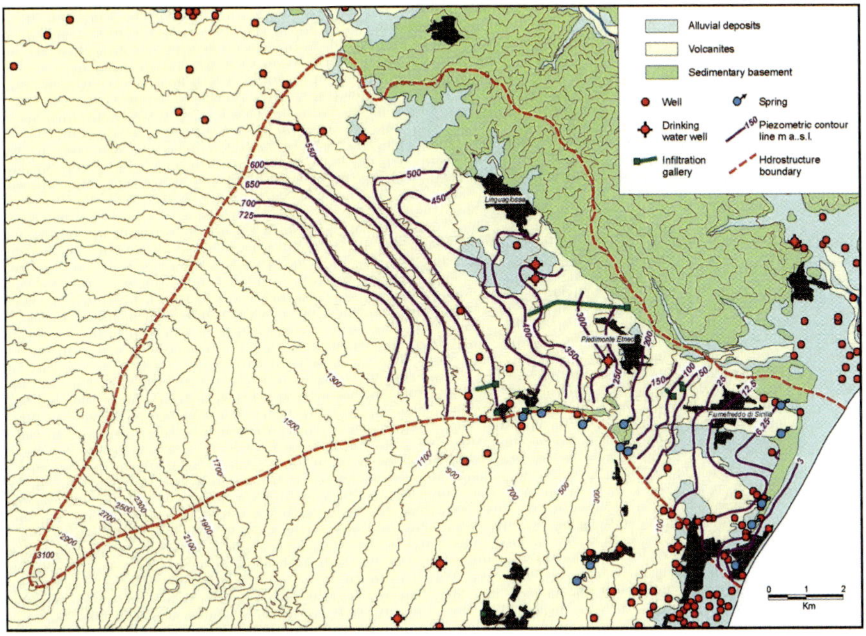

Figure 8. Hydrogeological map of the studied area

infiltration. The effects on the water levels in the piezometers were evident, with a rising temporal response after approximately 2 months from the peak of pluviometric values recorded in the piedmont zone (Figure 10). In various piezometers the levels vary from a minimum of approximately 1 m to a maximum of 20 m (Ferrara, 1999b).

By means of the correlation of groundwater levels, the conditions of the potentiometric surface in the period of maximum height (March 1996) and in that of minimal height (October 1995) have been reconstructed, thereby deducing for comparison the distribution of the variations determined by recharging in the entire piedmont area (Figure 11).

Precipitation at high altitudes of the volcano has contributed to such effects, with different times of transfer of the water flow: shorter along channels of higher permeability and longer in the greater part of the volcanic mass of lesser permeability, according to the described modality of circulation of waters within the aquifer. The tendency of water levels to lowering in the months immediately after the peak would indicate the progressive decline in the volume of regulating reserves stored in the aquifer, that lead up to the depletion phase; in this phase the groundwater flow is mainly fed by infiltration waters circulating in the least permeable part of the system.

In addition to the monitoring of the water levels in the piezometers of the piedmont area, the discharge measurements performed by the Regional Hydrographic Agency

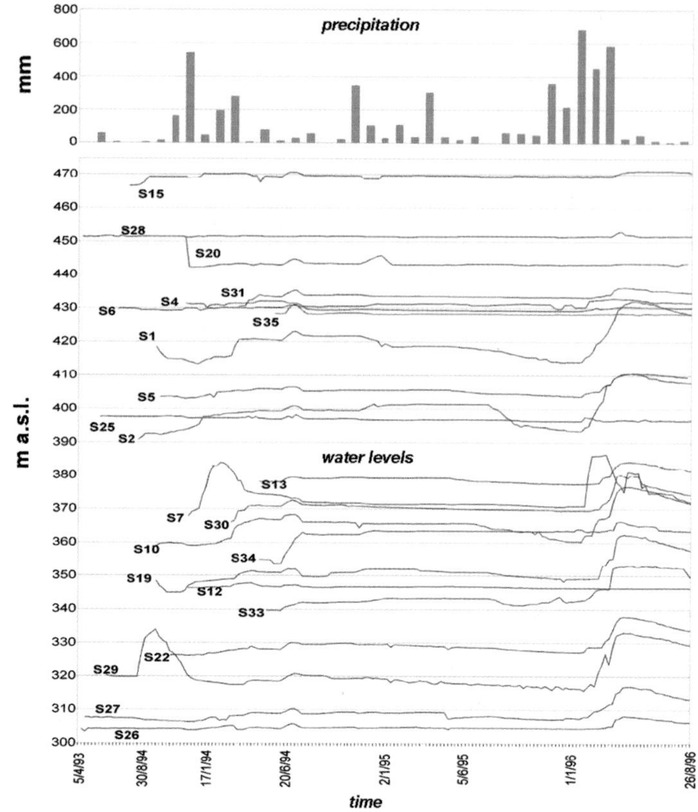

Figure 9. Fluctuations of water levels in piezometers and daily precipitations recorded at Linguaglossa station (April 1993–August 1996)

for approximately 40 years on springs in proximity of the Ionian coast, at the eastern end of the hydrostructure, have been analysed and evaluated.

The group of the Fiumefreddo springs, the most important for volume of discharged water, is situated at the contact point between the propagating ends of the lava flows and the impermeable sedimentary terrains of the basement, represented by the Pleistocene blue marly clays outcropping between the alluvial deposits of the coastal belt (Figure 12).

The intermittent measurements of discharge made between 1929 and 1964 indicated variable values in the spring outflow between a maximum of 2510 l/s (13.3.1950) and a minimum of 1492 l/s (10.6.1951). The monthly measurements carried out in successive years have highlighted cyclical variations in the outflow regime until 1986, with considerably different maximum and minimum values, attributable mainly to the variability of annual recharging of the aquifer. In none of these years, however, the discharge of springs at the end of the dry period approached depletion point.

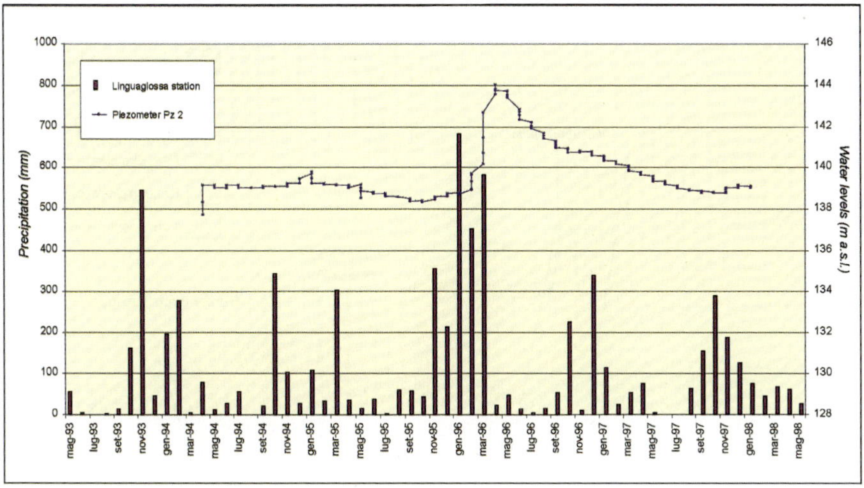

Figure 10. Effect of recharge on water levels during hydrologic year 1995–1996

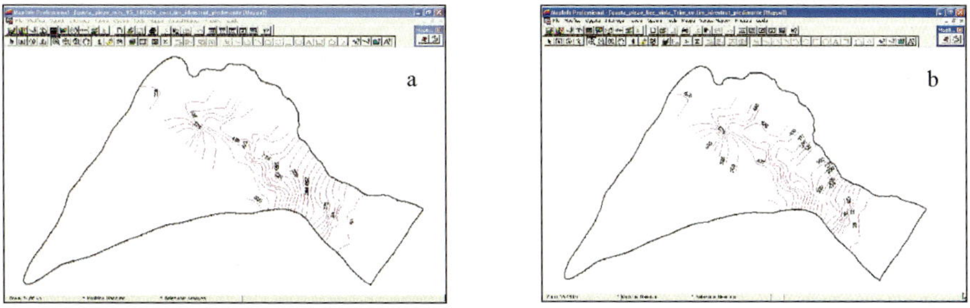

Figure 11. Potentiometric surface from October 1995(a) to March 1996(b)

From the hydrological year 1987–88, the discharge values have shown a constant decrease to the beginning of more consistent extractions from the aquifer in the upstream zone from the springs. The lowest values were measured in the three-year period 1989–91 during the summer months and early autumn, when withdrawals for agriculture were made in addition to the continuous extractions for drinking uses, whereas the water reserves stored in the aquifer approached depletion. The change in the trend of the pluviometric regime, begun in the hydrological year 1992–93, determined a gradual increase in the discharge values of springs, with a maximum in hydrological year 1995–96, with exceptionally abundant rainfall, followed in successive years by a gradual decrease, until returning to the previous conditions (Figure 13).

The analysis of the discharge values, correlated with those of monthly rainfall, has allowed acquiring important elements on the behaviour of the aquifer and, in particular, on the hydrodynamic process within it.

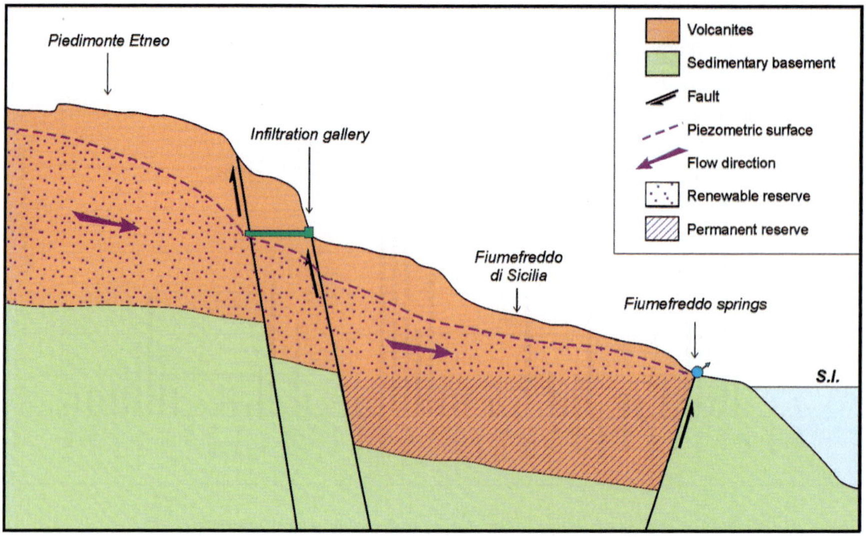

Figure 12. Schematic cross section of Fiumefreddo springs area

Given that it is a prevalently fractured hydrogeological system, the methodology in use considers the curve of annual outflow of the springs, in which two parts are generally distinguished: the line of filling (recharging) of the "reservoir" (Q = f (t) increasing) and the line of emptying (Q = f (t) decreasing), separated by the peak discharge. The second line can be divided in two further parts: the line of decrement and the line of depletion (Castany, 1968, Civita, 2004).

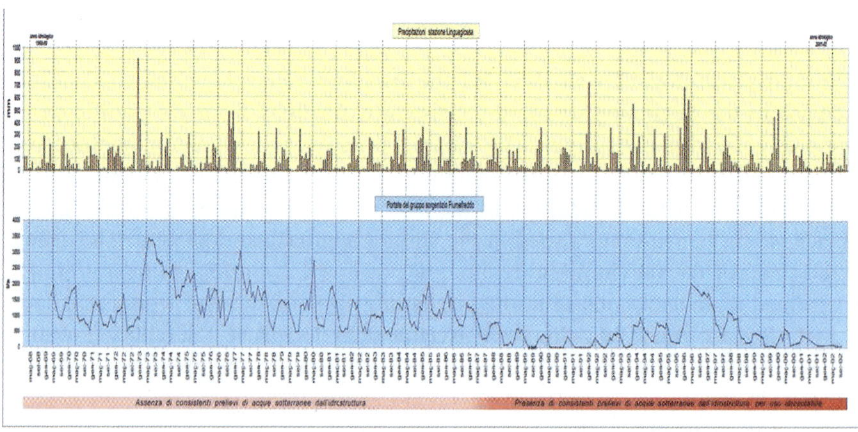

Figure 13. Comparison of Fiumefreddo springs discharge and rainfall at Linguaglossa station (May 1968–September 2002)

A model that considers the curve of complete emptying, from the peak of maximum discharge to the point of inversion of the function $Q = f(t)$ (beginning of new recharging), is the exponential model by Boussinesq (1904) and Maillet (1905):

$$Q_t = Q_0 \cdot e^{-\alpha \cdot t} \tag{1}$$

that is resolved for α (depletion coefficient):

$$\alpha = \frac{\log Q_0 - \log Q_t}{0,43429t} \tag{2}$$

and allows calculating the volume of storage (W_0) at time $t = 0$ integrating between 0 and ∞ the previous formula, from which:

$$W_0 = \frac{Q_0}{\alpha}.86400 \tag{3}$$

while the outflow volume at a time $t \neq 0$ is given by:

$$W_d = \frac{(Q_0 - Q_t).86400}{\alpha} \tag{4}$$

The depletion curves of the springs group have been obtained, by introducing into log-normal diagrams the discharge values measured in the period not influenced by precipitation in function of time,.

From the previous equations (3 and 4) the volumes stored in the aquifer at time $t = 0$ and those outflow values from the beginning of the emptying have therefore been calculated.

The discharge values in the period 1969–87, in which the spring regime was not influenced by consistent withdrawals from the aquifer, indicate the existence of equilibrium of the system, with limited decrements in the dry period and resumption on the arrival of following recharge. Such conditions are also highlighted by similar values of the depletion coefficients (Figure 14).

The intensive exploitation of the resources for drinking purposes from 1988 in the area upstream from the springs has determined substantial changes in the equilibrium of the system, clearly influencing the springs regime. This is evident, apart from the discharge values, by different and often anomalous coefficients of depletion compared to the previous period. A significant example of the changed conditions is given by the three hydrological years 1992–1995 (Figure 15).

From 1988 to 1995, the discharge of springs diminished on average by 70%, in spite of the absence of particularly remarkable pluviometric deficits in the area. In the hydrological year 1995-96, characterized by decisively higher precipitations than average and concentrated in a few weeks, the response of the aquifer was evident, in the same way as the discharge of the springs that reached a similar value to the mean value of the period 1969-87; likewise was observed for the depletion curve.

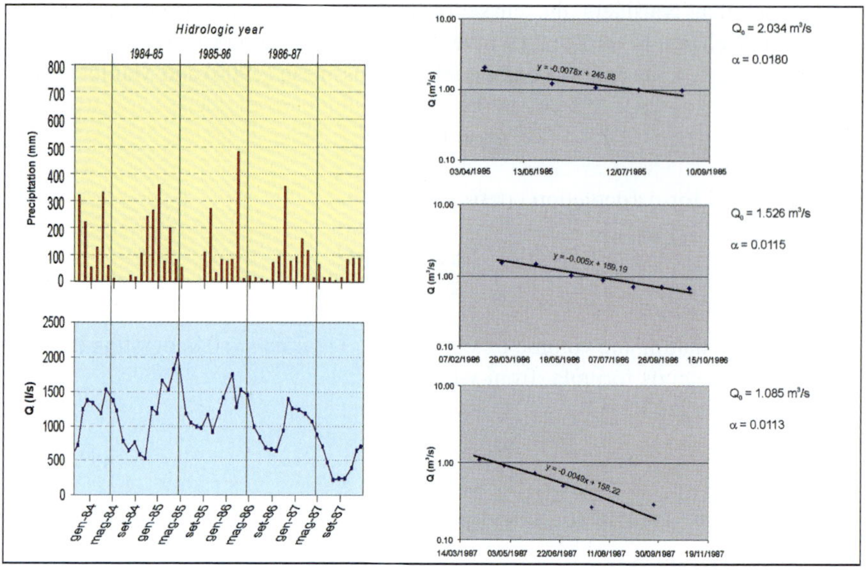

Figure 14. Depletion curves of springs discharge for the years 1984–1987

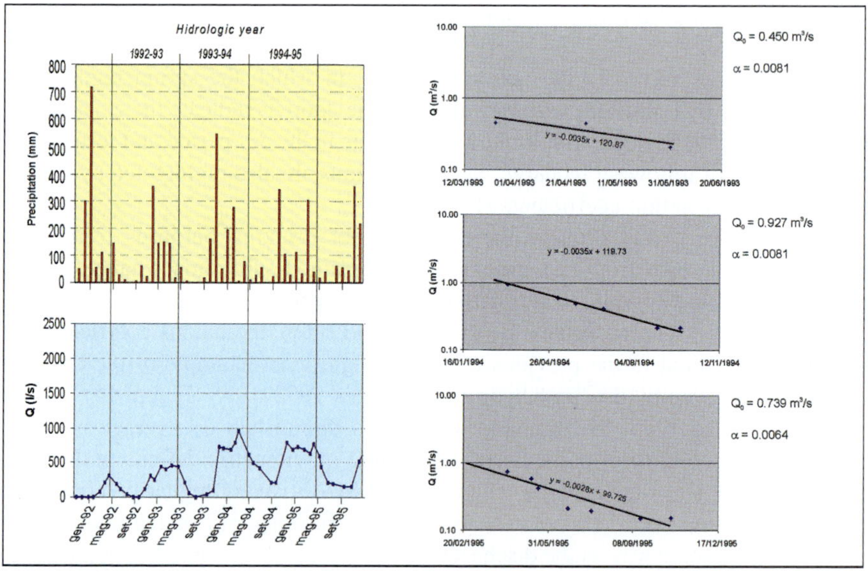

Figure 15. Depletion curves of springs discharge in the period 1992–1995

In the immediately following years, the system gradually returned to the previous conditions (Figure 16). In the first part of the hydrological year 1995–96, despite consistent extractions upstream, the discharge of springs reached, after approximately four months from the beginning of the rising piezometric levels (coinciding with the greater contribution of pluviometric inputs), the value of 2,02 m ^{3}/s in April.

Regarding the volume of effective infiltration in the same period, the stored renewable reserves in the system determined the rapid saturation of the more permeable parts of the aquifer and therefore the prompt reaction of water levels in the piezometers and the discharge of springs.

The successive inputs, feeding the flow in the less permeable zones of the aquifer, helped sustain the piezometric levels and the discharge of springs until the onset of recharging in the next hydrological year. This reveals a modest regulating ability of the aquifer and the existence of a groundwater circulation comparable to fast circuits with rapid recharging and equally fast emptying of the system, as the regime of the springs confirms.

Considering the mean storage capacity of the aquifer (W_{0m}), corresponding to recharging in the period 1969-2002 ($85, 30 \times 10^6\,m^3/yr$) and the mean value of the losses from the system (extractions and spring outflow) as emptying capacity (W_m), estimated for the same period at $79.20 \times 10^6\,m^3/yr$, a renewal rate ($T_{rin}$) of a little less than 100% (0,92) and a minimum time of renewal (t_{mr}) of 1,08 years have been evaluated. Such values confirm how all incoming water into the system departs in

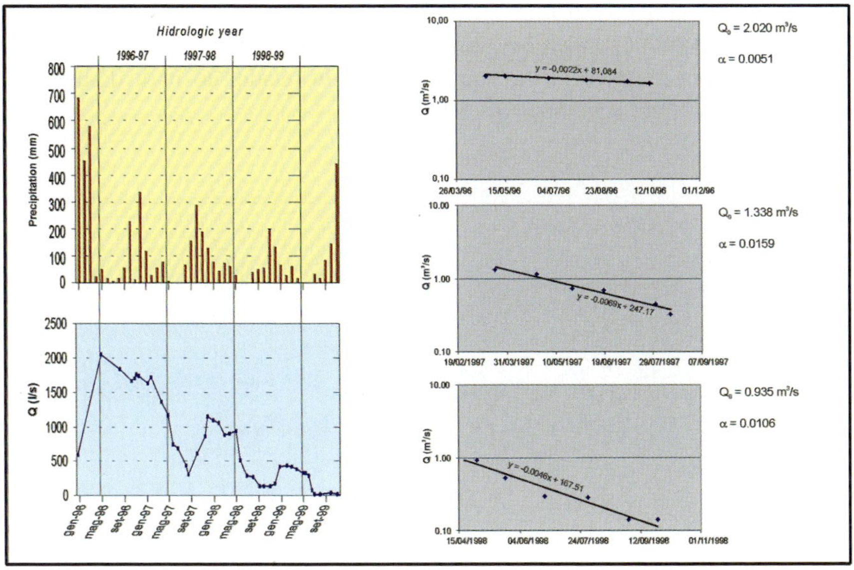

Figure 16. Depletion curves of springs discharge for the years 1996–1999

little more than a year and how the aquifer has therefore a poor regulating ability over the long term.

Similar evaluations have been carried out for the period in which the consistent withdrawals for drinking purposes began (1988–2002), considering a mean value of recharge equal to $87,29 \times 10^6\,m^3/yr$ and losses from the hydrostructure of $82,88 \times 10^6\,m^3/yr$. The renewal rate calculated is similar to the previous one (0,95), as also the renewal time (1,05 years), with the considerations that of it derive in case of many consecutive years with less precipitation than average.

3.5 Assessment of Groundwater Reserves

An estimate of average annual recharging of the aquifer determined by infiltration in the period 1969–2002 (Figure 17) indicates approximately $85.30 \times 10^6\,m^3/yr$ the amount of water resources theoretically available in the entire hydrostructure. In effect, this amount varies annually from a minimum of $33,30 \times 10^6\,m^3$ to a maximum of $171,90 \times 10^6\,m^3$, in relation to the variability of the height and distribution of precipitation of every single hydrological year, which considerably influences the infiltration of the aquifer system. The groundwater reserves of the aquifer has been estimated by means of the water balance with traditional methodologies, considering inflow (effective infiltration, reinfiltration of used waters) and outflow (extractions and water uses, springs discharge), in different situations, that is of long term average (1969–86), drought years (1987–90) and wet years (1995–96).

The results obtained for the period 1969–86 indicate a substantial equilibrium between the average volumes entering the aquifer and those flowing out, with negligible variations in the water reserves. Interesting results were obtained for periods 1987–90 and 1995–96, which have supplied useful indications for the evaluation of the groundwater stored in the aquifer system in relation to its hydrodynamic behaviour in different conditions of recharge.

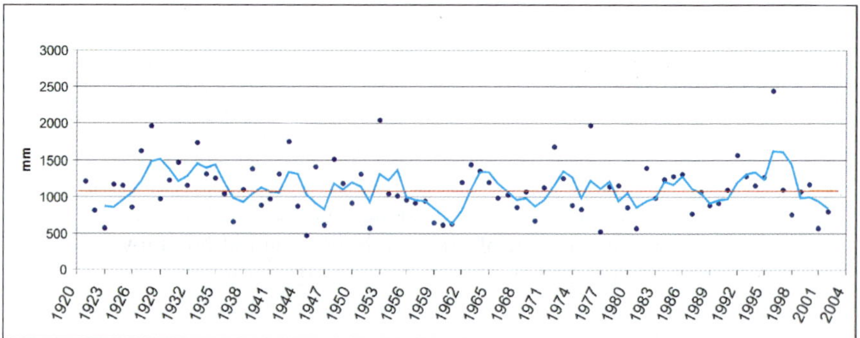

Figure 17. Precipitation values registered at Linguaglossa (yearly mean = red line; mobile mean (3 years) = blue line)

In the period 1987–1990, the lesser contributions to the hydrostructure due to less rainfall than the long term average for many consecutive years and differently distributed in the period of recharge, have determined a deficit of resources and with the constant incidence of considerable withdrawals for drinking use. Therefore, this deficit has been accumulated in this period thereby markedly affecting the yield of the springs, from which a reduction of more than 80% of the stored reserves in the aquifer has been estimated.

The exceptional pluviometric event of the hydrological year 1995-96 temporarily restored the water reserves of the aquifer, thereby producing obvious effects on the water levels and discharge of springs. From the positive variations of the piezometric surface, the increment in the saturated volume of the aquifer has been estimated and the corresponding volume of renewable reserves, using the mean value of the coefficient of storage deduced from pumping tests in wells in the area. In the first part of the period of recharging (October 1995–April 1996), the increase in the level of the piezometric surface was approximately 5 m on average, which diminished gradually to 3 m by the end of July. The volume of water stored in the aquifer was estimated at $116,84 \times 10^6\,m^3$, roughly 55% greater than the value of medium effective infiltration for the period 1969-86, equal to $75,38 \times 10^6\,m^3$.

The discharge of springs has shown a fast increase, reaching a peak of approximately $2\,m^3/s$, albeit having the same withdrawals upstream.

4. CONCLUSIONS

From the preliminary study of the hydrogeological structure and elaboration of the measurements of the piezometric levels and the springs discharge, it has been shown how renewable reserves equivalent to the mean value of effective infiltration can on average ensure the persistence of a groundwater outflow from the system, also given the consistent withdrawals from 1987–88. In such conditions, if the current amount of extractions remains unchanged, the reserves stored in the aquifer would theoretically be sufficient to avoid water crises in the supply system for drinking water and with respect to the Fiumefreddo River Natural Area Reserve in existence around the springs.

On the basis of the hydrodynamic behaviour of the aquifer, the margins of availability are however meagre and incapable of sustaining the groundwater outflow in the case of long term cycles of the aquifer recharge less than the mean value and irregularly distributed in the autumn-spring period, as occurred in the hydrological years from 1988 to 1994. In this period, although the rainfall have not been much less than the average, the amount of effective infiltration did not allow sustaining the outflow quantities from the hydrostructure in relation to the existing extractions. The volumes stored in the aquifer were nearly completely depleted and the reconstitution of the dynamic reserves was difficult.

The exceptional event of the hydrological year 1995–96, characterized by very abundant inputs concentrated in a few months, confirms this evaluation, as demonstrated by the behaviour of the piezometric levels and the springs discharge in the years immediately following such an event (1999–2002).

Therefore, for the purposes of forecasting and preventing the risk of water deficiency due to drought, by means of a correct management of the ground water resources, the above demonstrates the necessity for a careful and continuous observation of the aquifer behaviour, by means of a functional monitoring network of the water levels and springs discharge, in relation to the variability of the pluviometric data.

A correct setting up of a monitoring network and the relative interpretation of the data cannot however overlook the complete understanding of the aquifer system as concerns the lithostratigraphical, structural, hydrodynamic and permeability characteristics, without neglecting water chemical characteristics.

REFERENCES

Boussinesq, J. (1904) Recherches Theoriques sur l'Ecoulement des Nappes d'Eau Infiltrées dans le Soil et sur le Debit des Sources. *Journ. Math. Pures et Appliquées*, 10, pp. 5–78.

Branca, S. and Ferrara, V. (2001) An example of river pattern evolution produced during the lateral growth of a central polygenic volcano: the case of the Alcantara river system, Mt. Etna (Italy). *Catena*, 45, pp. 85–102.

Cassa per il Mezzogiorno. (1982) Indagini idrogeologiche e geofisiche per il reperimento di acque sotterranee per l'approvvigionamento del sistema V zona centro-orientale della Sicilia, C.M.P. S.p.A. (Unpublished technical report).

Castany, G. (1968) *Prospection et exploitation des eaux souterraines*, Dund, Paris.

Civita, M. (2004) *Idrogeologia applicata e ambientale*, C.E.A., Milano.

Ferrara, V. (1975) Idrogeologia del versante orientale dell'Etna. *3° Conv. Int. sulle Acque Sotterranee*, Palermo, pp. 91–144.

Ferrara, V. (1991) Modificazioni indotte dallo sfruttamento delle acque sotterranee nell'equilibrio idrodinamico e idrochimico dell'acquifero vulcanico dell'Etna. *Mem. Soc. Geol. It.*, 47, pp. 619–630.

Ferrara, V. (1999a) Vulnerabilità all'inquinamento degli acquiferi dell'area peloritana (Sicilia nord-orientale). *Studi sulla vulnerabilità degli acquiferi*, Pitagora, Bologna, 14, pp. 119.

Ferrara, V. (1999b) Strategie di gestione delle risorse idriche sotterranee a fini integrativi, sostitutivi e di emergenza nel settore nord-orientale dell'Etna. *3° Conv. Naz. sulla protezione e gestione delle acque sotterranee per il III millennio*, Parma, pp. 4.3–4.12.

Ferrara, V. (2001) *Carta idrogeologica del massiccio vulcanico dell'Etna*. C.N.R.–G.N.D.C.I., S.EL.CA., Firenze.

Ferrara, V. and Pennini, A. (1994) Caratteristiche idrogeologiche dell'acquifero vulcanico dell'Etna e metodologie per ottimizzare lo sfruttamento delle risorse idriche sotterranee. *Rencontre Int. des Jeunes Chercheurs en Géologie Appliquée*, EPFL-GEOLEP - Swiss Federal Inst. of Tech., Lausanne, pp. 97–102.

Ferrara, V., Maugeri, S. and Pappalardo, G. (1999) Gli acquiferi dell'area centro-orientale della Sicilia: risorse idriche, qualità delle acque e vulnerabilità all'inquinamento. *3° Conv. Naz. sulla protezione e gestione delle acque sotterranee per il III millennio*, Parma, pp. 1.89–1.97.

Ferrara, V. and Pappalardo, G. (2003) Intensive exploitation effects on the alluvial aquifer of the Catania plain (Eastern Sicily, Italy). *1st International Workshop "Aquifer Vulnerability and Risk"*, Salamanca, Guanajuato, Mexico, pp. 347–356.

Maillet, O.E. (1905) *Essais d'Hydraulique Souterraine et Fluviale*, Herfmann, Paris.

Regione Siciliana – Assessorato Lavori Pubblici (1994) Lavori di utilizzazione delle acque di Piedimonte Etneo per l'approvvigionamento idrico della città di Catania (Unpublished technical report).

CHAPTER 15

STUDY AND MONITORING OF SALT WATER INTRUSION IN THE COASTAL AREA BETWEEN MAZARA DEL VALLO AND MARSALA (SOUTH–WESTERN SICILY)

P. COSENTINO, P. CAPIZZI, G. FIANDACA, R. MARTORANA, P. MESSINA AND S. PELLERITO

Department of Chemistry and Physics of the Earth (CFTA), University of Palermo, Italy

Abstract: In this chapter the study of a coastal aquifer located in South-Western Sicily (between the towns of Marsala and Mazara del Vallo) is presented, carried out using geochemical, hydro-geological and geophysical techniques. The aquifer has been over-exploited to the point of being subject to intense and worrisome salt-water intrusion.

A preliminary chemical and physical characterization of the waters was carried out; this included measuring their conductivity and their chloride content. This allowed to detect the marine intrusion wedge in the coastal aquifer. A series of electromagnetic soundings, suitably calibrated by well logs, were effected in the whole area and allowed to create a 3D interpretative model of the resistivity distribution in the aquifer, thereby enabling to recognize the main intrusion directions and the pattern of the aquifer bed.

Furthermore an integrated geophysical 2D section was carried out along a line roughly perpendicular to the coast, in one of the zones that is particularly involved in the intrusion phenomenon. Field measures included ERT, IP, TDEM and seismic soundings, all of which were aimed at reconstructing a highly detailed geophysical section. The seismic soundings clearly show the lateral variation between the fresh and salt water, such as the overburden and the clayey bed of the aquifer.

The final target of this research is to propose an optimized management-model of underground resources. The lessons drawn from the use of different techniques for defining geophysical profiles suggest an integrated methodology to identify in detail the sea intrusion zone in aquifers. Therefore, the methodology used can be suitably extended and exported for studing and monitoring many similar Mediterranean coastal areas

Keywords: Salt water intrusion, groundwater, geophysical surveys

1. INTRODUCTION

Knowledge of groundwater quality, mainly referred to coastal aquifers with salt water intrusions, can be improved by defining standard monitoring and study procedures. These should be aimed at obtaining a detailed model of the aquifer, especially regarding

G. Rossi et al. (eds.), Methods and Tools for Drought Analysis and Management, 303–321.

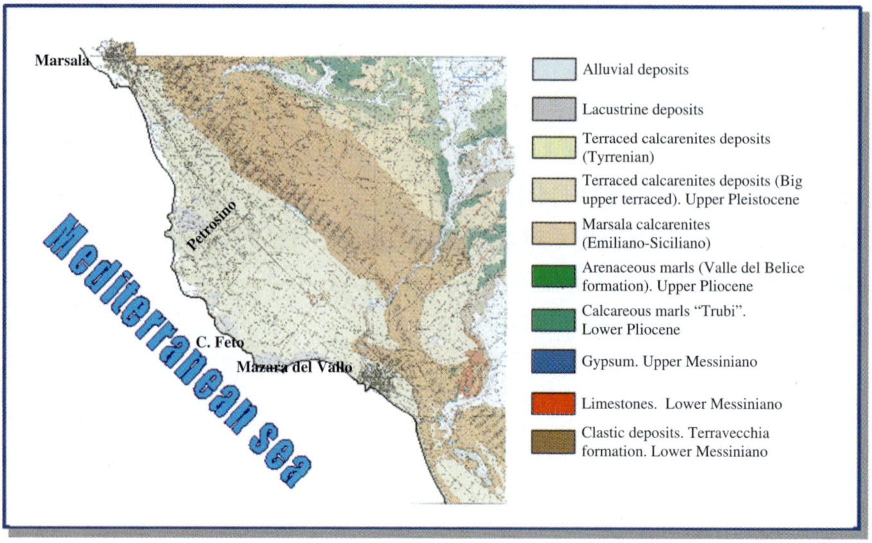

Figure 1. Geological sketch of the coastal area between Marsala and Mazara del Vallo (D'Angelo and Vernuccio, 1992; 1994)

salt water intrusion wedges. For this purpose a pilot project, which included a series of integrated studies using geophysical monitoring techniques (i.e. electrical, electromagnetic and seismic tomography, multi-parametric logs) was developed to define guidelines for the management of coastal aquifers in critical conditions.

The test-site is the coastal area between Mazara del Vallo and Marsala in South-Western Sicily (Figure 1). Here the coastal aquifer has been affected by sea water intrusion, caused by the over-exploitation of the local groundwater. Because of this harmful practice, the coastal aquifer has undergone a modification of the underground water flows which have in turn modified the natural equilibrium causing a serious intrusion of sea water. *Margi,* which were the most humid coastal areas, have disappeared. They are characterized by their unusual fauna and flora.

2. GEOLOGICAL, HYDROGEOLOGICAL AND GEOCHEMICAL SURVEYS

Preliminary surveys carried out in the zone (Cosentino et al., 2003), allowed us to define the hydrogeological characteristics of the aquifer and to characterize the preferential directions of the salt water intrusion. Further geochemical surveys allowed us to characterize the groundwaters according to the distribution of their chloride and carbonate contents.

The aquifer, which extends for approximately 150 km^2, is composed of pleistocenic sand and calcarenite deposits that overhang a clay-sand substratum. The hydrodynamic model (Figure 2) characterizes a strongly over-exploited aquifer with large zones affected by salt water intrusion.

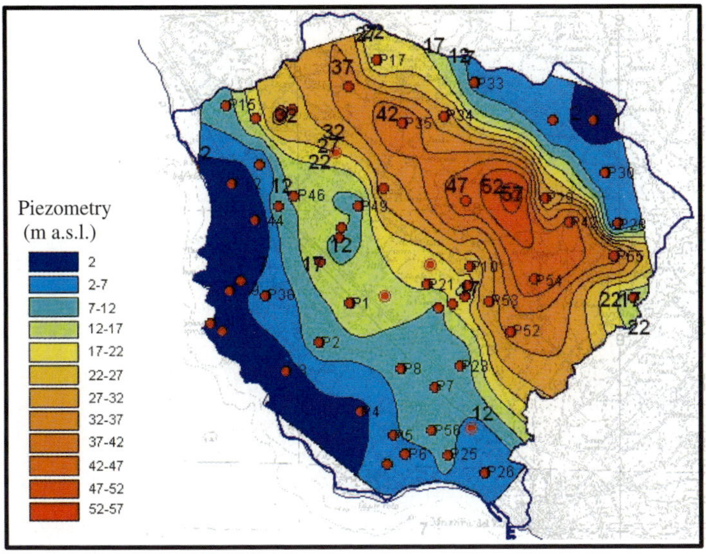

Figure 2. Piezometry map of the aquifer of Marsala – Mazara del Vallo

In natural conditions the near-surface groundwater heads towards the sea, coming up to the surface near the coast locally forming humid areas, the so-called *margi*. The pool water, substantially evaporated in the *margi*, and the sea water intrusion were limited to the contact areas between the *margi* and the sea. Part of the underground flow was drained off by small streams, today almost completely dry, whereas the

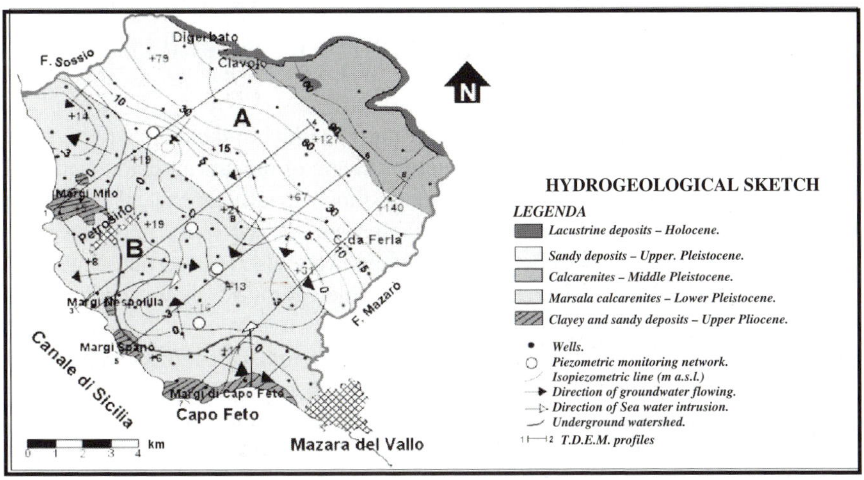

Figure 3. Hydrogeological sketch of the aquifer of Marsala – Mazara del Vallo

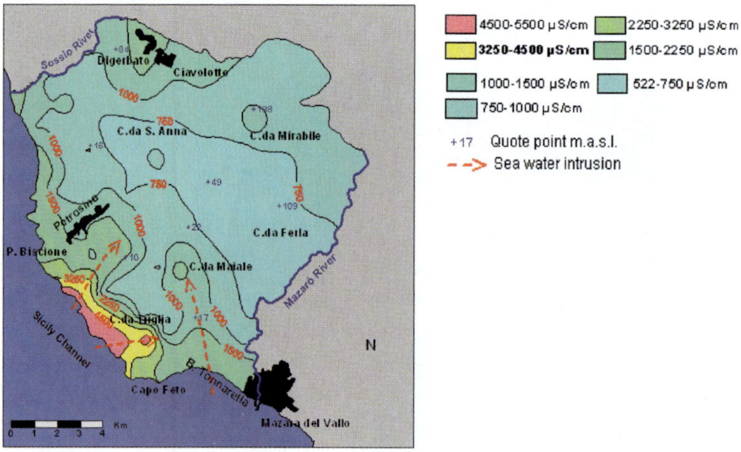

Figure 4. Map of conductivity of groundwater of the aquifer of Marsala – Mazara del Vallo

remaining flow recharged the superficial aquifer thereby feeding the *margi* and preventing the advance of the sea water wedge. The presence of the *margi* created a natural landscape with a delicate equilibrium: the superficial fresh water contrasted the sea water intrusion, so the expansion area of the *margi* was determined by the budget between evaporation and underground water contribution.

Hydrogeological surveys carried out in wells located in the zone allowed us to reconstruct the groundwater piezometry, shown in Figure 3.

Geochemical analyses of the water samples allowed us to elaborate a series of thematic geochemical maps. In particular, the groundwater conductivity map (Figure 4) allowed us to ascertain the main directions of the sea water intrusion.

3. 3D MODEL OF AQUIFER RESISTIVITY BY INTEGRATED GEOPHYSICAL SURVEYS

In order to obtain a three-dimensional interpretative model of the main structures of the aquifer and, in particular, to reconstruct the shape of the sea intrusion wedge, a series of integrated geophysical surveys, three-dimensional geophysical modelling of the aquifer and the localization of the main zones of sea water intrusion were carried out. Fifty electromagnetic TDEM surveys were also carried out, calibrated by four well-logs and two 2D electrical tomographies. Figure 5 shows the locations of the surveys.

TDEM surveys were located following some preferential directions (SSW-NNE) so as to make electro-stratigraphic sections that enhanced the zones with sea water intrusion (Fitterman and Stewart, 1986). The TEM-FAST 48 instrument (Figure 6) was used because of the following characteristics: precision, ease of handling and rapidity of acquisition. In the TEM-FAST 48 instrument a coil (transmitter loop) placed on the soil generates a series of electromagnetic impulses that are diffused

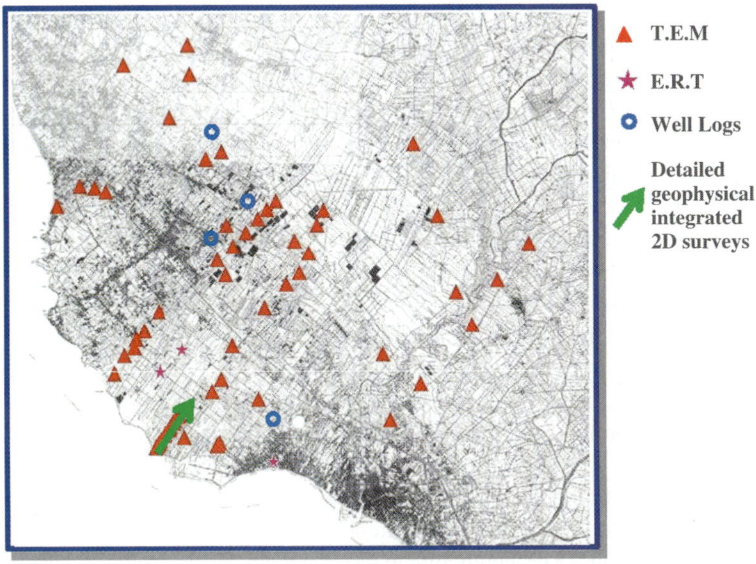

Figure 5. Location of the geophysical surveys carried out in the coastal area of Marsala – Mazara del Vallo

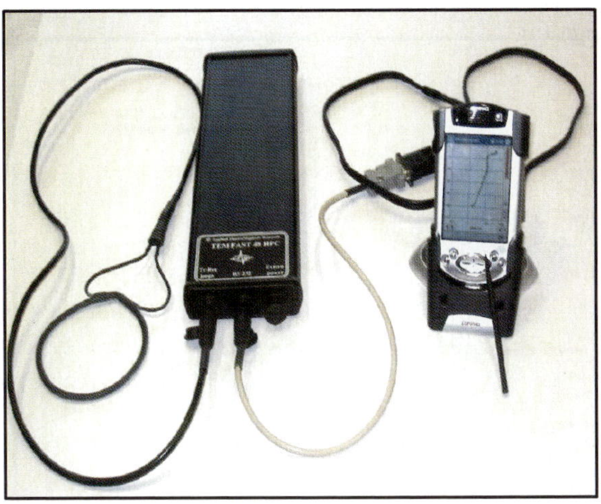

Figure 6. TEM-FAST 48 HPC instrument

in the subsoil inducing electrical eddy currents at different depths. These generate secondary electromagnetic fields. The analysis and the interpretation of induced signals, acquired by a receiver loop, allow us to recognize the electro-stratigraphic column of resistivity of the subsoil.

TDEM surveys were carried out by using the coincident loop configuration, with different coil lengths (12,5 m, 25 m and 50 m) in order to guarantee, case by case, a suitable penetration depth (Nabighian and Macnae, 1991). The data acquisition was made with a transmitting current of 3A. A time range varying from 2048 μs to 4096 μs was set.

TDEM measures were interpreted using TEM-RESEARCHER software, designed for the resolution of inverse problems in time domain electromagnetic soundings. Every sounding was interpreted based on a mono-dimensional model of subsoil with horizontal layers. At first, the data obtained from each sounding was interpreted in terms of a 1D model of depth varying resistivity (Figure 7). The analysis of the aforesaid models allowed us to estimate the height of the marly-clayey basement of the aquifer, corresponding to the roof of the "Marnoso-Arenacea della Valle del Belice" formation.

Subsequently electro-stratigraphic columns, obtained from the monodimensional interpretation of the TDEM soundings, were correlated (Figure 8) to obtain the assemblage of 8 resistivity sections (5 perpendicular to the coast line and 3 parallel). In order to obtain this assemblage, the re-interpretation of some of the curves of the apparent resistivity often had to be affected.

The resistivity sections (Figure 9) show a very irregular substrate and a large depression that coincides with the central zone of the investigated area. In this area,

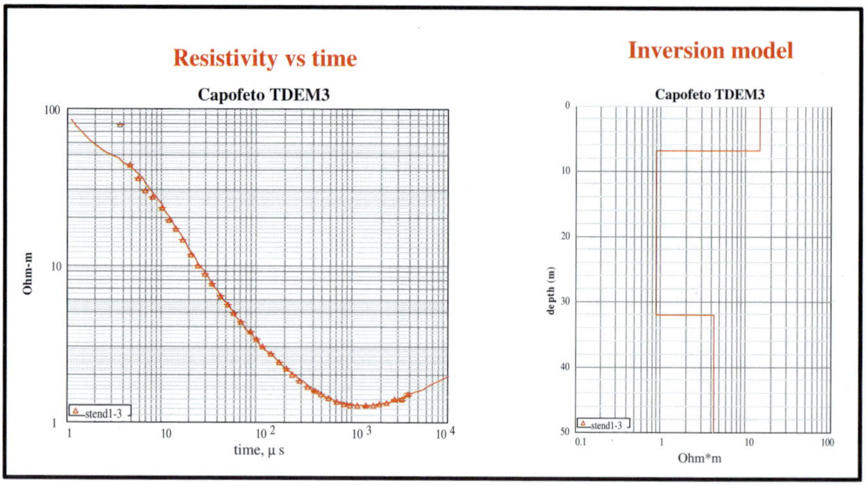

Figure 7. On the left, plot of the apparent resistivity versus time for a TDEM survey. On the right, interpretative electrostratigraphic model, showing pattern of resistivity versus depth

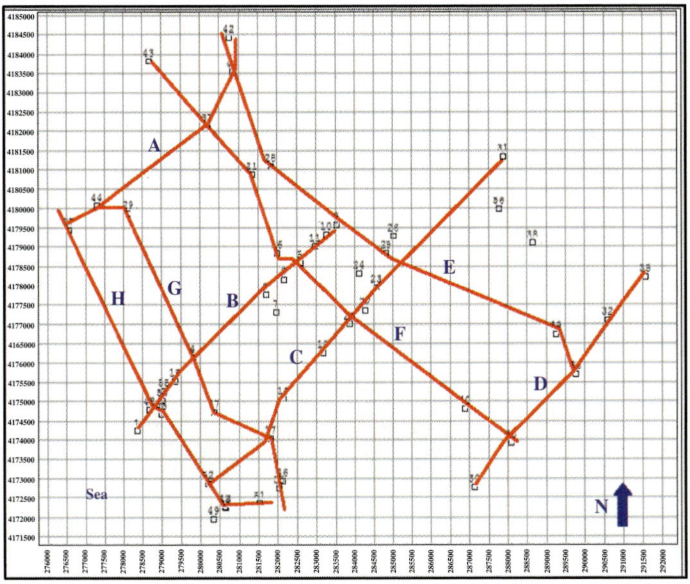

Figure 8. Location of TDEM surveys carried out in the area between Marsala and Mazara del Vallo. Red lines indicate the traces of the electro-stratigraphic sections, shown in Fig. 9

in correspondence with *Borgata Adragna*, the depression reaches a depth of – 30 m whereas near *Baglio Barberi* it reaches a depth of – 40 m.

From the locality called *Case Campanella*, and proceeding inland, the substrate reaches heights greater than 100 m near the northern side of the investigated area.

Finally the entire set of 1D TDEM inverting data was elaborated again to obtain a three-dimensional model of the aquifer resistivity (Fitterman and Hoekstra, 1984). This graphical 3D-plotting (Figure 10) allowed us to reveal and define the main directions of the sea water intrusion with more accuracy, by comparing these results with those shown by the isopiezometric lines of the hydrogeological map.

The analysis of the results highlighted the presence of extremely low values of resistivity along a strip of coast one kilometer wide. Evidently in this zone the sea intrusion is more intense than in the more internal areas. Moreover, the substantial agreement between the TDEM and the hydrogeological results was ascertained. In Figure 11 a perspective representation of an aerial photo of the zone is shown. In it, the resistivity map and the main lines of the sea water intrusion overlap in the coastal zones called margi. Considerations on the geometry of the aquifer obtained from electromagnetic surveys were strengthened by the data obtained from well logs. Moreover, they allowed us to identify some narrow layers having a high sand and clay content. These layers can limit vertical water flows by partially confining the aquifer in some areas.

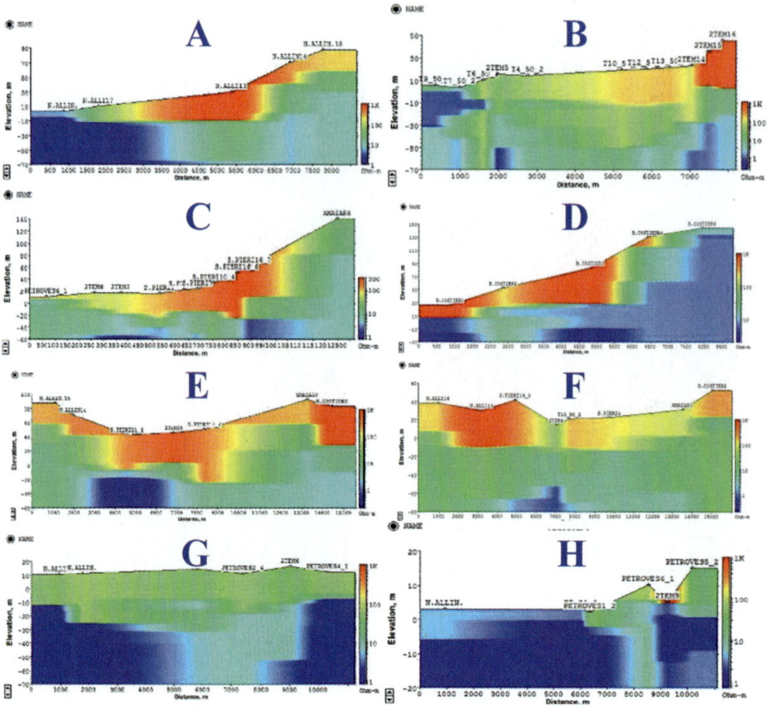

Figure 9. Electro-stratigraphic sections obtained from the profiles indicated in Figure 8

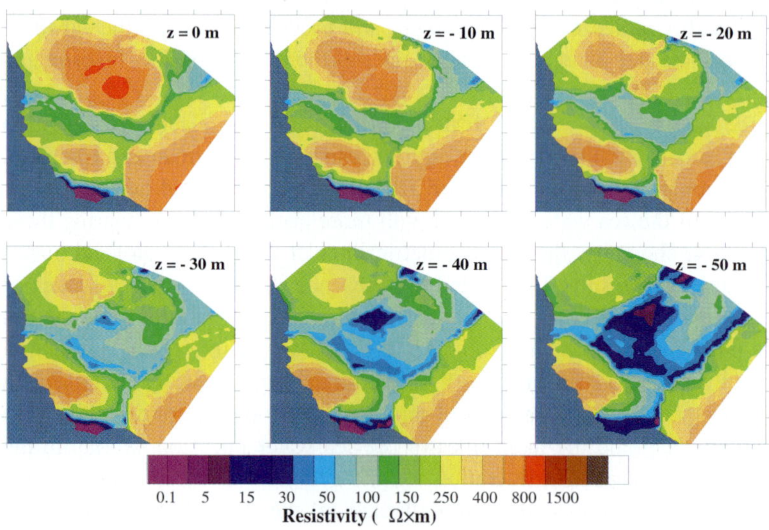

Figure 10. Electro-stratigraphic horizontal slices obtained by means of the combined interpretation of TEM-FAST surveys and their 3D interpolation

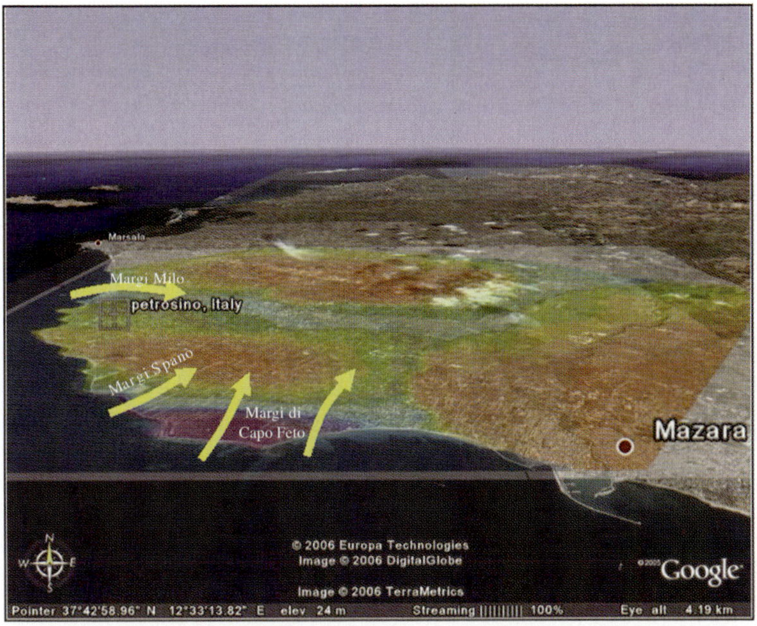

Figure 11. Overlapping of a resistivity horizontal slice on the aerial photo. The yellow arrows show the main directions of sea water intrusion (from Google ©)

4. METHODOLOGICAL STUDIES OF ELECTRIC RESISTIVITY TOMOGRAPHY (ERT)

Previous considerations on the general pattern of the sea intrusion wedge convinced us of the necessity to effect further surveys aimed at detailing the reconstruction of the aquifer in those zones affected by the intrusion. In fact, a geo-electrical investigation of the resistivity distribution in the subsoil would be useful to obtain a detailed reconstruction of the underwater structures. Then a geoelectrical methodology of data acquisition can be useful if optimized for the study of coastal aquifers interested by sea intrusion phenomena. Therefore a geoelectrical methodological study was carried out, by simulating synthetic models with electric parameters compatible with sea intrusion cases (Figure 12). In the chosen models the lateral and vertical resistivity variations in the subsoil were essentially determined by the pattern of the zones on the border between the fresh and salt water. Models simulate a sea intrusion (resistivity equal to $0.2\,\Omega \times \text{m}$) in coastal aquifers ($80\,\Omega \times \text{m}$) limited at the bottom by a clay basement ($2\,\Omega \times \text{m}$) and at the top by a resistive overburden ($150\,\Omega \times \text{m}$). The lateral passage between the fresh and salt water can be sharp or smooth, depending on the considered model.

The theoretical answers in terms of apparent resistivity values were evaluated assuming the execution of pseudosections using classic arrays (Wenner, Wenner-Schlumberger and dipole-dipole) and multielectrodic arrays (Linear Grid). This

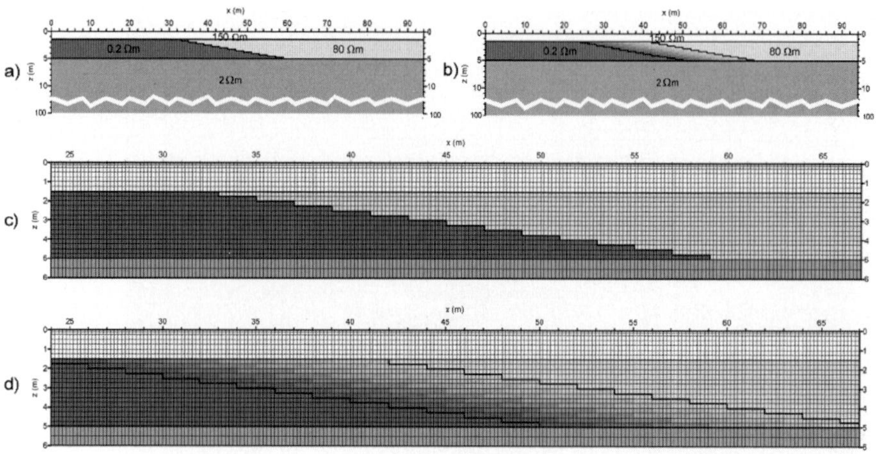

Figure 12. a) Model of an aquifer characterized by a sea intrusion wedge, with a lateral passage from the salt water intrusion zone to the fresh water zone. The aquifer is delimited at the bottom by a conductive basement and it is covered by a resistive superficial layer. b) the model is similar to the previous one, but with a transition zone between salt and fresh water. c) and d) Zooming of respectively a) and b) that show the choice of geometry and size of the model blocks

latter array (Fiandaca et al., 2005) allowed us to carry out a high number of measures for every current dipole, thereby reducing the acquisition time. The optimization of the number and the position of the current dipoles, allowed us to obtain results that were comparable, if not better, to classic arrays.

The linear Grid array is a 2D derivation of the resistivity grid array (Cosentino et al., 1999; Cosentino and Martorana, 2001) that concerns the use of a high number of potential dipoles for every current dipole considered. In fact a limited number of current dipoles was used in the same profile in contrast with the high number of potential dipoles. The array (Figure 13) was optimized to allow us to use it with a multichannel resistivity-meter.

Simulations on synthetic models were carried out considering 48 aligned and equal spaced electrodes and 21 different current dipoles, using two current electrodes every time and the remaining 46 for potential measures. The choice of the number and the positions of the current electrodes was finalized to obtain a good sampling density in the investigated zone.

Synthetic apparent resistivity data, with an added noise of different percentages (2% and 5%), were inverted with RES2DINV software (Loke and Barker, 1996), using the same optimized set of inversion parameters. In this way the inverted models, obtained from the theoretical pseudosections, were compared with the initial synthetic models using several comparison parameters to estimate the compatibility between these latter and the interpretative model. A quantitative evaluation of the resolution of each tested array was therefore obtained when applied to these hydrogeological problems. From a comparison of the results (Figure 14) it can be deduced that the best results were obtained using dipole-dipole and linear grid arrays.

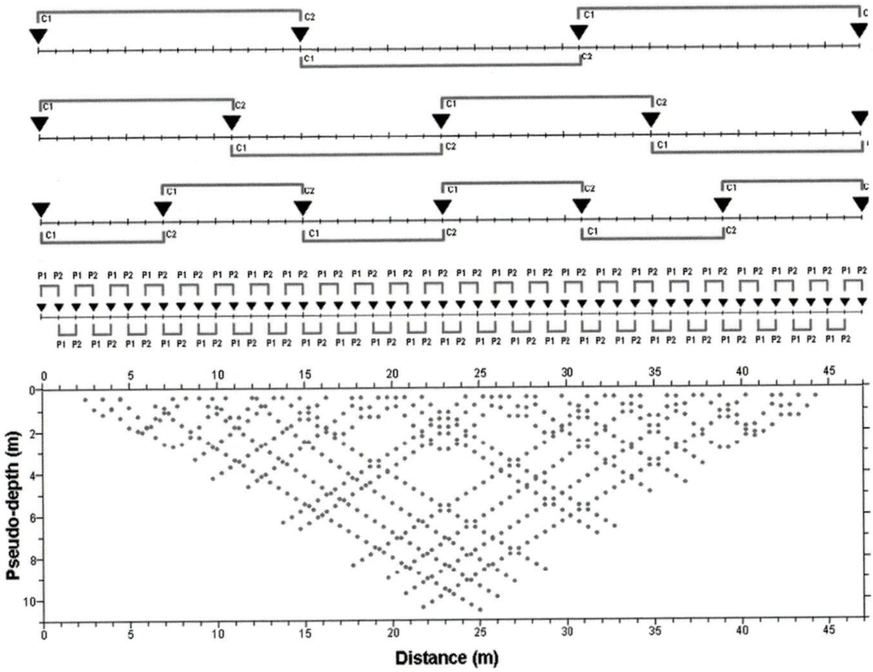

Figure 13. Top: outline of the Linear Grid array. For every current dipole C1-C2, potential measures for each adjacent electrode dipoles P1-P2 are carried out. Bottom: reference points for measures in the resulting pseudosection

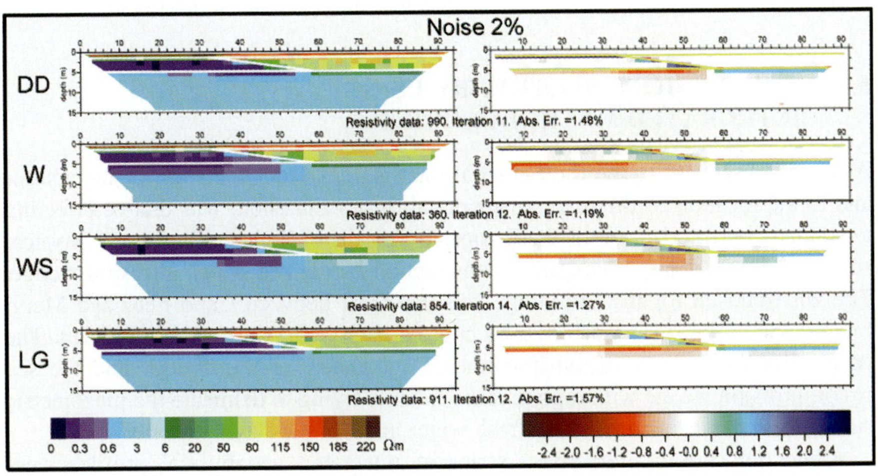

Figure 14. Inversion models (on the left) and correspondent patterns of the correlation parameters (on the right) of the model of sea intrusion shown in Figure 13. Simulations were executed with dipole-dipole (DD), Wenner (W), Wenner-Schlumberger (WS) and Linear Grid (RG) arrays. Results are referred to data with 2% noise

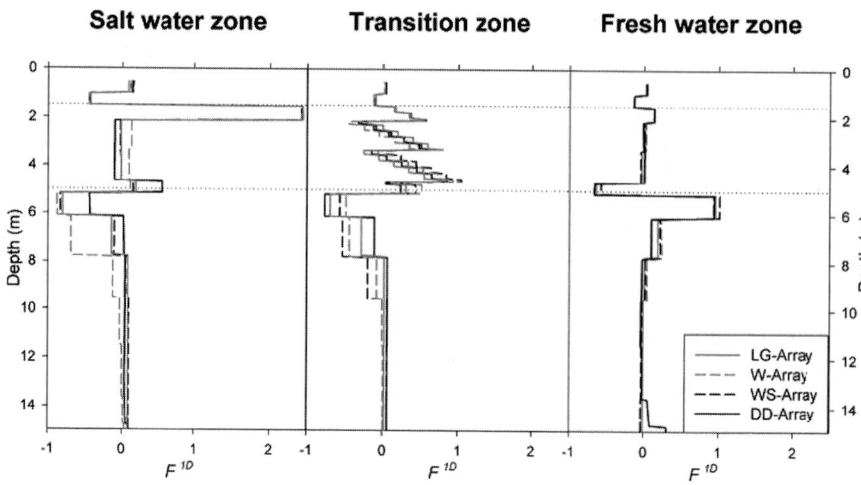

Figure 15. 1D misfit parameter relative to three different zones of the inversion model: salt water intrusion (on the left), to transition zone (at middle) and fresh water (on the right)

In order to estimate the approximation degree a 1D misfit parameter was used. In Figure 15 its patterns in simulations with errors of 2% are presented, the interpretative models are subdivided into three sections corresponding to three different zones of the profile: salt water intrusion, transition and fresh water. In the intrusion zone the worst results were obtained when using the Wenner array that wasn't able to locate the exact depth of the basement. On the contrary both the dipole-dipole and the linear grid recognized the contact between the clay basement and the aquifer well.

5. GEOPHYSICAL SECTION BY USING INTEGRATED METHODOLOGIES

A phase subsequent to the three-dimensional geophysical modelling of the aquifer and to the location of the main zones of salt water intrusion, was that of effecting a detailed characterization of these zones by means of high resolution geophysical surveys, aimed at the geometric reconstruction of the sea water intrusion wedge. The zone chosen for the soundings was the coast between Capo Feto and Margi Spanò, where the survey profile was approximately 1300 m long (Figure 16). The purpose of this was to reconstruct the 2D geometry of the aquifer and the sea water intrusion wedge with high precision, while trying to delineate the piezometric surface, the passage from salt to fresh water and the silt-clay substrate.

Along the chosen line AB, a series of integrated geophysical profiles were executed, using the following methodologies:

1. TEM-FAST electromagnetic surveys;
2. 2D resistivity tomographies;
3. 2D induced polarization tomographies;

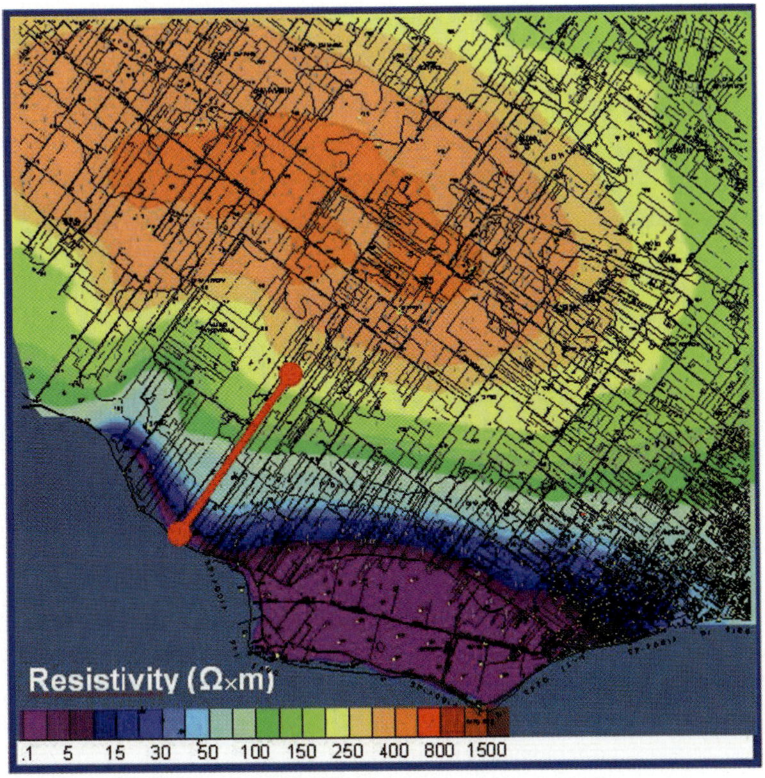

Figure 16. Slice of resistivity of aquifer at depth $z = -10\,m$ in the area of Capo Feto. The red line indicates the trace of detailed geophysical sections

4. refraction seismic profiles.

The interpretation of the acquired data with the aforesaid techniques was effected in a combined way, with the aim of linking together the interpretative models concerning different but specific parameters of the subsoil (i.e. electrical resistivity, chargeability, seismic wave velocity) as best as possible.

Along profile AB a series of TEM-FAST surveys was carried out with a *coincident loop* configuration and a square coil with side equal to 50 m, with the aim of obtaining an investigation depth greater than 40 m. The plot of the pseudo-section, from the apparent resistivity data, supplied a qualitative pattern of the subsoil conductivity. However the near-surface zone wasn't well detailed because of the limits imposed by the size of the coil.

The results of the surveys were interpreted with monodimensional layered models, whereas the results of the inversion models were interpolated to obtain a two-dimensional vertical electromagnetic section of resistivity (Figure 17). The section showed a resistive overburden over a lateral passage from very low resistivities (blue zone) to higher values (green zone). This variation was clearly interpretable as the transition zone from salt to fresh water.

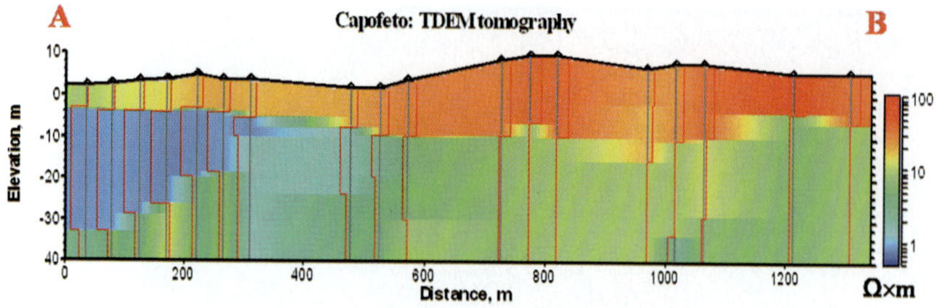

Figure 17. Capo Feto. Tomographic 2D section obtained from TEM-FAST surveys

The resistivity and induced polarization measures were carried out by placing a series of aligned electrodes in the ground spaced 2 m apart, and by connecting them to a Syscal Pro resistivity-meter, that allowed for the automatic injection of the current into the ground and the measurement of the electric potential at the surface. Measures of apparent resistivity or induced polarization were carried out using the same arrays, the interpretation of which enabled us to obtain vertical sections of resistivity and induced polarization.

Five resistivity tomographies were carried out along the AB line using the previously optimized linear grid array, for a total of 336 electrodes, spaced 2 m apart. The resulting vertical electrical section is shown in Figure 18. Furthermore, these

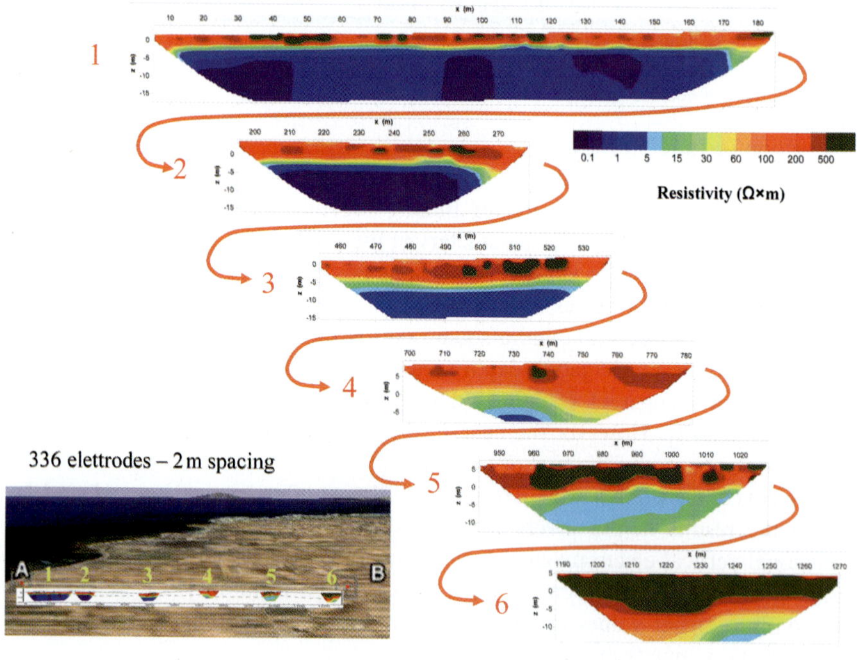

Figure 18. Capo Feto. Tomographic 2D sections obtained from resistivity surveys

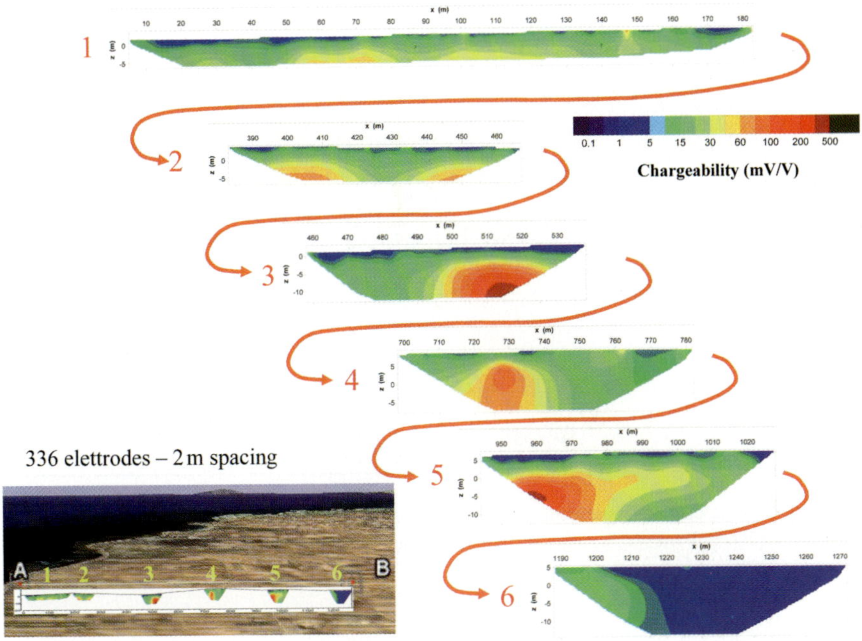

Figure 19. Capo Feto. Tomographic 2D sections obtained from I.P. surveys

sections also show a resistive cover overhanging a conductive zone in which a lateral increase of resistivity occurs when the distance from the coastal zone increases.

Measures of induced polarization were carried out with the same arrays. The obtained chargeability sections are shown in Figure 19. These are more difficult to interpret, since chargeability is not directly comparable with the presence of water, but is more linked to the clayey content. Unfortunately the obtained investigation depth didn't allow us to obtain information on the clayey basement.

Lastly, seismic measures were carried out using a series of aligned geophones, spaced 3 m apart, connected to an ABEM Terraloc MK6 seismograph by a multi-channel cable. Every section was obtained using 48 geophones and carrying out lateral and intermediate shots at different offsets for a total number of 7 shots for each profile. In this way several dromocrones were obtained, i.e. graphic times of arrival of waves / source-geophone distances (Figure 20), while the velocity of the seismic waves in the ground were deduced from these values . The interpretative seismic sections show an irregular cover, characterized by low values of seismic velocity (1000–1300 m/s) that cover a formation characterized by higher velocity (1950–2150 m/s).

The integrated analysis of the geophysical data regarding the same area but carried out using several methodologies, allowed us to correlate different inversion models. The result is the correct location and delimitation of the water table, the salt/fresh water transition zone, the lythological structures of the aquifer and the substrate.

The overlap and the comparison of the interpreted models obtained using different techniques (Figure 21) allowed us to highlight substantial analogies in the results

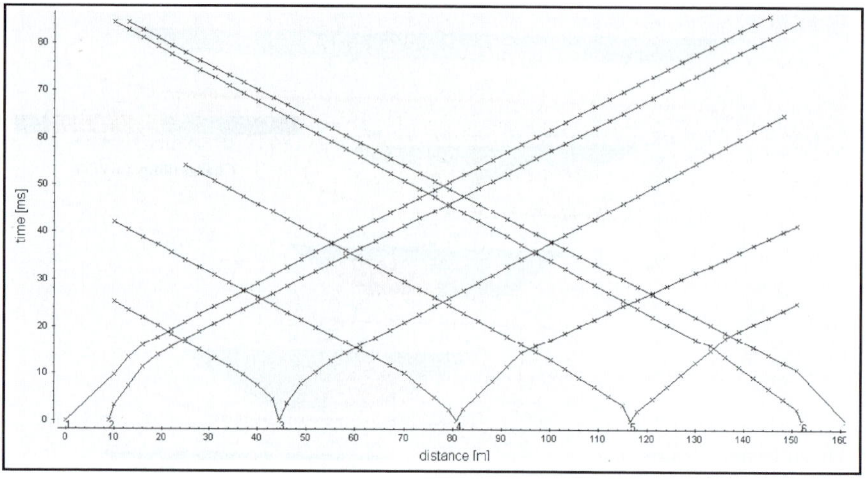

Figure 20. Seismic data are represented as a travel-time plotting, from which seismic velocities of the subsoil are deduced. For every seismic section seven shot points at different offsets were used

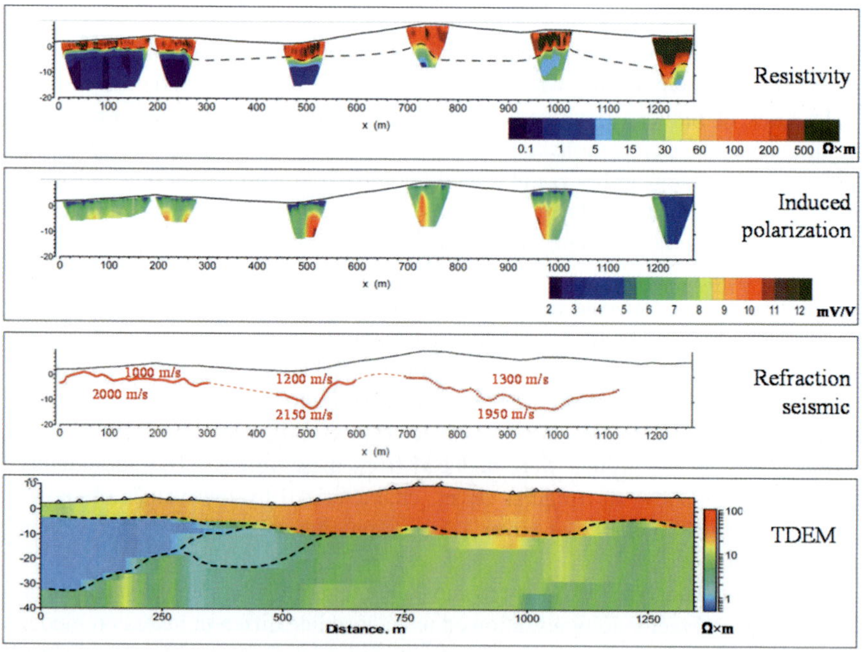

Figure 21. Capo Feto. Comparison of interpretative 2D models obtained from different geophysical methodologies

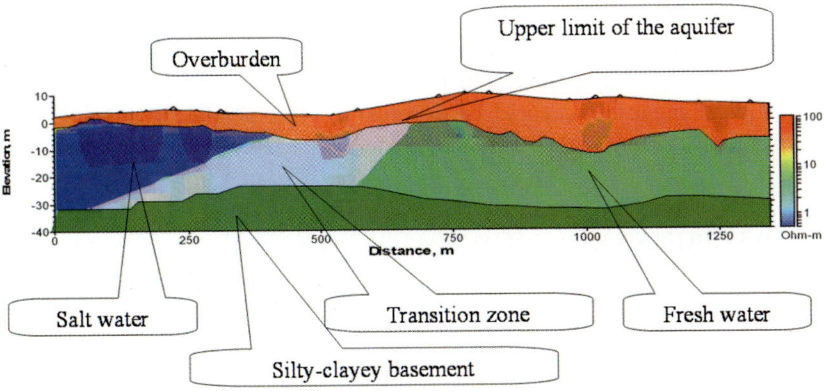

Figure 22. Capo Feto. Interpretative section obtained from integrated geophysical inversions

but also to underline some differences to be referred both to the analysis of the various parameters (i.e. resistivity, chargeability or seismic velocity) and to the different resolutions and depths of investigation of these techniques.

The combined interpretation of the different kinds of geophysical surveys allowed us to elaborate an interpretative section (Figure 22) in which the aquifer cover and the water table are characterized in good detail, while the sea intrusion zone, the transition zone and the fresh water zone are delimited with great precision. Finally upper limit of the siltous-clayey substrate is clearly evident.

6. CONCLUSIONS

The results of the geophysical surveys, together with the implementation of the geophysical monitoring techniques, allowed us to check, implement and partially support the previous hydrological balances of the coastal aquifer of Marsala-Mazara Del Vallo (Cosentino et al., 2003).

The integration of the selected geophysical methodologies (Geoelectrics, low-frequency electromagnetics – TDEM, Refraction Seismics and geophysical Logs) allowed decreasing the uncertainty limits of the various inversions, by means of a superposition of the various geometrical models obtained. However, TDEM and Geoelectrical investigations would be the basic tools for such kind of studies.

Some uncertainties do still remain, related to the real exploitation of the area and to the comparison of such utilisation with that both officially and unofficially declared. Furthermore, although the uncertainties regarding the exploitation heavily influence the balance, the geophysical methodology implemented and used allowed us to acquire fundamental information, the utility of which will be complete when the input/output data from the aquifer is more reliable and certain.

The methodological results obtained will allow us to draw some guidelines that could be applied to the study of coastal aquifers subject to intrusion risks. In particular, the proposed new geoelectrical array, which involves a remarkable reduction of the acquisition time without any qualitative loss, will be suggested not only in the phase of the preliminary study but also for the time-lapse monitoring periodical controls.

The mentioned guidelines, that will include methodologies to be used as well as technological suggestions, could be exported to other similar, hydrogeological cases typical of many coastal regions of the Mediterranean.

ACKNOWLEDGMENTS

We thank very much the Servizio Idrografico of the Sicilian Region, that awarded the grant for this study, and in particular Ing. Giuseppe Geraci and Ing. Teodoro Vega, who encouraged us to follow this line of research and gave us also particular technological and technician supports for carrying out the field activities.

Furthermore, we would thank the Group of Colleagues and Experts of the SEDEMED Project, as well as the Project Committee. Their support during the SEDEMED Meetings was very appreciated and stimulating.

Prof. G. Ranieri and his group of the University of Cagliari helped us during the acquisition of the long geoelectrical profiles, furnishing also a fast SYSCAL Resistivity-Meter, including the essential long multi-channel cables.

Finally, we have to thank many students and technicians of our Department. Their help during field and laboratory activities was precious.

REFERENCES

Cosentino, P.L., Deiana, R., Martorana, R., Pellerito, S. and Vega T. (2003) Geochemical and geophysical study of salt intrusion in the South-Western coast of Sicily. *Tecnología De La Intrusión de agua de mar en acuíferos costeros: Países Mediterráneos. IGME. Madrid 2003. ISBN. 84-7840-470-8*, 187–195.

Cosentino, P. and Martorana, R. (2001) The resistivity grid applied to wall structures: first results. *Proceedings of the 7th Meeting of EEGS, European Section*, Birmingham, ARCH02.

Cosentino, P., Martorana, R. and Terranova, L.M. (1999) The resistivity grid to optimize tomographic 3D imaging. *5th Meeting of EAEG, European Section*, Budapest, Hungary, Em12.

D'Angelo, U. and Vernuccio, S. (1992) Carta geologica del Foglio 617 "Marsala" scala 1:50.000. *Bollettino Società Geologica Italiana*, Vol. 113, Roma.

D'Angelo, U. and Vernuccio, S. (1994) Note illustrative della carta geologica Marsala (F°. 617 scala 1:50.000). *Bollettino Società Geologica Italiana*, Vol. 113, Roma, 55–67.

Fiandaca, G., Martorana, R. and Cosentino, P.L. (2005) Use of the linear grid array in 2D resistivity tomography. *Near Surface 2005 - 11th European Meeting of of Environmental and Engineering Geophysics*, Palermo, ISBN 90-737-8191-4, A023.

Fitterman, D.V. and Hoekstra, P. (1984) Mapping of saltwater intrusion with transient electromagnetic soundings, In:Nielsen, D.M. and Curl, M. (Eds.) *Proceedings of the National Water Well Assn. Conference on surface and borehole geophysical methods in subsurface investigations*, San Antonio: National Water Well Assn., 429–454.

Fitterman, D.V. and Stewart, M.T. (1986) Transient electromagnetic sounding for groundwater. *Geophysics* 51, 995–1005.

Loke, M.H. and Barker, R.D. (1996) Practical techniques for 3D resistivity surveys and data inversion. *Geophysical Prospecting*, 44, 499–523.

Nabighian, M.N. and Macnae, J.C. 1991 Time domain electromagnetic prospecting methods. *Electromagnetic methods in Applied Geophysics*, Vol. 2A, Tulsa, pp. 427–520.

PART V

DROUGHT IMPACTS AND MITIGATION MEASURES

CHAPTER 16

GUIDELINES FOR PLANNING AND IMPLEMENTING DROUGHT MITIGATION MEASURES

G. ROSSI, L. CASTIGLIONE AND B. BONACCORSO

Department of Civil and Environmental Engineering, University of Catania, Italy

Abstract: The severe drought events recently occurred all over the world have increased awareness of the harmful impacts they cause. Since drought, unlike other natural disasters, evolves in a long period of time, an effective reduction of related impacts is possible by implementing appropriate mitigation measures planned in advance, according to the indications provided by monitoring systems. In this chapter, the importance of a legislation for a correct drought management, based on a proactive approach, namely including both long-term actions oriented to reduce vulnerability of water supply systems and short-term actions to be implemented during droughts, is discussed. Moreover, the use of multicriteria analysis to compare different mitigation measures and to find the preferable combination of long and short-term actions is recommended. Finally, a framework for a timely implementation of drought mitigation measures and the adoption of an effective monitoring system for the evaluation of drought risk is presented

Keywords: Drought mitigation measures, drought management, proactive approach, drought impacts, drought contingency plans

1. INTRODUCTION

Drought, unlike other natural disasters such as floods or earthquakes, does not occur abruptly, but it evolves over a long period of time (Rossi, 2003). Such feature makes possible an effective mitigation of drought impacts, if a timely and reliable drought monitoring system is in operation (Cancelliere et al., 2006) and an appropriate plan, including the necessary actions to reduce the most severe damages caused by drought (in economic, social and environmental terms), has been prepared (Rossi, 2000).

In particular, the key issue for implementing an efficient drought management strategy consists in identifying in advance measures to mitigate drought impacts on the water supply systems, the productive sectors and the environment. To this end, the formulation of Guidelines for the definition of drought mitigation measures and for their appropriate use, in relation to the different drought conditions, can be extremely helpful.

Some concepts related to the definition of drought mitigation measures, as well as the necessary legislative and institutional framework for their implementation are

G. Rossi et al. (eds.), Methods and Tools for Drought Analysis and Management, 325–347.

discussed in this chapter. After a brief overview of basic concepts, drought types and main impacts (paragraph 2), different classifications of drought mitigation measures proposed at international level in the last decades are presented (paragraph 3). Then, the Spanish legal framework for drought management is analyzed, since it can be considered one of the most advanced legislations in Europe about drought issue. Also the legislative framework to cope with drought in Italy is summarized and some improvements for an advanced legislation oriented to prevent severe water shortages due to drought are proposed (paragraph 4). Finally, the main criteria to be adopted as Guidelines for the definition of the drought mitigation interventions (paragraph 5) and for their implementation (paragraph 6) are presented in some details.

2. DROUGHT PHENOMENON AND IMPACTS

A general consensus exists in distinguishing among drought, aridity and desertification concepts (Yevjevich et al., 1983). Unlike *aridity*, which refers to a permanent climatic condition with very low annual or seasonal rainfall, and the term *desertification*, used to indicate a long term and somehow irreversible process of decrease or destruction of biological soil potential, fostered by several factors such as climate, soil properties and above all human activities, the term *drought* refers to a casual (random) natural condition of a consistent reduction of water availability with respect to the normal values, spanning over a significant period of time and affecting a wide region.

According to the different components of the natural hydrologic cycle affected by a drought event, it is possible to distinguish among: meteorological, agricultural or hydrological drought (see Figure 1). In particular, a meteorological drought indicates a condition of reduction of precipitation with respect to normal values, consequent to precipitation variability probably caused by earth processes (such as geophysical and oceanographic interactions), interactions with the biosphere and maybe by sunlight energy fluctuations.

As a direct consequence of meteorological drought, a soil moisture deficit occurs, depending on the entity of the meteorological drought transformed by the water storage effect in the soil. Subsequently, when the previous deficit affects surface water bodies (rivers) and underground bodies (aquifers), an hydrological drought, as a surface and/or groundwater flow decrease with respect to the normal values, occurs. Finally, drought can have effects on water supply systems leading to water shortages. The latter is sometimes defined as operational drought, and in relation with the environmental, economical and social system features it can have economic and intangible impacts. Both the water availability reduction and its impacts depend, besides the importance of the drought event, on the efficiency of the mitigation measures adopted by the water supply and social-economic systems managers.

Even if a detailed classification of drought impacts is a difficult task, the main impacts can be distinguished in three categories: economical, environmental and social. In each of them there are different impacts, that are connected with the affected sector. In Table 1 a tentative list of main drought impacts is reported.

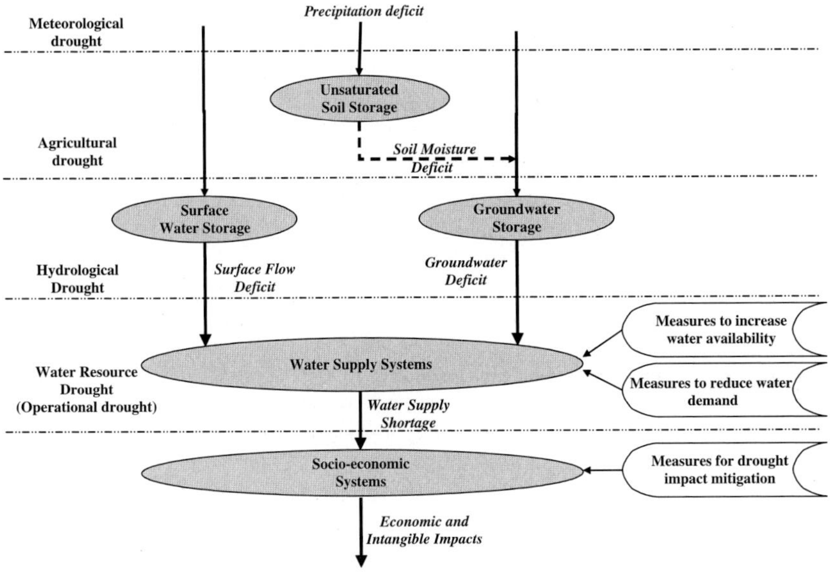

Figure 1. Drought phenomenon and role of drought mitigation measures (from Rossi, 2005)

Table 1. Drought impacts (from Yevjevich et al., 1978 and Rossi, 2004)

	Economic impacts
1)	Economic damage to agricultural production (crop reduction, damage in cultivations, insects epidemic, plants diseases)
2)	Economic damage to forest production (decrease of forest growth, woods fires, trees diseases)
3)	Economic damage to foremilk products and beef (reduction of pasture productivity, forced reduction of stock-farm, closing or reduction of public farm for pasture, increase of thefts, pasture fires)
4)	Economic damage to fishing (damage to river habitat and fishes caused by reduced flows)
5)	Economic loss to industries connected with agricultural production (food industries, industries producting fertilizing, ecc.)
6)	Economic damage to industries struck by hydroelectric energy reduction
7)	Unemployment caused by production decrease
8)	Economic damage to reduced navigability of streams, rivers and canals
9)	Damage to tourism sector due to the reduced water availability in water supply and/or water bodies
10)	Economic loss to entertaining (customers reduction, etc.)
11)	Economic damage to producers and tradesmen of amusing equipments
12)	Pressure on financial institutions (more risks in lending, capitals decrease etc.)
13)	Loss in public and local management revenue (because of reduction of taxes and taxes for hunting or fishing license, etc.)
14)	Income reduction for water firms due to reduced water delivery

(Continued)

Table 1. (Continued)

	Economic impacts
15)	Additional costs deriving from integrative water resources use
16)	Costs in emergency measures to improve resources and decrease demands (additional costs for water transport and removal, costs of advertising to reduce water use, etc.)
	Environmental impacts
1)	Lack of feed and drinking water
2)	Increase of salt concentration (in streams, underground layers, irrigated areas)
3)	Loss in natural and artificial lakes (fishes, landscapes, etc.)
4)	Damages to river life (flora, fauna)
5)	Damage to air quality (for example polluting dust)
6)	Damage to landscape quality (soil erosion, dust, reduced vegetation coverage etc.)
	Social impacts
1)	Inconveniences due to water system rationing
2)	Risks for health connected with increase of pollution concentration and discontinuous water system
3)	Impacts on way of living (unemployment, reduced saving capability, difficulty in personal care, reuse of water at home, street and cars washing prohibition, doubt on future, decrease of fest and amusing, loss of property)
4)	Iniquity in drought impacts and mitigation measures distribution
5)	Risks on public security due to more frequent fires (forests, pasture)
6)	Abandon of activities and emigration (in extreme cases)

3. DROUGHT MITIGATION MEASURES: A REVIEW OF CLASSIFICATIONS

3.1 General Remarks

From the drought impacts reported in Table 1 it can be inferred that a severe drought can be considered as a natural disaster due to its consequences. However it should be pointed out that the entity of the impacts mainly depends on the vulnerability of the different involved sectors. Indeed, the risk of water shortage for a water supply system is connected not only with the severity of a drought event, but also with the structural, managing, administrative and economical actions, adopted for the different elements of the water system in a preventive phase as well as in emergency conditions, when the phenomenon is underway.

In literature, several classifications of drought mitigation measures are available. The review that follows, reports synthetically the main classifications adopted in some important international meetings, and the classifications discussed in some monographs specifically dedicated to drought characteristics and drought management.

3.2 Classifications Based on the Purpose of the Measures

A consolidated and still valuable classification of drought mitigation measures has been proposed in the Conference "Drought Research Needs", held at the Colorado State University, Fort Collins, USA on December 1997 (Yevjevich et al., 1978), as a conceptual starting point to identify the main research subjects, that are necessary for an efficient strategy to cope with drought. This classification distinguishes three categories of measures: measures oriented to increase water supply, measures oriented to reduce water demands and measures oriented to minimize drought impacts (see Table 2).

The measures of the first category essentially include interventions that enable to improve water supply during drought through a better use of the existing water system resources, the utilization of new water supply sources and the adoption of proper managing rules in the use of water resources.

The demand-oriented measures aim to reduce the water use in order to meet water requirements regardless of reduced water supplies. Such measures include legal restrictions, rationing (also based on a rational and fair reallocation of limited water resources), economic incentives, as well as adoption of water recycle and water saving techniques.

The third category deals with the measures oriented to minimize drought impacts, and refers essentially to a reduction of damages or other negative consequences

Table 2. Consolidated classification of drought mitigation measures (from Yevjevich et al., 1978)

Water supply increase		
Existing supplies	New supplies	Complex supplies
– Surface storage – Subsurface storage – Interbasin transfer – Water conservation	– Emergency use of lakes – Saltwater conversion – Fossil waters – Weather modification	– Conveyance grids – Conjunctive water use – Snow and ice management
Water demand reduction		
Active strategies		Reactive strategies
– Legal restriction and public pressures – Economic incentives		– User recycling systems – User production adjustments
Drought impact minimization		
Forecasting	Risk sharing	Damages reduction
– Forecast and warning – Follow–up forecast and warning	– Insurance – Individual protection – Disaster aid	– Drought resistant crops – Agricultural techniques adjustment – Urban vegetation adjustment

caused by a drought event, through a risk spreading. It includes drought forecasting also based on monitoring systems, public campaigns to increase awareness of drought events, insurance, change in agricultural practices, etc.

3.3 Classifications Based on the Implementation Approach

In the volume "Coping with Droughts" (Yevjevich et al., 1983), which collects the main contributions presented at NATO Advanced Study Institute of Lisbon on 1980, the definition of mitigation measures is inserted in an organic and multidisciplinary framework of the various problems (drought identification, impacts, reduction of drought risk). In particular, two different approaches to drought phenomenon are distinguished: reactive and proactive approach. The reactive measures are defined as those measures that are basically implemented once a drought occurs and the visible impacts are already underway. These reactive measures include also the alternative of doing nothing, usually applied on the basis of the fact that individuals or organizations have enough resilience to stand impacts, and eventually includes a recovery after the end of the drought. The proactive (or preventive) measures are defined as all the measures, conceived or prepared in advance, which may help in the alleviation of drought consequences. The difference between the proactive and reactive measures is mainly in the approach: planning activity versus improvisation of various ad hoc measures. A very similar classification has been proposed by Werick (1993) and Whipple (1994), who individuate two types of response: tactical and strategic. Tactical measures are the actions implemented to cope with water shortage problems after a drought has started, and when it is too late to build new facilities. Strategic measures are actions planned in advance, which also include realization and/or improvement of hydraulic infrastructures and modification of existing laws and institutions.

3.4 Classifications Based on the Time-Frame

Dziegielewski (2000) proposed a net distinction between long-term measures aimed to improve drought preparedness, and short-term measures oriented to mitigate drought events when they already started, with emphasis on water supply problems.

Generally, drought mitigation through long-term measures entails a set of structural and non-structural adjustments to an existing water supply system, aimed at protecting the system from adverse effects of future drought by reducing its vulnerability to drought as risk of water shortage. Among the proposed long-term mitigation measures, three main groups can be distinguished: increase of storage facilities capacity, integrated managing of water resources in a wide area and improvement of water use efficiency.

Besides the long-term mitigation measures, the author proposes the implementation of short-term measures that refer to the capability of facing an ongoing drought. Such measures include planned actions before a drought beginning (included in contingency or emergency plans), and measures individuated when a drought is perceived as an heavy event. They mainly entail actions to improve water supply through new sources and to reduce water demand.

After the individuation of possible short-term and long-term measures to face expected shortages in water supply during drought periods, the choice among the different actions types is not easy. Although, long-term interventions could be more appropriate in a system for which emergency measures are frequently applied, on the other hand, if the risk of drought damages during the planning period is low or moderate, the best strategy can be to relay upon the short-term measures. Moreover, whereas public water supply agencies tend to emphasize the importance of long-term drought mitigation measures that would limit the need for drought response measures and minimize the chances of water crisis, the environmentalists tend to favour short-term drought responses measures, to reduce the need for additional hydraulic infrastructures and withdrawals to the detriment of in-stream water uses. Thus, only through an adequate understanding of the different roles of long-term preventive measures and short-term actions, it is possible to carry out an efficient drought mitigation.

To this end, the minimization of the interventions cost could be a criterion to select an appropriate combination of long term and short term measures.

3.5 Classification Proposed by the Water Scarcity Group

Recently, general recommendations on the mitigation measures for EU countries have been provided by the Water Scarcity Drafting Group (2006). Such a Group has been established by the initiative of Water Directors of EU, after drought events occurred in many Mediterranean countries during 2003, with the aim to develop, on behalf of the European Commission, an improved effort about water deficiencies due to drought.

The document presented in the Water Directors meeting (2 June 2006) identifies measures and procedures to be adopted in Europe and in the Mediterranean countries to provide and to share information and actions to face water deficiency problems. The document underlines the need for shifting from an approach based on the water crisis management to a new policy oriented to manage drought risk, by using adequate monitoring systems. Moreover, the development of specific plans oriented to drought risk planning and management is considered a key point for a successful mitigation of drought impacts. A very important element for an efficient drought management strategy is identified in the development of a drought watch system able to support a timely decision on the drought mitigation measures and a coordination among national, regional and local levels.

Drought plans should aim at reducing drought vulnerability of water supply systems as well as at improving their resilience to water shortage conditions. Therefore, an efficient drought plan must include short-term and long-term

Table 3. List of drought mitigation measures proposed by EU Water Scarcity Group, 2006

Category	Drought mitigation measure
• Water demand management	– Voluntary or mandatory rationing – Allocation of priorities – Pricing changes – Educational public campaigns – Incentives for municipal water saving – Incentives for agricultural water saving – Incentives for industrial water recycle – Alternatives to water consuming activities
• Changes in water resources management	– Water conjunctive use – Exchanges among water agencies (water banking) – Long–term changes in reservoir release rules – Changes in reservoir operational rules under drought conditions – Institutional changes – Legal changes – Operational coordination between water systems
• Water resources increase	– Use of groundwater as strategic reserve for drought – Reuse of treated wastewater – Desalination – Reallocation of resources – Water importation by barge – Use of lower quality water for specific use – Groundwater overexploitation – New reservoirs in the water supply systems – New water supplies interconnection

measures, whose optimal combination has to be selected according to appropriate criteria. A synthesis of the main measures included in the document is shown in Table 3.

4. LEGISLATIVE FRAMEWORK

4.1 The Example of Spain

An efficient drought mitigation requires a specific legislation which identifies drought mitigation measures and regulates the competences among public institutions involved in water resources management. Spanish legislation conceived to cope with drought is likely the most advanced legal framework in this field in Europe (Rossi, 2004).

Table 4 shows the division of the competences among the different organizations which have responsibility for facing with drought effects. It is worth underlying the essential role of the River Basin Authorities whose task is to prepare Drought Special Plans to face drought events in emergency conditions foreseeing the necessary solutions to satisfy water demands. In these Special Plans a particular

Table 4. Competences sharing for drought management in Spanish legislation

Institution	Legislative act	Competences
Ministry for Environment	Law 10/2001 (art. 27)	To establish hydrologic indicators networks to forecast drought events in the territory of the Autonomous Communities
Hydrographic Confederations (River Basin Authorities)	Law 10/2001 (art. 27)	To develop Drought Management Plans including measures to be adopted in alert or emergency conditions (after consultation of the "Water Council" which includes stakeholders and approval by the Ministry for Environment)
	Legislative Decree 1/2001 (art. 55)	To determine the releases from reservoirs and withdrawal from aquifers, limit water rights during drought and allow temporary exchange of water-rights among the users
Municipal water supply organisations	Law 10/2001 (art. 27)	To develop Drought Contingency Plans for Municipalities of more of 20.000 inhabitants
National government	Legislative Decree 1/2001 (art. 58)	To adopt the necessary measures for coping with extreme droughts and groundwater over-exploitation or similar necessities (after consultation of Hydrographic Confederation)
Ministry for Agriculture	Law 87/1978 and Min.Agr. order 1999	To regulate insurance system for drought damages in rainfed agriculture and pastures within the natural calamities recovery actions

attention is devoted to the coordination with municipal drought contingency plans with reference to water supply, mandatory rationing, management strategies modifications, relaxation of environmental constraints, as well as implementation of more restrictive measures as water availability decreases.

In addition to the contents of Table 4, it should be pointed out the role of the Permanent Management Office for drought situations, with the task of coordinating measures to be adopted in agricultural sector in drought situations (as provided by the Ministry of Agriculture order 6/9/1999).

4.2 The Italian Legislative Framework

Differently from Spain, the Italian legislative framework to cope with drought is not adequate, because: (i) the necessity of a proactive approach to face efficiently drought consequences does not seem to be widely shared, (ii) a clear distinction between long term and short term measures is lacking, and (iii) the assignment of competences among management organizations, Regions, State and Civil Protection Agency is ambiguous.

The legislative basis of the measures that can be adopted to cope with drought is reported in Table 5. Besides, the state of realization of the plans required by each law is indicated, as well as the eventual confirmation of the contents of the plans in the recent Legislative Decree 152/2006.

Even if the law 183/1989 does not explicitly refer to drought, it can provide a basis for an efficient drought mitigation activity based on the indications of the Basin Plan, that could include strategic actions to reduce drought vulnerability of water supply systems (long term measures). Nonetheless, the few approved basin plans generally do not deal explicitly with drought risk.

Law 225/1992 may be considered the basis for defining the interventions to be adopted when a natural calamity already occurred and has been perceived (short term or emergency measures). However the application of such a law to drought has been limited to public calamity declaration and to the appointment of commissioners for water emergency.

The directives for the individuation of drought risk areas included in D.P.C.M. 47/96 suggest to evaluate water deficiency risk in municipal sector by using the tool of the balance of water supplies and demands. However, in the preparation of the OTU plans, generally, no particular attention has been given to water resources variability; furthermore the drafting of drought management plans (or water emergency plans) has not been indicated among the tasks of the management agencies.

Finally, specific references to water shortage caused by drought and desertification are also included in the Legislative Decree 152/1999, which also establishes that Regions and Basin Authorities have to verify the presence of areas which are vulnerable to drought and desertification within their own territory, and eventually provide Water Protection Plans (WPP) including specific measures for such areas. The latter represents a planning tool to reach environmental quality levels of the basin's water resources by ensuring an effective water supply. Different Regions have already approved Water Protection Plans, but other have only started the preliminary survey about their territory. A not exhaustive list of regions that have approved WPP is: Marche, Piedmont, Lazio, Liguria, Lombardia, Emilia Romagna and Veneto, within the end of 2004, Tuscany and Sardinia in 2005, and Valle d' Aosta in 2006. Despite the above mentioned legislative prescriptions, the expected results have not always been achieved. In general, Water Protection Plans approved by Regions deal marginally with drought and desertification. A positive example is represented by Emilia Romagna that in its plan has identified vulnerable areas, both on the basis of previous studies carried out by the Basin Authority, and on the basis of spatial distribution of drought index SPI (Regione Emilia Romagna, 2004). This region has also defined protective actions in four categories: A) Soil protection; B) Sustainable water resources management; C) Impact reduction of productive activities and D) Reorganization of the territory. Moreover, criteria for drought management programs in municipal, agricultural and industrial fields are defined.

Table 5. Direct and indirect legislative indications in Italy for coping with drought impacts

Legislative Act	Technical Tool	Measures	Territorial Unit	Realization state	Legislative Decree 152/2006
L. 183/1989	Basin Plan (no indication on drought risk)	Long term measures to reduce vulnerability to drought of supply system	River Basin	Not approved plans of national basins	Confirmation of the contents included in Basin Plan
L. 225/1992	Forecast and prevention Plans for natural calamity (no indication on drought risk)	Short term measures	Region or district	Plans do not include drought risk	—
L. 36/1994 and DPCM 47/1996	Identification of drought risk areas (only for municipal water supply)	Short term and long term measures	Optimal Territorial Units (OTU)	Plans do not contain identification of drought risk areas	—
Leg.Decr.152/1999 and Decision CIPE 21/12/1999	Individuation of areas vulnerable to drought and desertification and definition of relative protection measures	Long term measures	Hydrograpic district	Protection plans contain only in a few cases the individuation of vulnerable areas and protection measures	Confirmation of protection plan contents

A further positive example of Water Protection Plans refers to Veneto that, thanks also to the Regional Program for coping with desertification (June 2000), has identified the areas at the foothills as areas vulnerable to desertification, because of the water resources exploitation and of the frequent wild fires, and has included the rationing of groundwater supply as protection measure for this area (Regione Veneto, 2004).

Also, Sardinia Region (Regione Sardegna, 2004) has identified areas sensible to desertification, based on a previous study of the Regional Agrometeorological Service (SAR, 2004) on desertification vulnerability, evaluated through four indices (Soil Quality, Climate Quality, Vegetation Quality and Territory Management Quality). The Water Protection Plan does not provide specific mitigation measures, even if a lot of actions have positive effects on risk reduction (for example alert level connected with water volume in the reservoirs). Moreover, new investigations have been planned for the realization of a GIS oriented to monitor areas affected by drought and desertification, in order to enact specific laws to protect such areas (these activities have been started within the ERSAT-SAR Project).

4.3 A Proposal for a New Legislative and Institutional Framework for Drought Management

As already mentioned in the previous paragraph, the actual Italian legislation for facing drought and water shortage risk is far to be adequate. The recent Legislative Decree 152/2006 on environmental protection has not improved the indications given by previous laws.

In a recent proposal (Rossi, 2004) is particularly stressed the need for a legislative act providing a comprehensive framework for sharing competences among the organizations responsible, at different levels, for developing the plans aimed at improving drought preparedness, at mitigating drought impacts and at implementing the planned courses of actions.

The act, to be issued by the Ministry of Environment, should include:

- a list of long term and short term drought mitigation measures;
- a clear assignment of competences among involved institutions;
- guidelines containing general criteria for drought preparedness and mitigation, on which the specific directives at a regional level should be based (see section 5).

In Figure 2 a proposal of competences sharing for the implementation of interventions to prevent and mitigate water shortages due to drought, is schematically represented together with the links among different agencies. Clearly, all the planning documents should foresee a wide information and public sharing about the required interventions to carry out, as required by the EU Directive 2000/60. Moreover, to avoid unacceptable delays in the implementation of the plans, an efficient mechanism of incentives and sanctions should be applied.

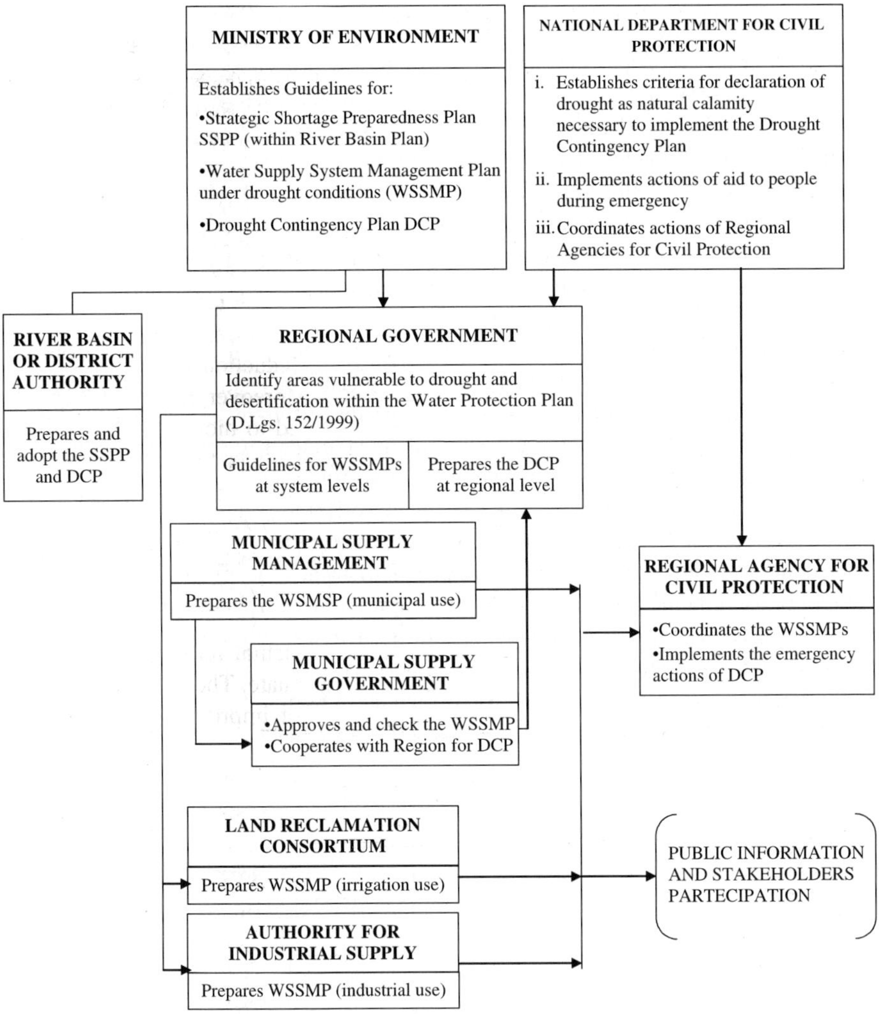

Figure 2. A proposal of competence sharing for drought management (by Rossi, 2004)

5. GUIDELINES FOR DEFINING DROUGHT MITIGATION MEASURES

Guidelines for drought mitigation should comprise the main criteria to be adopted, namely: the application of a proactive approach, the identification of the optimal combination of short term and long term measures and the individuation of adequate tools for the comparison and the ranking of the alternatives. Each single phase is discussed in details in what follows.

5.1 Application of a Proactive Approach

A first criterion considers the necessity to adopt a proactive approach instead of the reactive approach currently adopted to face water shortage risk. The reactive approach is in fact inadequate because it identifies emergency measures to limit water deficiency damages caused by drought once the phenomenon is started and perceived, often implying a waste of financial resources. On the contrary, to face shortage risk a proactive approach seems to be more adequate. This approach consists essentially of two different phases: preparedness of plans that allow to avoid and/or reduce water emergency consequences and implementation of such plans before, during and after a drought event.

As illustrated in Figure 3, after a preliminary assessment of water resources availability to meet long term needs, the first phase consists in the evaluation of the water shortage risk through an analysis of the different elements of a water supply system, by identifying the more vulnerable elements to drought phenomenon through economic, social and environmental investigations.

After the analysis of drought impacts on the different sectors, long term interventions are defined within planning documents (in particular River Basin Plan or Hydrographic District Management Plan). Such plans are not specifically oriented to drought management, but they can play an important role by reducing vulnerability of water supply system through water resources increase and/or demand (or losses) reduction. On the contrary, in this phase other two plans refer directly to drought fight: Water Supply System Management Plan and Drought Contingency Plan. The former should include measures to be adopted by the management organization of

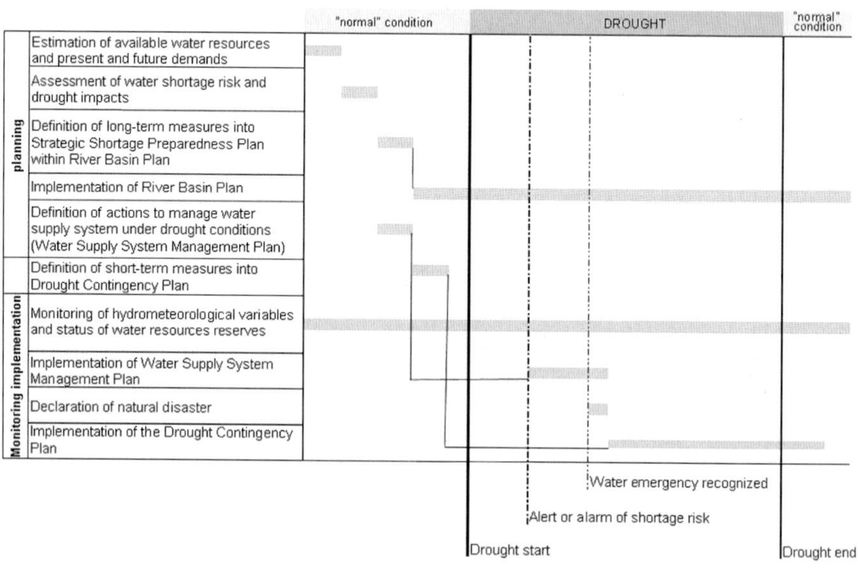

Figure 3. Main steps of a proactive approach in drought response

each supply system to avoid the begin of a real water emergency. The Contingency Plan should include short term measures to be adopted when drought causes heavy impacts (water emergency).

The second phase of a proactive approach entails a continuous monitoring of hydrometeorological variables and of the status of water reserves in order to identify possible water crisis situations and to apply the necessary measures before a real water emergency occurs. Nevertheless, if it is not possible to avoid a water crisis that appears as a natural public calamity (after a government declaration), the Drought Contingency Plan is implemented until the establishment of normal conditions. It is evident that a proactive approach, even if more complex, is more efficient than the traditional approach, since it allows to define in advance drought mitigation measures (both long term and short term) improving the interventions quality.

5.2 Adoption of Long-Term and Short-Term Measures

A second fundamental criterion to pursue an efficient mitigation consists of adopting long term and short term measures. According to a well consolidated classification, we can make the following distinction:

- Long term measures are oriented to reduce vulnerability of water supply system to drought, and they are included into the general water resources planning (Strategic Shortage Preparedness Plan into River Basin Plan or Hydrographic District Management Plan); moreover, such interventions could refer to drought event characterized by a fixed exceeding probability, namely a "design drought".
- Short term measures are oriented to reduce the damages of a drought event after its onset, and to mitigate effects of drought that overcomes a "design drought" (considered in first type actions), and they are included into Plans to face emergency situations prepared in advance (Drought Contingency Plans). Such plans would be implemented when monitoring systems indicate drought conditions.

In Table 6 and Table 7 classifications of respectively long term and short term measures in different sectors (municipal, agricultural, etc.), divided into the three categories explained in section 3.2, are reported.

5.3 Comparison and Ranking of the Alternatives

Among the different measures to be adopted the most appropriate combination of long term and short term actions has to be determined with reference to the vulnerability of the specific water supply system and to the drought severity. Given the high number and the different types of mitigation measures, it is necessary to adopt a proper evaluation procedure for the choice of the best combination.

A possible procedure that can be used is based on the joint application of water systems simulation models and multicriteria analysis (Rossi et al., 2005). Once that short term and long term interventions are identified and specific criteria to evaluate economic, environmental and social impacts are defined, the proposed

Table 6. Long term drought mitigation measures (Rossi, 2000 modified)

Category	Long-term actions	Affected sectors			
Demand reduction	Economic incentives for water saving	U	A	I	R
	Agronomic techniques for reducing water consumption		A		
	Dry crops instead of irrigated crops		A		
	Dual distribution network for urban use	U			
	Water recycling in industries			I	
Water supply increase	Conveyance networks for bi-directional exchanges	U	A	I	
	Reuse of treated wastewater		A	I	R
	Inter-basin and within-basin water transfers	U	A	I	R
	Construction of new reservoirs or increase of storage volume of existing reservoirs	U	A	I	
	Construction of farm ponds		A		
	Desalination of brackish or saline waters	U	A		R
	Control of seepage and evaporation losses	U	A	I	
Impacts minimization	Education activities for improving drought preparedness and/or permanent water saving	U	A	I	
	Reallocation of water resources based on water quality requirements	U	A	I	R
	Development of early warning systems	U	A	I	R
	Implementation of a Drought Contingency Plan	U	A	I	R
	Insurance programs		A	I	

U = urban; A = agricultural; I = industrial; R = recreational

Table 7. Short term drought mitigation measures (Rossi, 2000 modified)

Category	Short-term actions	Affected sectors			
Demand reduction	Public information campaign for water saving	U	A	I	R
	Restriction in some urban water uses (i.e. car washing, gardening, etc.)	U			
	Restriction of irrigation of annual crops		A		
	Pricing	U	A	I	R
	Mandatory rationing	U	A	I	R
Water supply increase	Improvement of existing water systems efficiency (leak detection programs, new operating rules, etc.)	U	A	I	
	Use of additional sources of low quality or high exploitation cost	U	A	I	R
	Over exploitation of aquifers or use of groundwater reserves	U	A	I	
	Increased diversion by relaxing ecological or recreational use constraints	U	A	I	R
Impacts minimization	Temporary reallocation of water resources	U	A	I	R
	Public aids to compensate income losses	U	A	I	
	Tax reduction or delay of payment deadline	U	A	I	
	Public aids for crops insurance		A		

U = urban; A = agricultural; I = industrial; R = recreational methodology enables the selection of the preferable alternatives according to the adopted criteria, by taking into account the opinions expressed by the involved stakeholders. Apart from the particular adopted technique, the application of AMC requires the definition of different alternatives and the evaluation of the impacts on each sector on the basis of prefixed criteria.

In Figure 4, a procedure for comparing and ranking mitigation alternatives is shown (Rossi et al., 2006). First simulation models are used to evaluate water system vulnerability to drought both in the current configuration and in future scenarios, through the calculation of proper performance indices. Then, according to the system vulnerability in the different considered cases, possible short term and long term drought mitigation interventions are individuated, giving the priority to those presenting an higher level of applicability with reference to economic, legislative and institutional constraints.

The next phase refers to impacts evaluation on the basis of prefixed economic, environmental and social criteria; also in this phase simulation models can be applied for a preliminary assessment of the effects of different alternatives. The choice of the criteria depends on drought system vulnerability as well as on the strategies of management and government at different levels, and on the policies adopted during previous drought. Once the criteria are defined and the impacts for each case are determined, a rank of alternatives can be carried out by using one of the different techniques proposed in literature and finally, the preferable alternatives can be selected by considering the different points of view expressed by the stakeholders.

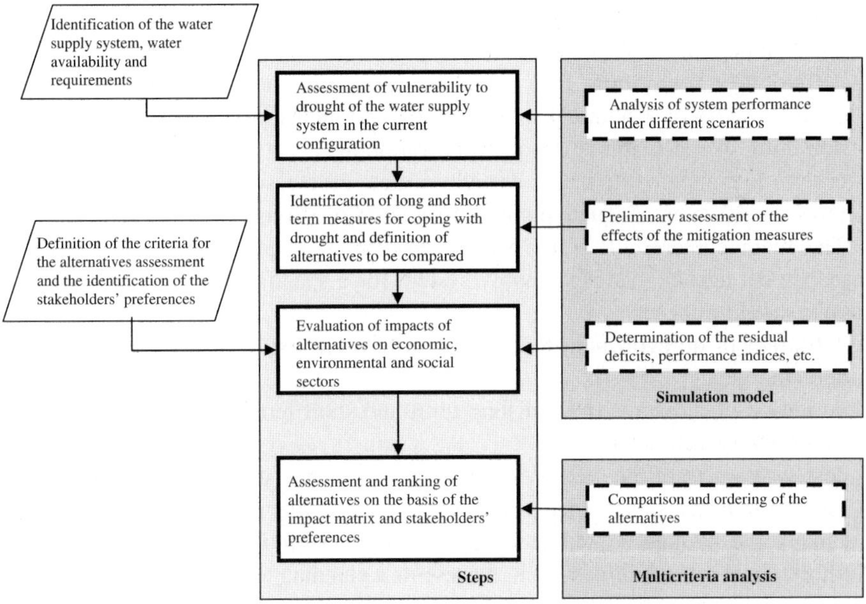

Figure 4. Comparison and ranking of drought mitigation measures (from Rossi et al., 2005)

6. FRAMEWORK FOR A TIMELY IMPLEMENTATION OF MEASURES

6.1 Planning Tools for Coping with Drought Risk

Drought preparedness and mitigation actions should be carried out by attempting to answer to some key questions, such as: How do we know when there is a drought? Which institution is in charge to manage drought related problems? What type of measures have to be implemented and when? What type of tools can be adopted to assess the effectiveness of the implemented measures?

A comprehensive solution could be represented by an integrated drought management, consisting of the following steps: planning, monitoring, implementation of planned measures, management of emergency situations not foreseen in advance and recovery of drought damages.

Once again such a strategy, which perfectly matches the requirements of a proactive approach is possible provided that an adequate institutional and legislative framework, which clearly defines tasks and duties among involved institutions, has been established.

According to the proposal discussed in the paragraph 4.3, three main tools have been individuated within the drought management planning process: Strategic Shortage Preparedness Plan, Water Supply System Management Plan and Drought Contingency Plan. A few detailed contents follow.

Strategic Shortage Preparedness Plan, whose institutions in charge are Basin or Hydrographic District Authorities, should provide:
- criteria to identify drought vulnerable areas;
- appropriate long term interventions for particular water supply systems;
- definition of the priority in water allocation under shortage conditions among different users (e.g. municipal, agricultural and industrial);
- definition of acceptable levels of water rationing;
- criteria to compare alternative drought mitigation measures;
- tools to ensure an adequate public information on drought problems.

The main contents of Water Supply System Management Plan prepared by management agencies of each supply system for a specific sector or multipurpose supply should refer to:
- definition of drought indicators for three different risk levels: pre-alarm, alarm and emergency;
- definition of measures (mix of long term and short term) to be implemented to avoid emergency conditions;
- cost assessment of the measures and indication of the source of financing;
- tools to improve stakeholders' participation and public awareness.

Finally the Drought Contingency Plan, whose institutions in charge are Basin Authority or Regional Government, represents a planning tool which identifies short term interventions to be adopted in drought conditions. In particular, it provides:
- drought indicators for calamity declaration;

- indications on the establishment of an Institutional Committee for coping with drought (e.g. Drought Task Force);
- list of short term drought mitigation measures (to increase water supply, to reduce water demands and to minimize drought impacts) with relative costs;
- special directives for coordinating actions among National Government, Regional Government and water supply agencies;
- tools to improve stakeholders' participation and public awareness;
- list of actions to recover drought damages.

The described drought management planning process is illustrated in Figure 5.

The planning process has to be dynamic in order to succeed in its purposes; therefore a review of the plans is necessary either periodically (i.e. every 5 years) and after each drought period, by evaluating: the performance and suitability of drought indicators to define threshold levels, the real effects of long term and short term measures and the effective implementation of the plan.

6.2 Selection and Implementation of Drought Mitigation Measures

A key point for an efficient drought prevention and mitigation is also represented by the way of selecting and implementing of different interventions on the basis of the priority of water allocation among the various uses, the indications provided by drought monitoring systems and the method adopted to assess drought risk.

At this end, it might be useful to recall some of the results obtained within the research program Medroplan (Mediterranean drought Preparedness and Mitigation Planning), collected in a draft version of the Drought Management Guidelines (MEDROPLAN, 2006).

First of all, the choice of drought management interventions has to consider two different priorities: the first refers to ensure adequate supplies of domestic water

Figure 5. Drought management planning process

available for public health, safety and welfare; the second is oriented to minimize negative drought effects on the economy, the environment and the social well-being.

Drought monitoring systems should provide information on meteorological conditions, on surface water and groundwater availability, and values of indices for drought monitoring and forecasting. Besides, it would be useful to use not only hydrological indices, but also indicators that describe social, economic and environmental drought consequences.

The assessment of drought risk should be carried out with reference to threshold levels, for instance alert, alarm and emergency, corresponding to different group of interventions. Figure 6 shows an example of thresholds and corresponding measures.

Alert scenario is declared when the monitoring shows the initial stage of a drought, which corresponds to moderate risk (i.e. 10 %) of consuming all water stored in the system and not being able to meet water demands. The objective of this phase is to be prepared in case the drought gets worse. This implies to ensure public acceptance of measures to be taken if drought intensity increases by raising awareness of the possibility of social impacts due to drought. The kinds of measures that are taken in alert situation are generally of indirect nature, implemented voluntarily by stakeholders and usually of low cost.

The alarm scenario is declared when the monitoring shows that drought is occurring and will probably have impacts in the future if measures are not taken immediately. In this case, there is a significant probability (i.e. greater than 30 %) of forthcoming water deficits. The objective in the alarm situation is to overcome

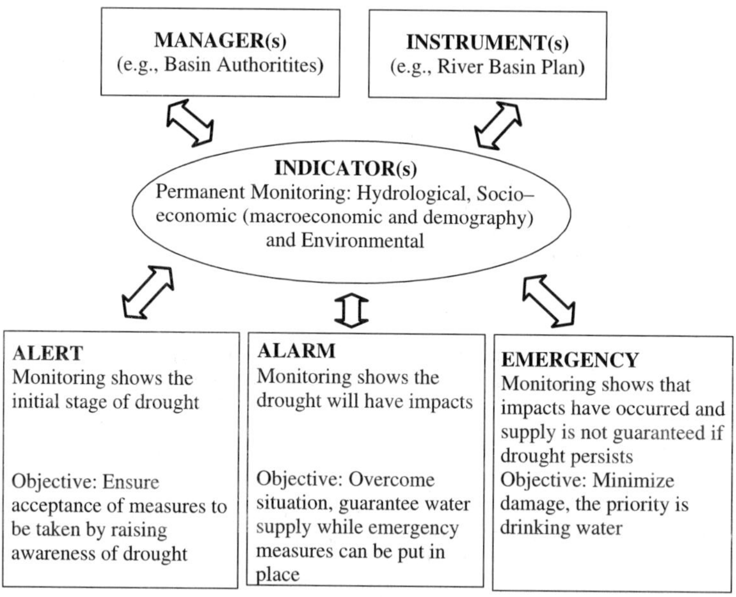

Figure 6. Threshold levels and objectives of the actions to be taken (from Medroplan, 2006)

drought avoiding the emergency situation by enacting water conservation policies and mobilizing additional water supplies. The kind of measures that are taken in the alarm situation are generally of direct nature, coercive and generally of low/medium implementation cost, although they may have significant impacts on stakeholders' economy. Most measures are not structural, and are directed to specific water use groups. Demand management measures include partial restrictions for water uses that do not affect drinking water or water exchange between uses with limitation of users' rights and priorities in normal condition.

The last scenario is relative to an emergency situation. It is declared when drought indicators show that impacts have occurred and supply is not guaranteed if drought persists. The objective is to mitigate impacts and minimize damages. The priority is to satisfy the minimum requirements for drinking water and crops. Measures adopted in emergency condition are of high economic and social cost, and they should be direct and restrictive. The nature of exceptional measures could be non structural, such as water restrictions for all users, subsides and low-interest loans.

7. CONCLUSIONS

Drought impacts mitigation represents one of the most challenging issue in water resources management, which could be successfully carried out by developing an integrated strategy consisting of planning, monitoring, implementation of planned and emergency measures and recovery of drought damages. Such a strategy, however, strictly relies upon a legislative framework providing appropriate planning tools for an efficient drought management, but also establishing a precise assignment of competences.

In this chapter a proposal for a legislative act aimed at regulating forecast and prevention activity of water shortage risk caused by drought has been presented. In addition to specific planning contents and a clear distinction of institutional competences, the proposed act should also include precise deadlines, compatible with technical and managing requirements, and sanctions in case of default by the involved institutions.

Secondly, a correct drought management should be based on the adoption of both long term and short term mitigation measures. Among the several possible actions, the most suitable combination of long term and short term measures for the considered water supply system has to be identified. In particular, for the choice of the best combination of interventions, the use of multicriteria analysis can be very helpful, since it enables to identify and select the preferable alternatives according to predefined criteria, by taking into account different points of view expressed by the involved groups of interest.

A further criterion for a correct drought mitigation is the activation of a dynamic planning process consisting of the following tools: Strategic Shortage Preparedness Plan, Water Supply System Management Plan and Drought Contingency Plan.

The final step within an integrated drought management is represented by the way of selecting and implementing different interventions. For this purpose, three main elements have to be considered: the priority of water allocation among the various uses, the indications provided by drought monitoring systems and the method adopted to assess drought risk. With reference to the latter, the individuation of risk levels (i.e. alert, alarm and emergency), associated with different groups of interventions, could be a valuable tool for decision makers.

In conclusion, it should be pointed out that an integrated drought management strategy is a key step toward the reduction of the most adverse drought effects, since it makes possible to shift from a reactive to a proactive approach, which is widely recognized to be the more appropriate way for a successful drought mitigation.

ACKNOWLEDGEMENTS

This research has been carried out on behalf of UIR-Sicilia (Contract Rep. 112 of 02/11/2004) within the project Sedemed II, funded by the European Program INTERREG III B – MEDOCC, and partly funded by EU Project "Mediterranean Drought Preparedness and Mitigation Planning" (MEDROPLAN, EuropeAid MEDA, contract no. ME8/AIDCO/2001/0515/59770-P027).

REFERENCES

Cancelliere, A., Di Mauro, G., Bonaccorso, B. and Rossi, G. (2006) Drought forecasting using the Standardized Precipitation Index. *Water Resources Management* (accepted for publication).

Dziegielewski, B. (2000) Drought preparedness and mitigation for public water supplies. In: D.A. Wilhite (Ed.) *Drought: a Global Assessment* Vol. II, Routledge, London, pp. 194–205.

MEDROPLAN (2006) Drought Management Guidelines (Draft) (edited by A. Iglesias et al.)

Regione Emilia Romagna (2004) Piano di Tutela delle Acque, http://www.ermesambiente.it/PianoTutelaAcque.

Regione Sardegna (2004) Piano di Tutela delle Acque, http://www.regione.sardegna.it.

Regione Veneto (2004) Piano di Tutela delle Acque, http://www.regione.veneto.it/ Territorio+ed+Ambiente/ Ambiente/ Acqua/Ciclo-Acqua/ Piano+di+ Tutele+delle+Acque. htm.

Rossi, G. (2000) Drought mitigation measures: a comprehensive framework. In: J.V. Vogt and F. Somma (Eds.) *Drought and Drought Mitigation in Europe*, Kluwer Academic Publishers, Dordrecht, pp. 233–246.

Rossi, G. (2003) Requisites for a drought watch system. In: G. Rossi et al. (Eds.) *Tools for Drought Mitigation in Mediterranean Regions,* Kluwer Academic Publishers, Dordrecht, pp. 147–157.

Rossi, G. (2004) Prevenzione e mitigazione delle carenze idriche dovute a siccità. *L'Acqua*. 4: pp. 9–22.

Rossi, G. (2005) Siccità: dalla gestione dell'emergenza alla gestione del rischio di deficienza idrica. *L'Acqua*. 3: 117–127.

Rossi, G., Cancelliere, A. and Giuliano, G. (2005) Case study: multicriteria assessment of drought mitigation measures. *Journal of Water Resources Planning and Management*. 131(6): pp. 449–457.

Rossi, G., Cancelliere, A. and Giuliano, G. (2006) Role of decision support and multi-criteria methods for the assessment of drought mitigation measures. In: J. Andreu, G. Rossi, F. Vagliasindi and A. Vela (Eds.) *Drought Management and Planning for Water Resources*, Taylor & Francis Group, Boca Raton, pp. 203–240.

Servizio Agrometeorologico Regionale per la Sardegna (SAR) (2004) Carta delle aree sensibili alla desertificazione, a cura di Motroni, A., Canu, S., Bianco, G. and Loj, G.

Water Scarcity Drafting Group (2006) *Water Scarcity Management in the context of WFD*, Salzburg, June 1–2, 2006.

Werick, W.J. (1993) National study of water management during drought: results-oriented water resources management. In: *Water Management in the 90's*, ASCE New York, pp. 445–450.

Whipple, W. (1994) *New Perspectives in Water Supply*, Lewis Pubblication, Boca Raton, Florida.

Yevjevich, V., Hall, W.A. and Salas, J.D. (1978) *Drought Research Needs*. Water Resources Pubblication, Fort Collins, Colorado.

Yevjevich, V., Da Cunha, L. and Vlachos, E. (1983) *Coping with Droughts*. Water Resources Pubblication, Littleton, Colorado.

CHAPTER 17

DROUGHT IMPACTS IN AGRICULTURE: WATER CONSERVATION AND WATER SAVING PRACTICES AND MANAGEMENT

L. S. PEREIRA
Institute of Agronomy, Technical University of Lisbon, Portugal

Abstract: This chapter revises the measures and practices that can be applied by the agriculture sector, rainfed and irrigated agriculture, to reduce both water demand and water consumption to mitigate the impacts of droughts. The analysis focuses on water conservation and water saving practices and management measures, which may lead to reduce the demand, control supplies and to improve the efficient use of water and water productivity

Keywords: Rainfed agriculture, irrigated agriculture, irrigation methods, deficit irrigation

1. WATER CONSERVATION AND WATER SAVING: CONCEPTUAL APPROACH

The terms water conservation and water saving are generally associated with the management of water resources under scarcity. However, these terms are often used with different meanings in accordance with the scientific and technical disciplines or the water user sector considered. Very often, both terms are used as synonyms. *Water conservation* refers to every policy, managerial measure, or user practice that aim at conserving or preserving water resources, as well as combating the degradation of water resource quality. *Water saving* aims at limiting or controlling water demand and use for any specific purpose, including the avoidance of water waste and the misuse of water. In practice both perspectives are complementary and inter-related; however, these terms should not be used as synonyms.

When water scarcity is due to drought, water conservation and saving require policies and practices that may be common with other water scarcity regimes (Pereira *et al.*, 2002a). However, coping with droughts requires a distinction between *preparedness (or proactive) and emergency (or reactive) measures*, the first consisting in preparing for the application of the response measures during and after drought according to the respective severity and impacts.

Preparedness measures correspond to *risk management*, i.e. to the development of those measures that protect agricultural and other vulnerable sectors from the

G. Rossi et al. (eds.), Methods and Tools for Drought Analysis and Management, 349–373.

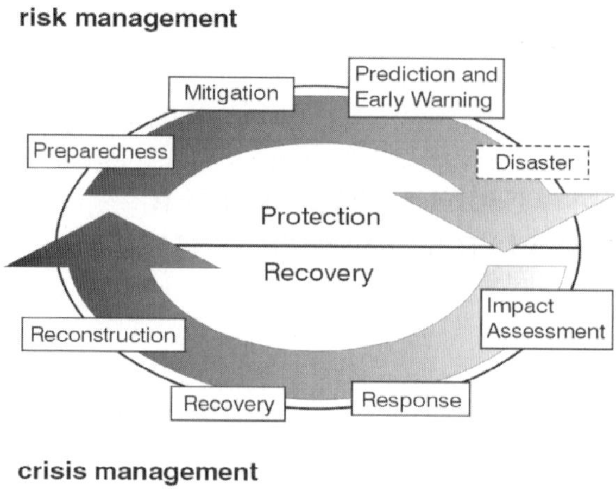

Figure 1. Cycle of drought disaster management (source: National Drought Management Center, University of Nebraska, USA)

impacts of drought (Figure 1). They include the preparation of measures, particularly the mitigation ones, in relation to the foreseen problems to be dealt if a drought occurs, i.e. the establishment of drought contingency plans. They also include the observation of weather and global circulation anomalies such as the North Atlantic Oscillations (NAO), which is ideally performed by a drought observatory or a Drought Watch System (Rossi, 2003), as well as the exploration of weather data and models that may produce prevision or prediction of drought occurrences (Paulo *et al.*, 2005; Paulo and Pereira, 2007). Also included the follow-up of the drought events to characterize the respective severity and its evolution as well as drought end. Mitigation measures are implemented during drought and their characteristics depend upon the severity of drought and mainly, of drought impacts.

Reactive measures also correspond to the so called *crisis management* that aims at the recovery of the drought affected system (Figure 1). Included are the impact assessment, and the response, recovery and reconstruction measures aimed at re-establishing the lost balances observable after a drought has ended.

Impacts of drought in agriculture depend upon the crops and crop patterns, the period when drought occurs in relation to the crop cycles, and on the water availability that may compensate for the lack of water. They also depend on drought severity and duration. A rainfed annual crop may achieve production if minimum water requirements are satisfied by scarce but opportune rains, or may be lost if rains fall out of the critical periods. A rainfed perennial crop may be less affected if the rooting system is able to explore water far deep, but suffers if the soil is shallow and no soil water reserves are available. Irrigation may overcome the drought effects if groundwater is used as main source or if the surface reservoir is well refilled before drought becomes severe; irrigation will be quite limited if water is lacking. Moreover, deficit irrigation strategies may not be implemented if yield losses due to water stress affect yields beyond the economic limits, which depend on many factors including the variation of prices of commodities.

The variety of situations that may occur in practice do not allow to affirm that measures and practices for water conservation and saving will produce effective control of impacts. However, while water conservation creates better conditions to use the limited available water resources and water savings contribute to an improved use of the resource, both of them tend to mitigate the impacts of drought. Their effectiveness depends upon the local circumstances and the ability and means available for farmers and managers to cope with drought. A review is presented in the following sections and selected bibliography is included.

2. WATER CONSERVATION AND WATER SAVING TO COPE WITH DROUGHT

Water conservation corresponds to a variety of measures and practices aimed at better using the available resources or at increasing them out of the normal conditions (Table 1). *Water conservation measures for drought preparedness* include:

1. development and effective implementation of drought watch systems as a main component of the meteorological and hydrological information systems,
2. predicting drought initiation, evolution and ending, basically prediction of transitions among drought classes of severity,
3. storage and regulation reservoirs to mitigate the effects of the diminishing availability of the resources during drought, as well as improved conditions for operation, maintenance, and management of water supply systems, mainly for controlling operational losses and for improved flexibility in operation,
4. controlling and planning for larger groundwater withdrawals aiming at augmenting the water availability during drought periods,
5. establishment of water allocation policies to be enforced under drought, taking into consideration the social, economic and environmental uses of water,
6. planning for the augmentation of available water resources during drought, including waste-water re-use and the use of non conventional water resources;
7. development of water technologies and practices to be adopted by the end users that help in reducing the demand and controlling the water wastes,
8. development of institutional conditions for drought preparedness and management, including for timely application of drought mitigation measures,
9. establishing water pricing and financial incentives and penalties aiming at reducing water consumption and use and avoiding water wastage and misuse, including the control of water quality degradation by effluents and return flows,
10. enhancing public awareness of the economic, social and environmental value of water, thus favouring the adoption of drought mitigation measures.

When a drought occurs, *water conservation for drought mitigation* should be implemented. Then, measures and practices include:

1. exploring the drought watch system to monitor the drought onset, development and termination, as well as to produce information for decision makers and water users,

Table 1. Water conservation measures and practices and their relative importance to cope with droughts (Source: Pereira *et al.*, 2002a)

Water conservation measures and practices	Relative importance
• Meteorological and hydrological information networks to support planning and real time operation and management of reservoirs and irrigation systems (drought watch systems)	H
• Storage and regulation reservoirs for improved availability of water resources	H
• Land and water use planning and management	L
• Enforcing water quality management measures and practices	L
• Improved conditions for operation, maintenance, and management of water supply systems	H
• Maintenance of required discharges for ecological purposes in natural streams and water bodies	L
• Controlling ground-water withdrawals, recharge and contamination	L
• Enforcement of water allocation policies focusing on the prevalent problems	H
• Augmentation of available water resources through the re-use of treated waste-waters, drainage and low quality waters, for specified uses	H
• Exploring non-conventional water sources by households, and farmers	H
• Adoption of water technologies and practices by the end users for resource conservation	H
• Development of soil and water conservation practices in rainfed and irrigated agriculture	H
• Development of soil and crop management practices for restoring the soil quality,	L
• Combating soil and water salinization	L
• Development of participative institutions to water management, including legal and regulatory measures	M
• Application of water pricing and financial incentives that favour efficient water uses, treatment and re-use	H
• Adoption of penalties for water wasting and misuse, and degradation of the resource	H
• Enhancing public awareness of the economic, social and environmental value of water, including nature conservation	H

H = high to very high importance; M = important but not having first priority; L = important but having low priority.

2. implementing changes in reservoir and ground-water management rules,
3. enforcing drought oriented water allocation and delivery policies, and
4. adoption of farm water storage and soil water conservation practices, which cannot help during drought but must be in place before the drought starts.

Water conservation must be complemented with water-saving programmes, which are essentially reactive. *Water saving for drought mitigation* concerns measures and practices summarized in Table 2.

Coping with drought requires that measures, practices and policies of water conservation and water saving be effectively applied in relation to the respective problems.

Table 2. Water saving measures and practices and their relative importance to cope with droughts (Source: Pereira *et al.*, 2002a)

Water saving measures and practices	Relative importance
• Control of evaporation losses from reservoirs	H
• Control of leaks from canals and conduits	H
• Control of spills from canals and other hydraulic infrastructures	H
• Exploring information systems (drought watch systems) for decision makers, water system management and improved water uses	H
• Implementing reservoir and ground-water management rules	H
• Enforcing water scarcity oriented water allocation and delivery policies	H
• Adoption of reduced demand crops, cropping patterns (drought tolerant crop patterns), cultivation practices, and irrigation techniques	H
• Cropping and irrigation practices oriented to control non-point source pollution by agro-chemicals, fertilisers and erosion sediments	L
• Adoption of deficit irrigation practices	H
• Adoption of irrigation and drainage practices favouring salinity management	L
• Adoption of farm water storage and soil water conservation practices	H
• Use of inferior quality water for irrigation	H
• Water price policies in relation to the used water volumes, the specific uses, and the productivity of water use	H
• Incentives for reducing water demand and consumption	L/M
• Penalties for excessive water uses as well as for low quality effluents and return flows	H
• Education and campaigns for adoption by end-users of water saving tools and practices.	H

H = high to very high importance; M = important but not having first priority; L = important but having low priority.

3. WATER CONSERVATION IN DRYLAND AGRICULTURE

Dryland agriculture, also called rainfed agriculture, is crop production without irrigation. Crop water requirements are satisfied by the rainfall occurring during the crop season. Dryland agriculture is traditionally practised in water scarce areas when rainfall during the rainy season is sufficient for the crop to develop and produce. Yields are commonly poorer than those obtained thanks to irrigated agriculture and vary in a large range from one year to the other, following the trends in the temporal variation of precipitation. The use of water conservation practices is usually required for successful crop production under water scarcity conditions.

Water conservation in dryland agriculture may assume quite different facets and include the selection of crops that are less affected by rain water deficits, the adoption of cropping practices that favour the ability of crops to escape water stress, and soil management practices that help to conserve water in the soil for crop use. A recent review has been produced by Unger and Howell (1999).

Typically, the term water conservation in dryland agriculture is used for technologies which aim to improve the amount of water infiltrated and stored in the soil. However it includes other technologies that improve the water productivity of the crops, i.e. the yield quantity produced per unit quantity of water. In fact, by increasing water productivity the water use is improved and the resulting demand for water may be better controlled and, in certain cases, the demand for other water uses may be better satisfied.

In dryland agriculture, crops and crop varieties are selected taking into consideration their tolerance to the water stress conditions that characterise the environments where they are cultivated. In general, these crops correspond to centuries of domestication of plants native to these environments, but new varieties have been introduced in the last decades following scientific plant breeding and improvement programmes. The most common food crops are wheat, barley and millet among cereals, and beans, cowpea and chickpea as legumes (pulses), as well as sunflower and safflower.

A summary of the mechanisms of crop resistance to water stress according to the nature of physiological and morphological processes and characteristics involved is presented by Pereira *et al.* (2002a). The term drought resistance is commonly used by plant breeders and physiologists. It is also used here to describe the water stress mechanisms. These resistance mechanisms refer to three main groups: *drought escape, drought avoidance* and *drought tolerance*.

Drought escape corresponds to plant physiological characteristics typical of arid environments that make plants escape from prolonged periods of water stress, as is the case for plants that develop very fast and have a quite short biological cycle, or that have morphological characteristics that favour reduced impacts of insufficient water availability. *Drought avoidance* consists of plant characteristics that provide for reduced transpiration and improved conditions to extract water from the soil. These characteristics increase the resistance of water vapour transfers from the plant to the atmosphere and therefore reduce transpiration. However, since photosynthesis is affected, yields are reduced. *Drought tolerance* consists of plant characteristics that make them able to cope with water deficits and include adjustments in cell turgor, capabilities to retain water in the tissues, maintaining photosynthesis despite transpiration being reduced, and controlling the senescence and abscission of leaves, as is the case for many shrubs in arid areas. Again, as transpiration is reduced photosynthesis is also decreased and plant yields are small. However, these traits are useful in plant improvement programmes.

Despite progress in plant breeding research, including genetic manipulation, results are still somewhat disappointing. The number of genes and characters with potential for altering the plant responses to water stress is too much large, thus making genetic manipulation difficult and slowly responsive. This is contrary to the success of plant improvement programmes focusing on resistances to given diseases or salinity, at least for some species. Benefits of plant responses to water stress are commonly contradictory to crop yield responses, i.e. a water stress resistant variety is able to yield regularly with less water use than other common varieties of the same crop but yields less than a more sensitive variety. This fact makes it difficult to get them adopted by farmers. In addition, crop responses to water stress are influenced by other environmental factors such as wind and temperature, and by the cropping conditions (Wilkinson, 2000).

Due to limitations of plant breeding programmes, research is also oriented to evaluate water use efficiency, i.e. the harvestable yield per unit water consumed, and the water productivity (WP), i.e. the yield per unit water used of crops commonly used in arid and semi-arid zones. These experiments often combine the assessment of water yield performances with the evaluation of improved crop management techniques and practices such as fertilising and soil tillage practices or relative to planting dates and water conservation. Yield responses to supplemental irrigation are also considered when irrigation water may be available during the critical periods for crop water stress. Maximizing the water productivity is generally not acceptable since achieving higher WP may be opposed to higher land productivity and therefore to higher incomes to the farmers. However, under conditions of drought, when the main limiting factor is not land but water, then this objective may be acceptable and allow for extending the area irrigated with the same amount of water (Pereira *et al.*, 2002b).

Crop management. Water conservation in dryland agriculture mainly refers to crop management techniques and to soil management practices. Techniques for crop management to cope with water scarcity and drought are summarized in Table 3 and relate to three main approaches:

A *Techniques to manage crop risk,* which concern crop management techniques designed to minimise the risks of crop failure and to increase the chances for beneficial crop yield using the available rainfall.

B *Controlling the effects of water stress* by adopting techniques and practices that reduce the impacts of water deficits on crop development and yield:

C *Water conservation cropping techniques,* designed to increase the available soil water, to control soil water losses by evaporation, to minimise the transpiration by weeds and their competition for water, as well as to reduce crop transpiration when water stress is extreme. These water conservation techniques highly relate to soil management.

Soil management practices. They refer to tillage and land-forming practices that favour rainfall infiltration into the soil, water storage in the soil zone explored by roots, capture of runoff to infiltrate the soil, control of evaporation losses from the soil and weeds, extraction of water by plant roots, and crop emergence and development (Table 3).

These practices have long been known to have positive impacts on water conservation in dryland farming. However, results of any soil management technology depend upon the soil physical and chemical characteristics, the land-forms and geomorphology, the climate and the kind of implements used. All these factors interact, creating variable responses in terms of crop yields. Thus, when a technique is to be introduced in a given environment and it is substantially different from the traditional and well-proved practices adopted by local farmers, it is advisable to perform appropriate testing before it is widely adopted. However, the principles

Table 3. Crop and soil management for coping with limited water availability (Source: Pereira *et al.*, 2002a)

Techniques	Benefits
Crop risk management	
• Selection of crop patterns taking into consideration the available season rainfall and the crop water productivity	• Lesser water stress effects, including under drought
• Choice of water stress resistant crop varieties when drought is likely to occur	• Lesser impacts on yields
• Use of short cycle crops/varieties	• Lesser crop water requirements
• Planting after the onset of the rainy season	• Ensure crop establishment
• Early seeding	• Avoidance of terminal stress
• Early cutting of forage crops	• Avoiding the degradation of the crop
• Grazing drought damaged fields	• Alternative use when grain yield is lost
• Supplemental irrigation of dryland crops at critical crop growth stages	• Avoid loosing the crop yield when a drought occurs
Controlling effects of water stress	
• Include a tilled bare soil fallow in the rotation	• Increase in soil moisture
• Adopt large plant spacing of tree crops, large row spacing for annual crops, and low density seeding for cereals in drought prone areas	• High exploitable soil volume by tree or plant, increased available soil water
• Reduced fertiliser rates	• Adaptation to reduced yield potential and minimising salt effects
Water conservation cropping techniques	
• Conservation tillage	• Control of soil evaporation losses
• Adequate seed placement	• Prevention of rapid soil drying around the seed
• Seed placement in the furrow bottom and planting in soil depressions	• Place root systems where soil moisture is better conserved
• Pre-emergence weed control	• Alleviating competition for water, avoiding herbicide effects on crop
• Early defoliation	• Decrease crop transpiration surfaces and use leaves for animal feeding
• Windbreaks	• Decrease wind impacts on evaporation,
• Anti-transpirants and reflectants	• Reduction of plant transpiration
Soil management for increasing surface water retention and runoff control	
• Soil surface tillage for increased surface roughness	• Ponding of rainfall excess, larger time opportunity for infiltration
• Tillage for contour (and graded) furrows and ridges	• Runoff and erosion control, storage in furrows, increased time for infiltration
• Residues and crop mulching	• Runoff retardation and higher infiltration
• Furrow dikes	• Rain water storage in furrow basins/pits and increased infiltration amounts
• Bed surface cultivation	• Runoff control and increased infiltration

Soil management for increasing soil infiltration	
• Organic matter for improving aggregation	• Improved soil aggregates and increased infiltration
• Conservation tillage	• Preserve soil aggregates and infiltration rates
• Mulches, crop residues	• Soil surface protection, better aggregates and higher infiltration rates
• Traffic control	• Less soil compaction and improved water penetration in the cropped area
• Chemicals for aggregates	• Favours maintaining soil aggregation and infiltration rates
Soil management for increased soil water storage	
• Loosening tillage	• Increased soil porosity, soil water transmission and retention
• Subsoiling to open natural or plough made hardpans	• Improved soil water transmission and storage, and increasing the soil depth exploitable by roots
• Deep tillage/profile modification in clay horizons	• Increased water penetration and soil depth exploitable by roots
• Chemical and physical treatments of salt-affected soils	• Increased infiltration and availability of soil water
Soil management for controlling soil evaporation	
• Crop residues and mulching	• Decrease energy available on soil surface for evaporation
• Shallow tillage	• Control soil water fluxes to the soil surface
• Chemical surfactants	• Decrease capillary rise
Land management for runoff control in sloping areas: terracing, contour ridges and strip cropping	• Reduced runoff, increased infiltration, and improved soil water storage

of soil management for water conservation are of general application, regardless of the size of the farm, the traction used, or the farming conditions.

Soil management practices for water conservation, are often common to the practices for soil conservation, i.e. they not only provide for augmenting the soil moisture availability for plant growth but they also contribute to the control of erosion and soil chemical degradation. As these practices produce changes in soil infiltration rates and amounts, soil water storage and runoff volumes, they may also produce relatively important changes in the hydrologic balance at the local field scale and, when widely adopted, at the basin scale. The soil management practices for water conservation can be grouped as follows:

A *Runoff control and improved water retention on the soil surface* to provide for a higher amount of rainwater that can infiltrate into the soil and a larger time opportunity for the infiltration to occur.

B *Improvement of soil infiltration rates*, which refer to a variety of practices that aim at increasing water penetration into the soil and maintaining high rates of infiltration. This limits the rain water that may run off or that evaporates.

C *Increasing the soil water storage capacity* by improving the soil water holding characteristics, increasing the depth of the soil root zone, or favouring soil water conditions for water extraction by plant roots.

D *Control of soil evaporation,* which may be achieved in different ways, mainly by using mulching.

E *Runoff control in sloping areas,* which correspond to well known erosion control measures, also provide for water conservation in sloping landscapes. Most of them are associated with modifications of the land forms, reducing both the slope lengths and the slope angles.

F *Water harvesting,* which are techniques used in arid lands to maximise the fraction of rainfall that is used for crops. They include: micro water-harvesting, micro-watersheds, runoff farming, and water spreading. However these techniques are generally not applied in view of drought mitigation.

4. WATER SAVINGS AND CONSERVATION IN IRRIGATED AGRICULTURE

4.1 General

Demand management for irrigation to cope with water scarcity consists of reducing crop irrigation requirements, thus adopting irrigation practices that lead to higher irrigation performances and water saving, controlling system water losses, and increasing yields and income per unit of water used. It includes practices and management decisions of agronomic, economic, and technical nature. The objectives of irrigation demand management can be summarised as follows:

- *Reduced water demand* through selection of low demand crop varieties or crop patterns, and adopting deficit irrigation, i.e. deliberately allowing crop stress due to under-irrigation, which is essentially an economic decision.
- *Water saving / conservation,* mainly by improving the irrigation systems, particularly the uniformity of water distribution and the application efficiency, reuse of water spills and runoff return flows, controlling evaporation from soil, and adopting soil management practices appropriate for augmenting the soil water reserve, which are technical considerations as dealt with in section 3.
- *Higher yields per unit of water,* which requires adopting best farming practices, i.e. practices well adapted to the prevailing environmental conditions, and avoiding crop stress at critical periods. These improvements result from a combination of agronomic and irrigation practices.
- *Higher farmer income,* which implies to farm for high quality products, and to select cash crops. This improvement is related mainly to economic decisions.

The agronomic aspects of irrigation demand management refer essentially to those described previously in section 3. They concern crop improvement relative to resistance to water stress and respective water productivity, cropping techniques that favour coping with lesser water availability, and soil management for water conservation. Economic decisions concern the decision making processes relative to the selection of crop patterns and farming practices that reduce the crop irrigation demand, and include the evaluation of the economic returns and feasibility of water saving and conservation practices. The technical aspects of irrigation demand management which concern the various practices within irrigation are dealt in this section.

Issues for irrigation demand management often refer only to irrigation scheduling, giving to irrigation methods a minor role. However, an integrated approach is required (Pereira, 1996; 1999). Irrigation scheduling is the farmers decision process relative to "when" to irrigate and "how much" water to apply at each irrigation. The irrigation method concerns "how" that desired water depth is applied to the field. The crop growth phase, its sensitivity to water stress, the climatic demand by the atmosphere, and the water availability in the soil determine when to apply an irrigation or, in other words, the frequency of irrigation. However, this frequency depends upon the irrigation method, i.e. on the water depths that are typically associated with the farm irrigation system. Therefore, both the irrigation method and the irrigation scheduling are inter-related.

Irrigation scheduling requires knowledge of crop water requirements and yield responses to water (cf. Allen *et al.*, 1998), the constraints specific to the irrigation method and respective on-farm delivery systems (cf. Pereira and Trout, 1999; Pereira *et al.*, 2002b), the limitations of the water supply system relative to the delivery schedules applied, and the financial and economic implications of the irrigation practice. Improving the irrigation method requires the consideration of the factors influencing the hydraulic processes, the water infiltration into the soil, and the uniformity of water application to the entire field. Therefore, irrigation demand management to cope with water scarcity is discussed here with respect to both the irrigation systems and scheduling.

4.2 Irrigation Systems

Several performance indicators are currently used in on-farm irrigation. The uniformity of water application to the entire field is commonly evaluated through the *distribution uniformity* (DU), which is the ratio between the average infiltrated water depth (mm) in the low quarter of the field and the average infiltrated water depth (mm) in the entire field (Burt *et al.*, 1997; Pereira, 1999). The distribution uniformity essentially depends upon the characteristics of the irrigation system and less on the farmer management. In other words, high DU can only be achieved when the farmers manage well the irrigation system and it is well designed and maintained, whilst poorly designed and/or maintained irrigation systems definitely lead to low DU (Pereira *et al.*, 2002b).

The main farm efficiency indicator is the *application efficiency* (AE), the ratio between the average water depth (mm) added to root zone storage and the average depth (mm) of water applied to the field. AE is a measure of the quality of irrigation management by the farmer and is strongly related to the appropriateness of decisions on when and how much water is applied. Due to the limitations imposed by the farm system characteristics, the application efficiency depends upon the distribution uniformity. In general, if DU values observed are high, the application efficiencies will also be good when irrigation scheduling is appropriate.

Useful relations between irrigation uniformity and crop yields have been made available, which show that attaining high DU is a pre-condition to achieve high application efficiencies, and therefore to obtain a good match between the amounts of water applied and the crop use requirements. Therefore, DU is an indicator that relates well to the system characteristics that favour water conservation and saving, as well as to higher water productivity. Thus improvements in performance of farm irrigation methods and systems are mainly discussed with regard to their effects on distribution uniformity as analysed by Pereira *et al.* (2002b).

Surface irrigation systems. Several surface irrigation methods are used in practice. The main ones are:

- *Basin irrigation*, which is the most commonly used irrigation system world-wide. Basin irrigation consists of applying water to levelled fields bounded by dikes, called basins. Two different types are considered, one for paddy rice irrigation, where ponded water is maintained during the crop season, and the other for other field crops, where the ponding time is short, just until the applied volume infiltrates. The latter can be divided into two categories: traditional basins, with small size and traditional levelling; and modern precision-levelled basins, which are laser levelled and have large sizes and regular shapes. For row crops, the basins are often furrowed with the crops planted on raised beds. For cereals and pastures, the land is commonly flat inside the basin. Basin irrigation is most practical when soil infiltration rates are moderate to low and soil water holding capacity is high, so large irrigations (> 50 mm) can be given. Inflow rates for basin irrigation have to be relatively high (> $2 \mathrm{l\,s^{-1}}$ per meter width) to achieve quick flooding of the basin and therefore provide for uniform time of opportunity for infiltration along the basin length. Basins must be precisely levelled for uniform water distribution along the irrigated field.
- *Furrow irrigation*: water is applied to small and regular channels, called furrows. There is a small discharge in each furrow to favour water infiltration while the water advances down the field. Furrow irrigation is primarily used for row crops. Fields must have a mild slope, and inflow discharges must be such that advance is not too fast, nor too slow, i.e. the time elapsed since inflow starts at the upstream end until the water arrives to the other end must be in equilibrium with the infiltration to avoid either excess runoff at the downstream end, or excess infiltration in the upstream zone. Efficient furrow irrigation nearly always requires irrigation times longer than advance times. Runoff at the downstream end, typically close to 40% of the applied water, should be collected, stored, and reused.
- *Border irrigation*: water is applied to short or long strips of land, diked on both sides and open at the downstream end. Water is applied at the upstream end and moves as a sheet down the border. Border irrigation is used primarily for close growing crops such as small grains, pastures, and fodder crops, and for orchards and vineyards. The method is best adapted to areas with low slopes, moderate soil infiltration rates, and large water supply rates. Borders are most common and practical on small slopes (<0.005 m m^{-1}) but they can be used on steeper

slopes if infiltration is moderately high and the crops are close growing. Design and management of very flat borders approximates conditions for level basins. Precise land levelling is required, and inflow rates should be neither erosive, nor producing too slow or too fast advance.

In traditional systems, the water control is carried out manually according to the ability of the irrigator. In small basins or borders and in short furrows, the irrigator cuts off the supply when the advance is completed. This practice induces large variations in the volumes of water applied at each irrigation, and it is difficult to control "how much" water is applied. Over-irrigation is then often practised. In modernised systems, some form of control of discharge and of automation is used. The fields are often precision levelled, and the supply time and the inflow rate can be known and controlled.

In surface irrigation, the distribution uniformity, DU, mainly depends upon the system variables controlling the water advance along the field and the way in which the water is distributed and infiltrates throughout the field (Pereira *et al.*, 2002b). These variables include the inflow discharge, field length and slope, uniformity of the land surface topography in relation to land levelling conditions, and the duration of water supply to the field. The application efficiency, in addition to these variables, is influenced by the soil water deficit at the time of irrigation. The farmer's skill plays a major role in controlling the time duration of water application, in applying the water at the appropriate soil water deficit and in maintaining the system in good operational conditions. However, his capability to achieve higher performances is limited by the system characteristics and, often, by the delivery rules relative to supply timing, duration and discharges, which are dictated by the canal network managers. Therefore, adopting improved scheduling and management rules is often beyond farmers decisions and ability because the off– and on-farm system limitations are often the prime constraint.

Improvements in surface irrigation systems aimed at reducing the water volumes applied and increasing the water productivity to cope with water scarcity are listed in Table 4. Techniques common to all methods refer to:

A *Land levelling,* which provides conditions for reducing the advance time and water volumes required to complete the advance, better water distribution uniformity (DU) and application efficiency (AE), as well as better conditions for adopting deficit irrigation and easier control of the required leaching fraction (LF).

B *Irrigation with anticipated cut-off,* i.e. cutting the inflow to basins or borders before the advance is completed, or to furrows before the downstream area is irrigated. This technique reduces water application volumes but the last quarter or half of the field is often under-irrigated. This practice is easily applicable but may have important impacts on yields and economic returns.

C *Field evaluation,* which consists of monitoring irrigation events in farmers' fields. It does not provide water savings by itself but produces detailed information to the farmers and extension staff that permit the selection and further adoption of corrective measures for controlling percolation and runoff, improving DU and

Table 4. Techniques for improving surface irrigation systems aimed to cope with limited water availability

Techniques	Benefits
Basin irrigation	
• Higher discharges, reduced widths and/or shorter lengths	• Fast advance time, reduced volumes applied, easier application of deficit irrigation and control of the leaching fraction
• Higher basin dikes	• To catch storm rainfall, providing for conjunctive use of irrigation and rainfall
• Corrugated basin irrigation for row crops	• Faster advance, improved emergence and rooting of the crops planted on the bed, easy adoption of row crops in rotation with rice
• Maintaining low water depths or just soil saturation in rice basins	• Lower seepage and percolation losses and better conditions to store any storm rainfall
Furrows and borders	
• Irrigation with alternate furrows	• Reduced water application to the entire field and deep rooting of the crops is favoured
• Reuse of tail water runoff	• Avoidance of runoff losses, increased systems efficiency and control of quality of return flows
• Closed furrows and borders	• Avoiding runoff at the downstream end
• Contour furrows	• Runoff and erosion control in sloping land
• Surge flow	• Faster advance, reduced percolation and runoff losses, higher performance, in addition to providing for system automation
• Continuously decreased inflow rates, (cablegation)	• Control of percolation and runoff losses by continuously adjusting flow rates to infiltration, providing for automation
On-farm water distribution	
• Gated pipes and layflat pipes	• Easier control of discharges, control of seepage and automation
• Buried pipes for basins and borders	• Easier control of discharges, control of seepage and automation
• Good construction of on-farm earth and lined canals	• Easier control of discharges to the fields and control of seepage operational losses
• Automation and remote control of farm systems	• Improved conditions for operation, easier application of improved irrigation scheduling, including in real time, and precise irrigation management techniques

AE, and adopting water saving practices, as well as improving the timeliness and duration of irrigation events leading to better system management and salinity control.

D *Improved design and modelling.* As for field evaluation, it does not produce water savings but provides for it through the selection of the best combination of field sizes, slope, inflow discharges and time of application that optimise conditions for controlling deep percolation, runoff, leaching fraction applications, and deficit irrigation. It requires support by extension services and should be based on data from field evaluations.

Surface irrigation systems are, in general, not able to apply small irrigation depths but only large ones. Because of system and delivery scheduling constraints, irrigation scheduling has to be simple. The use of easy irrigation calendars, or the adoption of scheduling charts produced by irrigation scheduling simulation models to take into consideration the actual climatic demand, are in general useful.

Sprinkler irrigation systems. The main ones are:

- *Set systems*: the sprinklers irrigate in a fixed position and can apply small to large water depths. Set systems include solid set or permanent systems as well as periodic-move systems, which are moved between irrigations, such as hand-move, wheel line laterals and hose-fed sprinklers. A wide range of sprinklers can be selected for a variety of crops and soils as well as for environmental conditions.
- *Travelling guns*: a high pressure sprinkler continuously travels when irrigating a rectangular field. The high application rates and the characteristics of the moving system make travelling guns unsuitable for applying very small or large depths, or to irrigate heavy soils and sensitive crops. In addition, these systems have a high energy requirement and may have low performances and high evaporation losses when operating under hot, arid and windy conditions.
- *Continuous move laterals,* where the sprinklers or sprayers operate while the lateral is moving in either a circular or a straight path, named center-pivot and linear move systems respectively. These systems are designed to apply small and frequent irrigation in very large fields.

In sprinkler irrigation, the irrigation uniformity essentially depends upon variables characterising the system such as the pressure, discharge, throw and application rate of the sprinklers, spacing between sprinklers and between the pipe laterals where sprinklers are mounted, and head losses along the pipe system network. These variables are set at the design phase and they are difficult to modify by the farmer. Thus DU is the performance indicator that characterises the irrigation system. The efficiency AE depends upon the same system variables as DU and, moreover, the duration and the frequency of the irrigation events. The AE indicator often follows DU and largely depends upon the irrigation scheduling applied. Thus, the irrigator can do little to improve the uniformity of irrigation and is constrained by the system characteristics for any improvements in irrigation performance (Pereira *et al.*, 2002b).

Field evaluations can provide good information to farmers on how to improve management and introduce limited changes in the system. In *set sprinklers*, the uniformity is often lower than that potentially attainable due to (i) excessive spacings between sprinklers and laterals, (ii) too much variation in pressure within the operating system, and (iii) insufficient or excessive pressure at the sprinklers. The main problems in *travelling guns* refer to (i) excessive distance between towpaths, (ii) inadequate pressure, (iii) asymmetric wet angle, and (iv) variable advance velocity. In *lateral moving systems*, more frequent causes for non-uniformity are (i) inadequate pressure distribution along the lateral, mainly when operating in non flat areas, (ii) application rates much above infiltration rates, (iii) use of an end gun sprinkler without an appropriate source of pressure, and (iv) excessive pressure in systems with sprayers. Wind effects are common to all systems. Problems reported above are generally due to poor design, or lack of design. In addition, very often, the farmers have a very poor knowledge of their own systems and do not receive extension support for system selection or to maintain and manage the systems.

The improvement of sprinkler irrigation systems to cope with water scarcity, as summarised in Table 5, mainly concerns a variety of practices aimed at reducing the volumes of water applied, avoiding runoff and limiting wind drift and evaporation losses, as well as to increase DU and AE and water productivity.

Other less specific measures contribute to improving water use in sprinkler systems. However, most of them apply only to large commercial farms having high technological support. This is the case for automation and remote control, which support enhanced management, including precise water application, real time irrigation scheduling and energy management strategies. It is also the case for fertigation - the application of fertilisers with the irrigation water – that improves the use of both the water and the fertilisers, and precise water application - when advanced information systems make it possible to apply differentiated water and fertiliser amounts allowing for differences in soil conditions and crop development.

Microirrigation systems. They apply water to individual plants or small groups of plants. Application rates are usually low to avoid water ponding and minimise the size of distribution tubing. The microirrigation systems in common use today can be classified as:

- *Drip irrigation,* where water is slowly applied through small emitter openings from plastic tubing. Drip tubing and emitters may be laid on the soil surface, buried, or suspended from trellises.
- *Microspray irrigation*, also known as micro-sprinkling, where water is sprayed over the soil surface. Microspray systems are mainly used for widely spaced plants such as fruit trees but in many places of the world they are used for closed space crops in small plots.
- *Bubbler systems*, that use small pipes and tubing to deliver a small stream of water to flood small basins adjacent to individual trees. Bubbler systems may be pressurised with flow emitters, or may operate under gravity pressure without emitters.

Table 5. Improvements in sprinkler irrigation systems aimed at reducing volumes of water applied and increasing water productivity to cope with limited available water

Objectives	Techniques
• *Optimising overlapping of sprinkler jets*	• Adopt appropriate sprinkler spacings • Adopt appropriate towpath spacings for travelling guns • Use sprinkler nozzles that suit available pressure • Enhanced alignment of continuous lateral move systems
• *Minimising discharge variations within the operation system*	• Design for pressure variations not to exceeding 20 % of the average sprinkler pressure • Use pressure regulators in sloping fields • Booster pumps for end gun sprinklers in moving laterals • Monitor and adjust pumping equipment
• *Minimising evaporation and wind drift losses*	• Irrigation during non windy periods • Smaller spacings in windy areas • Sprinklers with low jet angles for windy areas • Suspended spray heads or LEPA piping and heads instead of sprinklers on the top of moving laterals • Large sprinkler drops and application rates in windy areas • Orient set laterals and travelling gun towpaths perpendicular to prevailing wind direction • Avoid gun sprinklers under high winds and heavy soils
• *Maximising infiltration and avoiding runoff*	• Adopt application rates smaller than infiltration rate • Soil management practices that favour infiltration • Furrow dams in sprinkled row crops

Microirrigation uniformity depends upon various system variables such as pressure variation and emitter discharge variation, emitter flow characteristics (pressure – discharge relationship, susceptibility to variations in temperature, emitter orifice size in relation to susceptibility to clogging), emitter material and manufacturing variability, emitter spacings, head losses in the lateral tubing, pressure variation due to field slopes, and filtering characteristics. With the exception of maintenance, the farmer can do very little to achieve good distribution uniformity. As for sprinkler systems, DU is the indicator for the system performance and essentially depends on decisions taken at design stage. The application efficiency mainly depends upon the same system variables as DU and on management variables related to the irrigation frequency and time duration. Therefore, the farmer may improve the application efficiency when adopting appropriate irrigation schedules, but performances are limited by the system constraints.

Field evaluations also play an important role in guiding farmers, creating information for design of new systems, and for quality control of design and services. Results of field evaluation show that micro-irrigation performances are often lower then expected. Low DU is mainly due to inappropriate control of pressure, discharge variation within the operating set, insufficient filtration and filters maintenance, lack of pressure regulators, poor selection of emitters, and poor information on manufacturing specifications and characteristics.

Improvements in micro-irrigation systems to cope with water scarcity aim at achieving high uniformities of water distribution (DU) in the entire field, which is a pre-condition to achieve water savings, as well as for efficient water use and productivity. They are summarised in Table 6.

There is evidence on the need for good design and management when micro-irrigation systems are used under limited water availability. Since these systems apply water only in a part of the field and near the crop roots, they are able to provide

Table 6. Improvements in micro-irrigation systems aimed at reducing volumes of water applied and increasing the water productivity to cope with limited water availability

Techniques	Benefits
• Single drip line for a double crop row	• Reduced water use
• Microsprayers in high infiltration soils	• Avoid deep percolation losses which would be produced by drippers
• Use drippers in low infiltration sloping soils	• Avoid runoff losses which could be produced by microsprayers
• Adjust application duration and timing to soils and crops	• Control of percolation losses and salt distribution and accumulation in the soil
• Adopt pressure regulators in large sets and in sloping areas	• Provide for emitter uniformity by avoiding variations in pressure in the operating set
• Adopt self-compensating emitters in long and sloping laterals	• Provide for emitter uniformity by avoiding variations in pressure along each lateral
• Use appropriate filtering and filters locations	• Control emitter clogging and consequent non-uniformity of emitter discharges
• Frequent filter cleaning	• Control of emitter clogging
• Chemical treatment to combat emitter clogging	• Helps appropriate functioning of emitters
• Careful maintenance	• To fully use the beneficial characteristics of the system
• Automation	• Helps operation and the adoption of irrigation scheduling, mainly in real time
• Adopt fertigation and chemigation	• Enhances water, fertiliser and chemicals efficiency, and the control of weeds and soil diseases
• Improve system design for management under scarcity	• Provides for selecting emitters and system layout, and helps to adopt sub-optimal operation under limited water supply
• Field evaluation	• Identification of corrective measures to the system and, mainly to its management and maintenance

for higher production using less water than other irrigation systems. However, these systems are much more expensive than surface and sprinkler systems despite the fact that costs tend to become lower over time. This implies that their adoption has to be economically feasible, i.e., that yields must be high enough to produce high farm incomes. To achieve this, systems have to be designed for achieving the best results without system water losses, but not just to save water. Progresses in design are required to find solutions that may allow the systems to operate under optimal and sub-optimal conditions when water is not sufficient to fully satisfy the crop requirements.

4.3 Irrigation Scheduling and Deficit Irrigation

Research has provided a large variety of tools to support improved irrigation scheduling, i.e. the timeliness of irrigation and the adequateness of volumes applied. Irrigation scheduling techniques may be used with diverse objectives in the practice of farmers. The application of appropriate irrigation scheduling techniques permits them to optimise the timeliness and the volumes applied, thus controlling return flows, deep percolation, transport of fertilisers and agro-chemicals out of the root zone, and avoiding waterlogging in the parts of the field receiving excess water. Economic and environmental benefits are also obtained because better conditions are created for achieving the target yields.

When water is limited, and certainly insufficient to achieve maximum yields, some kind of reduced demand scheduling has to be used. This may be managed under very well controlled approaches as for deficit irrigation, or just to minimise the effects of water stress as for reduced irrigation when drought occurs. *Deficit irrigation* is an optimising strategy under which crops are deliberately allowed to sustain some degree of water deficit and yield reduction. *Reduced irrigation* is a remedial strategy where crops are irrigated using the minimal amounts of water available during the more critical growth periods in such a way that some yield may be achieved and, for perennial crops, future yields are not compromised. Therefore, for deficit irrigation the timeliness and volumes of water applied must be well scheduled in such a way that the economic returns of the irrigated crops are optimised; differently, for reduced irrigation only the water productivity may be optimal. However, deficit irrigation under saline conditions is difficult because making deficit irrigation and salinity control compatible may not be achievable. Then, reduced demand takes the form of *controlled saline irrigation* where leaching is achieved regularly through most seasons but not in years when water is scarce. Provisions for additional leaching have to be made as soon as more water is available.

The irrigation scheduling techniques may be based on: a) soil water indicators, including soil electrical resistance, soil water potential, soil water content, and remotely sensed soil moisture; b) crop indicators such as canopy temperature, changes in diameter of stems or fruits, sap flow, and remote sensing of crop stress; c) climatic indicators using information on pan evaporation, crop evapotranspiration and rainfall, remotely sensed ET; and d) soil water balance and irrigation calendars, mainly using models and diverse information technologies, including the Web. The irrigation scheduling techniques using climatic indicators are widely used. They require local or regional weather data and some kind of water balance approach. The degree of accuracy depends upon the nature, quality and spatialisation of weather data, as well as the water balance approach used. Preferably, evaporation and evapotranspiration data should be used as input of water balance models operating in real time or using historical data.

Appropriate irrigation scheduling can only be applied when the farmers have control on the timing and duration of irrigation. Pressurised conveyance and distribution systems are often managed on demand, i.e. farmers are in control of irrigation timings and duration. Therefore, irrigators supplied by these systems are free to select and adopt the irrigation schedules they consider more appropriate to their crops and farming practices. However, rigid delivery rules may be enforced by the system managers during periods of drought and limited water supply, which restrict farmers decisions.

The implementation of irrigation scheduling practices requires efforts that imply the involvement of water and agriculture management institutions, farmers organisations, water user associations, and others to stimulate farmers and the providers of the technologies to make use of them. Farmers do it when they understand the benefits, are informed, receive appropriate support and the communities are involved. Thus, before adopting new technologies for irrigation scheduling it is necessary that problems to be solved in a given location be clearly identified, objectives be set, tools and means be made available by the government, the market or any other institution. Then farmers apply an improved schedule that will increase water productivity, control environmental impacts, save water and increase agricultural incomes.

Deficit irrigation, as mentioned before, is an optimising strategy under which crops are deliberately allowed to sustain some degree of water deficit and yield reduction. The optimisation approach leads to economical viability; the allowed water deficits favour water saving, and the control of percolation and runoff return flows reduce losses of fertilisers and agro-chemicals. The adoption of deficit irrigation implies appropriate knowledge of crop ET, of crop responses to water deficits, including the identification of critical crop growth periods, and of the economic impacts of yield reduction strategies. Therefore, appropriate deficit irrigation requires some degree of technological development to support the application of irrigation scheduling techniques. These are built upon validated irrigation scheduling simulation models or based on extensive field trials. In this case, the optimal irrigation schedules are often based on the concepts of water-use efficiency and water productivity. Experimental studies often include the assessment of planting dates since the water productivity is influenced by the rainfall use by the crop, which varies with the planting date.

The irrigation scheduling strategies for deficit irrigation may be designed through using irrigation scheduling simulation models after these have been calibrated or

Table 7. Management of irrigation supply systems to cope with limited water supply

Management techniques	Benefits
Add to supply for drought mitigation	
• New sources of surface water, short distance water transfers	• Increase local water availability
• Increased groundwater pumping	• Adds to normal water sources
• Transfer of water rights	• Re-allocation of available water
• Use/reuse of low quality water for irrigation and landscape	• Alternative sources of water
• Conjunctive use	• Maximises use of available rainfall and water resource
• Reinforce the use of other non-conventional waters	• Prevents extreme scarcity
Improved reservoirs operation	
• Information systems, including remote sensing, GIS, models	• Information for optimised operation and management
• Hydrological forecasting and drought watch systems	• Improved assessment of supplies
• Upgrading monitoring	• Improved use of operation tools
• Application of optimisation, risk, and decision models	• Optimised management rules; and water allocation
Conveyance and distribution systems	
• Canal lining	• Avoidance of seepage losses
• Improved regulation and control	• Higher flexibility, better service and reduced operation losses
• Automation and remote control in canal management	• Improved delivery management and low operation losses
• Low pressure pipe distributors	• Reduced spills and leaks, higher flexibility, easier water metering
• Change from supply oriented to demand oriented delivery schedules	• Favours farmers to apply water saving irrigation management
• Intermediate storage (in canal, reservoirs, farm ponds)	• Increased flexibility and reduced operation losses
• Involve farmers in delivery schedules planning to cope with limited supply	• Allows farmers to adopt best management practices
• Adopt demand delivery scheduling in pressurised systems	• Higher flexibility for water savings at farm
• Water prices in relation to volumes of water diverted and times for use	• Induce farmers to save water and to irrigate by night (automation)
• Information systems	• Provides for optimised operation maintenance and management
• Application of optimisation methods to schedule deliveries	• Increased reliability and equity, and reduced farm demand
Maintenance and management	
• Effective systems maintenance	• Avoids spills and leaks and improves operation conditions
• Personnel training	• Allows to implement more demanding technologies
• Information to farmers and other users	• Knowledge on the system constraints and saving issues

validated for the local conditions. When calibrated using field trials data, those simulation models are particularly useful to extend field results to other areas and other soil types. However, it is also required that economic results are associated with water use and yield data to evaluate which reduced demand strategies are economically viable. Such economical analyses are often lacking and extrapolations from other studies are difficult because costs and benefits change with farming systems, labour, production factors and water prices, while benefits vary with the price of the harvestable yields and local markets. Summarising, there is a great uncertainty and risk in the decision process relative to deficit irrigation.

5. SUPPLY MANAGEMENT

The importance of supply management strategies to cope with water scarcity is well identified in literature and observed in practice (Table 7).

Supply management (Table 7) is generally considered under the perspective of enhancing reservoir and conveyance capabilities to provide higher reliability and flexibility of deliveries required for improved demand management, as well as to control the operational losses such as by seepage, leaks and discharges. Supply management is also considered under the perspective of systems operation, particularly related to delivery scheduling.

Training of personnel in operation, maintenance and management are required to enhance the quality of service, for the technological upgrading of the systems and to carry out water saving programmes. Training is also necessary to develop skills required to contact with the public and for the communication with users. Users and farmers information is of paramount importance to increase the awareness of the value of water and the importance of water savings. Information is required to involve farmers and other users in water saving and conservation programmes.

6. CONCLUSIONS

A variety of measures and practices can be applied in rainfed and irrigated agriculture to reduce both water demand and water consumption to mitigate the impacts of droughts. They refer to water conservation and water saving practices and management measures. Their application requires appropriate knowledge of drought processes and related information aimed at developing preparedness measures appropriate to the locally prevailing agricultural socio-economic and environmental conditions, and at the timely implementation of emergency measures. All are mitigation measures that may lead to reduce the demand, control supplies and to improve the efficient use of water and water productivity, so minimising drought impacts.

ACKNOWLEDGEMENTS

This research was funded by the INTERREG project SEDEMED I and II. The support of the Agricultural Engineering Research Center is also acknowledged.

REFERENCES

Allen, R.G., Pereira, L.S., Raes, D., Smith, M. (1998) *Crop Evapotranspiration. Guidelines for Computing Crop Water Requirements.* FAO Irrig. Drain. Pap. 56, FAO, Rome, 300p.

Burt, C.M., Clemmens, A.J., Strelkoff, T.S., Solomon, K.H., Bliesner, R.D., Hardy, L.A., Howell, T.A., Eisenhauer, D.E. (1997) Irrigation performance measures: efficiency and uniformity. *J. Irrig. Drain. Engng.* 123: 423–442.

Paulo, A.A., Pereira, L.S. (2007) Prediction of SPI drought class transitions using Markov chains. *Water Resources Management* DOI 10.1007/s11269-006-9129-9

Paulo, A.A., Ferreira, E., Coelho, C., Pereira, L.S. (2005) Drought class transition analysis through Markov and Loglinear models, an approach to early warning. *Agric. Water Manage.* 77: 59–81.

Pereira, L.S. (1996) Inter-relationships between irrigation scheduling methods and the on-farm irrigation systems. In: M. Smith *et al.* (Eds.) *Irrigation Scheduling: From Theory to Practice.* FAO Water Reports 8, ICID and FAO, Rome, pp. 91–104.

Pereira, L.S. (1999) Higher performances through combined improvements in irrigation methods and scheduling: a discussion. *Agric. Water Manag.* 40(2): 153–169.

Pereira, L.S., Trout, T.J. (1999) Irrigation methods. In: H.N. van Lier, L.S. Pereira and F.R. Steiner (Eds.) *CIGR Handbook of Agricultural Engineering*, vol. I: Land and Water Engineering, ASAE, St. Joseph, MI, pp. 297–379.

Pereira, L.S., Cordery, I., Iacovides, I. (2002a) *Coping with Water Scarcity.* UNESCO IHP VI, Tech. Doc. Hydrol. 58, UNESCO, Paris, (http://unesdoc.unesco.org/images/0012/001278/127846e.pdf).

Pereira, L.S., Oweis, T., Zairi, A. (2002b) Irrigation management under water scarcity. *Agric. Water Manag.* 57: 175–206.

Rossi, G. (2003) Requisites for a drought watch system. In: G. Rossi, A. Cancelliere, L.S. Pereira, T. Oweis, M. Shatanawi, and A. Zairi (Eds.) *Tools for Drought Mitigation Mediterranean Regions.* Kluwer, Dordrecht, pp. 147–157.

Unger, P.W., Howell, T.A. (1999) Agricultural water conservation – a global perspective. In: M.B. Kirkham (Ed.) *Water Use in Crop Production*, The Haworth Press, New York, pp. 1–36.

Wilkinson, R.E., ed. (2000) *Plant-Environment Interactions.* Marcel Dekker, New York.

FURTHER READING

Ayars, J.E., Phene, C.J., Hutmacher, R.B., Davis, K.R., Schoneman, R.A., Vail, S.S., Mead, R.M. (1999) Subsurface drip irrigation of row crops: a review of 15 years research at the Water Management Research Laboratory. *Agric. Water Manag.* 42: 1–27.

Campos, A.A., Pereira, L.S., Gonçalves, J.M., Fabião, M.S., Liu, Y., Li, Y.N., Mao, Z., Dong, B. (2003) Water saving in the Yellow River Basin, China. 1. Irrigation demand scheduling. *Agric. Engng. Intern.* Vol. V (www.cigr-ejournal.tamu.edu).

Depeweg, H. (1999) Off-farm Conveyance and distribution systems. In: H.N. van Lier, L.S. Pereira and F.R. Steiner (Eds.) *CIGR Handbook of Agricultural Engineering*, vol. I: Land and Water Engineering, ASAE, St. Joseph, MI, pp. 484–506.

El Amami, H., Zairi, A., Pereira, L.S., Machado, T., Slatni, A., Rodrigues, P.N. (2001) Deficit irrigation of cereals and horticultural crops. 2. Economic analysis. *Agr. Engng. Intern.* Vol. III, (www.cigr-ejournal.tamu.edu).

English, M., Raja, S.N. (1996) Perspectives on deficit irrigation. *Agric. Water Manag*. 32(1): 1–14.

Goussard, J. (1996) Interaction between water delivery and irrigation scheduling. In: M. Smith *et al.* (Eds.) *Irrigation Scheduling: From Theory to Practice*. FAO Water Reports 8, ICID and FAO, Rome, pp. 263–272.

Hatcho, N. (1998) Demand management by irrigation delivery scheduling. In: L.S. Pereira, and J.W. Gowing (Eds.) *Water and the Environment: Innovation Issues in Irrigation and Drainage*, E&FN Spon, London, pp. 239–246.

Horst, M.G., Shamutalov, Sh.Sh., Pereira, L.S., Gonçalves, J.M. (2005) Field assessment of the water saving potential with furrow irrigation in Fergana, Aral Sea Basin. *Agric. Water Manage.* 77: 210–231.

Horst, M.G., Shamutalov, Sh.Sh., Gonçalves, J.M., Pereira, L.S. (2006) Assessing impacts of surge-flow irrigation on water saving and productivity of cotton. *Agric. Water Manage.* (in press, doi:10.1016/j.agwat.2006.06.014)

Keller, J., Bliesner, R.D. (1990) *Sprinkler and Trickle Irrigation*. Van Nostrand Reinhold, New York.

Mantovani, E.C., Villalobos, F.J., Orgaz, F., Fereres, E. (1995) Modelling the effects of sprinkler irrigation uniformity on crop yield. *Agric. Water Manag.* 27: 243–257.

Maracchi, G. (2000) Agricultural drought – a practical approach to definition, assessment, and mitigation strategies. In: J.J. Vogt, and F. Somma (Eds.) *Drought and Drought Mitigation in Europe,* Kluwer, Dordrecht, pp. 63–75.

Minhas, P.S. (1996) Saline water management for irrigation in India. *Agric. Water Manage.* 38: 1–24.

Pereira, L.S. (2003) Irrigation demand management to cope with drought and water scarcity. In: G. Rossi, A. Cancelliere, L.S. Pereira, T. Oweis, M. Shatanawi, and A. Zairi (Eds.) *Tools for Drought Mitigation in Mediterranean Regions*, Kluwer, Dordrecht, pp. 19–33.

Pereira, L.S. (2005) Water and agriculture: facing environment and scarcity challenges. *Agr. Engng. Intern.* Vol. VII (www.cigr-ejournal.tamu.edu).

Pereira, L.S., Cai, L.G., Musy, A., Minhas, P.S., eds. (2003) *Water Savings in the Yellow River Basin. Issues and Decision Support Tools in Irrigation.* China Agriculture Press, Beijing, 290 p.

Pereira, L.S., Zairi, A., Mailhol, J.C. (2006) Irrigation de Surface. In: J.R. Tiercelin, and A. Vidal (Eds.) *Traité d'Irrigation, 2ème edition*, Lavoisier, Technique & Documentation, Paris (in press).

Rodrigues, P.N., Pereira, L.S., Zairi, A., El Amami, H., Teixeira, J.L., Slatni, A., Machado, T. (2001) Deficit irrigation of cereals and horticultural crops. 1 Simulation of strategies to cope with droughts. *Agr. Engng. Intern.* Vol. III, (www.cigr-ejournal.tamu.edu).

Rossi, G. (2003) An integrated approach to drought mitigation in Mediterranean regions. In: G. Rossi, A. Cancelliere, L.S. Pereira, T. Oweis, M. Shatanawi, and A. Zairi (Eds.) *Tools for Drought Mitigation Mediterranean Regions*, Kluwer, Dordrecht, pp. 3–18.

Reca, J., Roldán, J., Alcaide, M., López, R., Camacho, E. (2001) Optimisation model for water allocation in deficit irrigation systems: I. Description of the model. *Agric. Water Manag*. 48: 103–116.

Santos, F.L. (1996) Quality and maximum profit of industrial tomato as affected by distribution uniformity of drip irrigation system. *Irrig. Drain. Syst.* 10: 281–294.

Sarwar, A., Bastiaanssen, W.G.M. (2001) Long-term effects of irrigation water conservation on crop production and environment in semi-arid areas. *J. Irrig. Drain. Engng*. 127(6): 331–338.

Tarjuelo, J.M., de Juan, J.A. (1999) Crop water management. In: H.N. van Lier, L.S. Pereira and F.R. Steiner (Eds.) *CIGR Handbook of Agricultural Engineering*, vol. I: Land and Water Engineering, ASAE, St. Joseph, MI, pp. 380–429.

Tarjuelo, J.M., Martin de Santa Olalla, F., Pereira, L.S., Guest Editors (2005) *Land and Water Use: Environmental Management Tools and Practices*. Special Issue of *Agricultural Water Management*, vol. 77.

Vlachos, E., James, L.D. (1983) Drought impacts. In: V. Yevjevich, L.V. Cunha, and E. Vlachos (eds.) *Coping With Droughts*. Water Resources Publications, Littleton, CO, pp. 44–73.

Vogt, J.V., Somma, F., Eds. (2000) *Drought and Drought Mitigation in Europe*. Kluwer, Dordrecht.

Warrick, A.W., Yates, S.R. (1987) Crop yield as influenced by irrigation uniformity. In: D. Hillel (ed.) *Advances in Irrigation*, vol. 4,: Academic Press, Orlando, pp. 169–180.

Zairi, A., El Amami, H., Slatni, A., Pereira, L.S., Rodrigues, P.N., Machado, T. (2003) Coping with drought: deficit irrigation strategies for cereals and field horticultural crops in Central Tunisia. In: G. Rossi, A. Cancelliere, L.S. Pereira, T. Oweis, M. Shatanawi, and A. Zairi (eds.) *Tools for Drought Mitigation in Mediterranean Regions*. Kluwer, Dordrecht, pp. 181–201.

CHAPTER 18

ASSESSMENT OF WATER SHORTAGE IN URBAN AREAS

C. BRAGALLI, G. FRENI AND G. LA LOGGIA

Centre for Studies on Urban Hydraulics, Polytechnic of Milano, DIIAR – Sez. CIMI, Italy

Abstract: Recent history has demonstrated that extreme hydrological events such as floods and droughts can create additional stress on water supplies essential for human and ecosystem health. As stated several times by European Environmental Agency, the prudent and efficient use of water is thus an important issue in Europe and a number of policies and mechanisms have been used or have been formulated to ensure sustainable use of water in the long term. Urban uses are responsible of almost the 17% of the total European fresh water consumption and they are rapidly growing depending on the extension of urban areas and concentration of population in cities. The present chapter will discuss the phases in the development of a drought management plan starting from the estimation of water supplies and demands, the evaluation of drought risk and urban area vulnerability and, finally, planning procedures and possible mitigation measures

Keywords: Urban areas; risk assessment; mitigation measures

1. INTRODUCTION

The share of urban water demand varies considerably when different countries are taken into account. Mediterranean water resources are limited, fragile and threatened. They are already intensely utilised, especially in the South and East, and they are often badly administered. Natural input (renewable water resources) is very unequally shared between countries and populations: (72% in the North, 23% in the East, 5% in the South). Moreover, its distribution on a domestic level, which determines the degree of independence of each country as regards water resources, is also very uneven. This input is also subject to a very irregular time scale, and generally tends to be very low. Mediterranean water resources are particularly sensitive to drought. The impact of variable availability of natural resources can be exalted by incorrect management, leading to high leakages, and inefficient water usage, leading to high consumption connected to several human activities that could be carried out mobilising less resources.

G. Rossi et al. (eds.), Methods and Tools for Drought Analysis and Management, 375–398.

The scarcity and disparity of water resources are:

- exacerbated by the different levels of usability – and therefore mobilisation costs – and particularly environmentally sustainable usability.
- increased because only a part of natural water resources can be stored and utilised. Basin management is generally recommended but is not common practice; it is unsuitable for arid areas (with no functional basin), large karstic zones or highly fragmented basins.
- intensified by the threat and impact of human activity which disrupts water regimes and leads to a deterioration in water quality, and also by the vulnerable nature of some chronically over-utilised resources: salinisation of coastal aquifers (Spain, Israel, etc.), and even the disappearance of sources (Tunisia);
- made more complex by partitions between numerous countries (the Balkans, the Middle East, the Nile basin).

Water scarcity affects urban areas more than the other parts of a country. The reasons have to be found in the concentration of population and, generally, in water demanding users (industrial and commercial entities) and in the potential social impact of inhomogeneous water supply. The estimates of future water demand for urban use, over the next 30 years (Figure 1), differ among the European countries. Demand is expected to increase further in France, Greece, the Netherlands and the United Kingdom. Stable demand is forecast in Hungary (compared with 1990) and

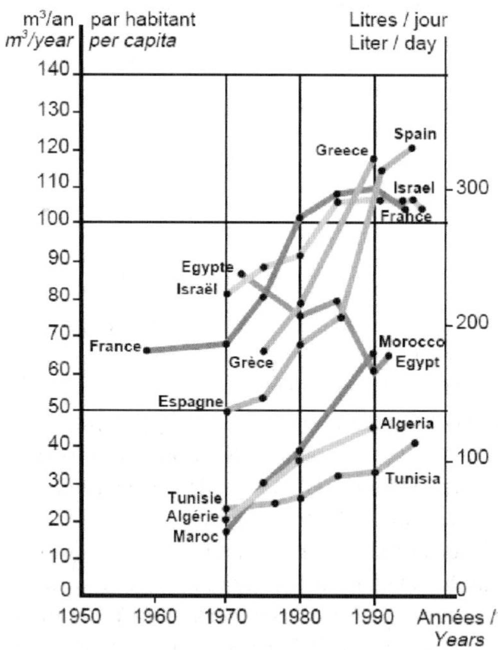

Figure 1. Changes in urban water demand per inhabitant in several countries over recent decades (Margat and Vallée, 1999)

a decrease is expected in Bulgaria and Italy. These trends have been estimated based on assumptions of an increase in population and changes in lifestyle (further penetration of water consuming appliances, changing habits, etc.), increase in water prices (e.g. Bulgaria) as well as the growth of public awareness leading to a more economic water use (Margat and Vallée, 1999).

The delicate equilibrium between urban water demand and water supply availability and the large impact that urban water scarcity can have on population highlights the need for applying efficient water supply and distribution to reduce the occurrence of scarcity events; moreover effective water usage limits the impact that water scarcity can have on population.

The following paragraphs will briefly describe the most common tools for urban water consumption estimation and forecasting, comparing future water demands with water resources availability. The balance between water supply and demand, as well as the status of water storage and distribution system, can be thus used for estimating the risk of urban water scarcity and then provide the basis for water management planning and efficient mitigation of water scarcity impact.

2. WATER DEMAND EVALUATION MODELS

In this context, the first issue to be addressed regards the definition of robust methods for water demand forecasting. Several methods of demand forecasting are often combined, even within a single utility. Several commonly used methods are discussed in this section, although there is usually significant overlap in their use. The amount of data economically available will usually determine the degree of sophistication in the method of forecasting applied. Rigorously, models should be divided into forecasting models and estimation models: this depends on their ability of reconstructing the evolution of water consumption data in time or simply evaluating some characteristics of the future consumption process such as its average, variance, etc.

2.1 Models for Water Demand Estimation

Estimation models evaluate and analyse water demand characteristics without reconstructing its evolution in time. Therefore these models do not allow to project future consumption time series on the basis given by present consumption data and system characteristics. Estimation model are nevertheless capable to define future water consumption in a probabilistic framework also analysing the behaviour of different user classes and possible modifications in water uses connected with technological improvements. These models can be classified as follows (American Water Works Association, 1992):

- **Per capita model.** As its name suggests, the per capita model simply calculates the total production or consumption per capita for a historical period and applies the current year per capita consumption (sometimes with a trend) to the population projections for future periods.

- **Extrapolation model.** Extrapolation models plot, in a scatter diagram, annual or monthly consumption related to time or to population and draw a line manually to capture the "slope" or relationship between the variables.
- **Disaggregate water uses model.** The billing systems in most large water utilities permit at least some disaggregation of total billings, usually by residential, commercial, industrial, and municipal or public facilities. Further breakdowns are often available for single family and multifamily residential, institutional, and irrigator accounts. The basic approach applied to disaggregated segments of total water consumption is usually to isolate cubic meters per day per unit for each segment, which is then applied to projections of the base units into future years. The base unit used could be utility accounts, employment, floor space, population, or any other unit of measure that makes sense for a particular segment.
- **Multiple regression model.** Multiple regression models simultaneously evaluate a combination of independent variables that can include population, households or dwelling units, household income, lot sizes, land use, employment, and various weather variables.
- **Land use models.** Land use models concentrate on current and projected uses of residential, commercial, industrial, and public lands within the ultimate boundaries of the water utility. Residential land use is frequently divided into two or more density classifications, and the non-residential categories are also segmented if there are different patterns of consumption.

2.2 Models For Water Demand Forecasting

Forecasting models have usually more complex structures if compared with the ones described in the previous paragraph. On the other hand, their complexity has to be compensated with high quality consumption data, including long and high resolution time series, able to describe not only the characteristics of consumption events but also their distribution in time. These models are able to describe main consumption process characteristics, such as average, variance and other statistics, but they can also allow for the definition of synthetic water consumption time series describing the future evolution of the process. Those models can be classified as follows (American Water Works Association, 1999):

- **Artificial neural network models.** Artificial Neural Network models (ANNs), as well as other parametric regressive models (Genetic Algorithms, Simulated Annihiling approaches), are based on the same philosophy of multiple regression ones even if the relationship between independent and dependent variables is not simply definable by equations, being dependent on the shape, extension and connections of the Neural Network.
- **Univariate forecasting models.** Univariate models have limited applicability to long term water demand forecasting but they are frequently used for short term analysis. Their specific limit is the consideration that the consumption process is homogeneous in time (it will perpetuate in time with the same characteristics measured in the past). Among these models, rectangular pulse approach is most

frequently adopted in practical applications. Rectangular pulse models differentiate each other for the characteristics of statistical distribution that are introduced even if all of them simulate water consumption as the superposition of elementary demand event with intensity, duration and frequency defined as random variables.

In the following paragraphs, the two most used approaches (Artificial Neural Networks and univariate rectangular pulse stochastic models) will be briefly discussed.

2.3 Artificial Neural Network Model

The Artificial Neural Network are mathematical models that theorize mental and cerebral activities attempting to take "massively" advantage of parallel local elaboration processes and the property of storing that it is supposed to exist in the human brain. These ones, then, are formed by a lot of linear elements, called neurons, organized in order to form a network that recalls the way in which the neurons are connected in biological brains. The most generally used artificial neural network in engineering applications is the Back Propagation Artificial Neural Network (Ashu et al. 2001) that consists of a lot of stratified calculation elements (neurons) tied among them in such a way that the exit of each element of a layer depends on the element of a layer, or level, that precedes them. For example in the Figure 2 it is shown a three level totally interconnected network: an input layer, an hidden layer, where the real elaboration takes place, and an output layer.

In order to "train" the network to carry out a specific function it is previously necessary to prepare a suitable set of "training data". The training phase is managed by the algorithm Error Back Propagation that, in force of the comparison of the incomes matrix to the network with the matrix of wanted values (target), calculates

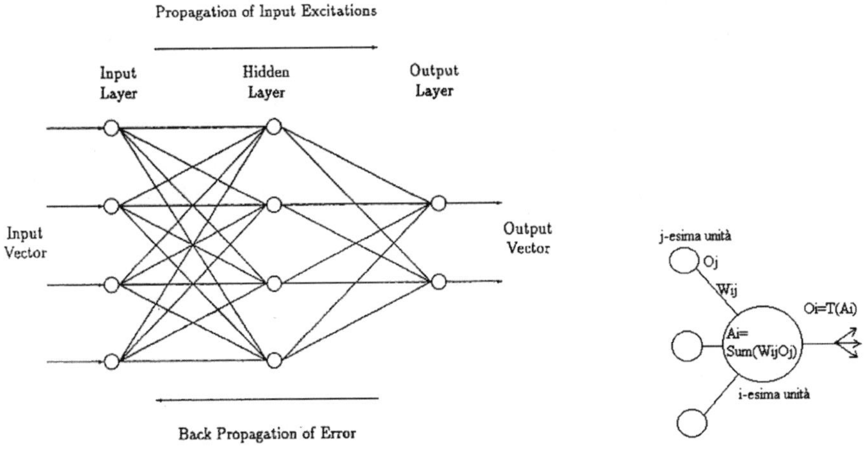

Figure 2. Structure of a Back Propagation Artificial Neural Network

the error E. That value is afterwards used to suitably "correct" the synaptic weights in such a way as to minimize, gradually, the error itself.

The calibration and validation of ANNs can be performed by the mean of automated calibrators that now are available in the most common mathematical application software. For water consumption analysis, most commonly, the developed ANNs use feed-forward back propagation algorithm in which every layer receives input only from the previous one. The algorithm of Levenberg-Marquardt (Levenberg, 1944; Marquardt, 1963) is often used for the training process and the performance, defined as the square of the average output error, is the parameter used to define the level reached by the training network after that all training cycles (epochs) are completed. The model prediction efficiency can be obtained through a "trial and error" procedure during the training phase making changes in the number of neurons present in the hidden layer and in synaptic connections initialization in order to minimize the error.

2.4 Poisson Model With Rectangular Impulses

Poisson rectangular pulses model has been firstly proposed and applied in order to describe rainfall temporal patterns (Rodriguez-Iturbe, 1986; Cowpertwait et al., 1996). Subsequently, it has been transferred to water demand modeling giving good results (Buchberger and Wu, 1995; Buchberger and Wells, 1996; Alvisi et al., 2003; Freni et al., 2004) and confirming the adaptability of such models to describe processes that come from the aggregation and superposition of single events. According to the hypothesis that the instantaneous consumption is regulated by a non-homogeneous Poisson process with rectangular pulses, it is supposed that the consumption intensity consists in a series of instantaneous pulses. Each pulse position is regulated by a Poisson process $N(t)$ with parameter λ that indicates the consumption event frequency. At every instant T_n, in which a consumption event begins, it is associated a couple $\mathbf{U}_n = (t_r^{(n)}, i_r^{(n)})$, being $t_r^{(n)}$ the event duration and $i_r^{(n)}$ its intensity. It is assumed throughout that the features of the events, $\mathbf{U}_n$, are identically distributed and mutually independent of the occurrence time T_n; $t_r^{(n)}$ and $i_r^{(n)}$ are also mutually independent so that the distribution of $\mathbf{U}_n$ is determined by the marginal distributions of $t_r^{(n)}$ and $i_r^{(n)}$.

A scheme of the model with rectangular impulses is shown in Figure 3. Commonly, both consumption event durations and intensities are considered to be well adapted by mutually independent exponential distributions. The model

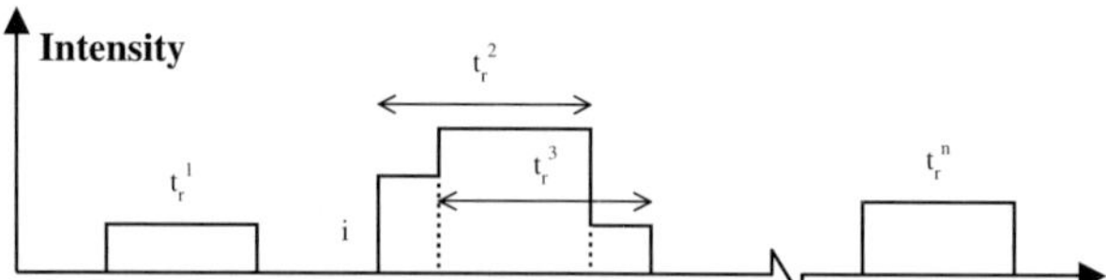

Figure 3. Scheme of the model with rectangular pulses

is thus univocally defined identifying three parameters: the frequency of the consumption event λ, the inverse of the consumption event duration expected value $\eta = 1/E[t_r]$ and the consumption event intensity expected value $\mu = E[i_r]$. Neglecting the theoretical description of the model that can be found in cited literature, in the following only the solving equation system is presented. Having three unknowns, three equations dealing with aggregated consumption average, variance and covariance are commonly adopted even if some examples are present in literature where additional equations are used for allowing parameters optimization.

Buchberger and Wu [1995], who have proposed the normal law for analyzing consumption intensities, have introduced an additional parameter: the variance σ^2 of the intensity distribution. Thus an additional equation is necessary for estimating model parameters. Guercio et al. [2001] have proposed the use of the correlation function at lag-1 (Eq. 4). In the following, the solving equations are reported only for the latter case because, as demonstrated by Fontanazza et al. (2006), they better adapt to water scarcity conditions:

$$E[Y] = \frac{T\lambda\mu}{\eta} \tag{1}$$

$$Var[Y] = \frac{2\lambda}{\eta^3}\left(\mu^2+\sigma^2\right)\left(T\eta - 1 + e^{-T\eta}\right) \tag{2}$$

$$Cov[T_i, Y_{i-2}] = \frac{\lambda}{\eta^3}\left(\mu^2+\sigma^2\right)\left(1-e^{-T\eta}\right)^2 e^{-Tn(k-2)} \tag{3}$$

$$\rho_Y(1) = \frac{Cov[Y_i, Y_{i-1}]}{Var[Y_i]} = \frac{1}{2}\frac{\left(1-e^{-\eta T}\right)^2}{\left(\eta T - 1 + e^{-\eta T}\right)} \tag{4}$$

The covariance equation (3) is commonly specified in the case that k = 3.

Referring back to literature for a more detailed description of calibration and validation procedures for such models, it is worth to give a brief rèsumè of most common calibration methods:

Most commonly parameters are estimated by the moments method comparing statistical moments of the population and of the measured samples; the method has a quite simple application but it can have the inconvenient of generating a stochastic model able to represent the average behavior of the consumption patterns and loosing the ability of representing consumption variance especially at high temporal and spatial aggregation scale.

Another calibration method is based on maximum likelihood concept which allows for fitting the empirical frequency distribution of consumption events by the mean of statistical distribution, minimizing defined objective function; this method can give results similar to the moment method if applied to zero and first order moments of the disaggregated data, but it can be adopted introducing also statistics of aggregated consumption data.

3. URBAN DROUGHT RISK

3.1 Introduction

Water shortage risk connected to drought events affecting a certain urban area is the result of the exposure to a hydrological scenario characterized by drought severity, and of the vulnerability of the water supply sector and of the territorial framework, determined by a combination of economic, environmental and social factors. The characteristics of the Risk and Vulnerability Assessment techniques include hazard focus, geographical and sector focus, subject and purpose of the vulnerability assessment, and input and output parameters. Regarding the natural hazard focus, there is a tradition of concentrating on earthquake and flood hazards: unfortunately, the drought phenomena has received so far too little attention, while development occurring in rich and poor countries alike has laid the basis for increasing vulnerability (Bender, 2002).

The concept of implementing mitigation measures to reduce natural hazard impacts has been an important component of emergency management. Therefore incorporating mitigation into planning efforts is a relatively new concept for droughts. One way to identify appropriate drought mitigation actions seems to be an overall risk analysis as part of drought planning: a combination of drought risk assessment and drought risk management. This approach makes sense, given that risk evaluation, mitigation and planning for drought are so closely linked. Although considered a natural hazard, droughts differ from the other natural hazards in several ways: it is a slow-onset hazard, often described as a creeping phenomenon, making it difficult to determine when a drought begins; droughts do not have a universally accepted definition, causing confusion about whether a drought exists and its severity; although drought is not as physically destructive as most natural disasters, it can affect vast areas and cause a wide range of economic, environmental and social impacts (Hayes et al., 2004).

3.2 Characterization of The Territorial Area

The **territorial area** that includes the urban zones served through the **water supply sector** can be characterized as shown in Figure 4, identifying three **sections**:

- **hydrologic scenario:** reduction of surface flow and groundwater storage due to precipitation deficit**;**
- **water supply sector**: water resource, water supply infrastructure, water supply demand, water losses;
- **territorial framework**: location of urban areas, location of vulnerable users (schools, hospitals, rest homes, location of the productive users, such as zootechnical activities, industries using drinkable water for production process, tourism).

Each sector can be divided into different **components** (Table 1).

In particular, water supply infrastructure **WSI** is constituted by all the necessary water systems connecting resources to water demanding users. Therefore it is

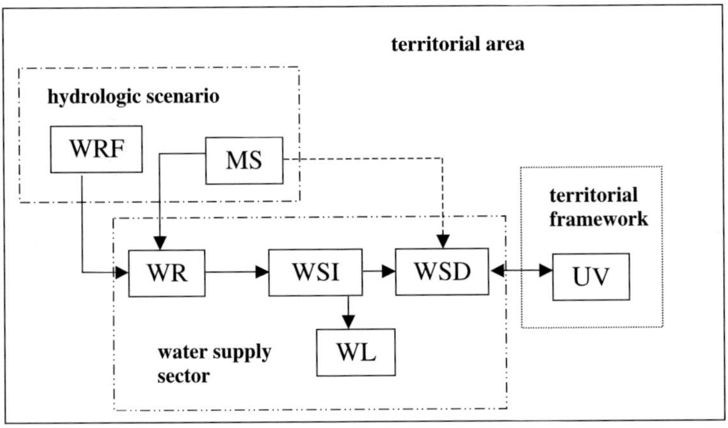

Figure 4. Characterization of the territorial area: sections and components of every section

essential to identify causes and effects, and to establish what sectors, social components and ecosystems are mostly at risk, finding out the appropriated mitigation measures in order to reduce the vulnerability when water resource drought occurs.

3.3 Definitions

Water shortage is the situation in which water demand from consumers in the urban area cannot be completely satisfied: for this reason a **water supply deficit** results. According to the definitions given in MEDiterranean DROught preparedness and mitigation PLANning project (MEDROPLAN, 2005), a water supply deficit can be caused either by drought or by human actions such as population growth, water misuse and inequality in the access to water. If water deficit is absent and distribution network does not succeed in following water demand, water shortage must be imputed to inefficiencies in water supply infrastructure, operations and identified service levels. A water shortage condition is an event having social, environmental and economic impact on the urban area; these impacts can affect

Table 1. Sections and components in urban area

SECTIONS	COMPONENTS	
hydrologic scenario	Water resources features	WRF
	Meteoclimatic situation	MS
water supply sector	Water resources	WR
	Water supply infrastructure	WSI
	Water supply demand	WSD
	Water losses	WL
territorial framework	Users vulnerability	UV

land use, water uses, economic development and, in the worst cases, public order, population growth and social and cultural assessment.

A **water shortage risk event** is an event that produces a water supply crisis identified by the water supply deficit, the result of which is an impact on the social, environmental and economic component of a part or of the whole territorial framework. A **composed water shortage risk event** is a sequence of independent risk of shortage events.

A **scenario** is the state or condition assumed by the components of the area that includes the urban area served through the water supply sector, that is consequent to an initial water shortage condition, from which a risk event or a composed risk event occurs:

scenario = initial condition + risk event (or composed risk event) of water shortage

In United Nation International Strategy for Disaster Reduction (UNISDR) project, **risk** is defined as the probability of harmful consequences, or expected losses (deaths, injuries, property, livelihoods, economic activity disrupted or environment damaged) resulting from interactions between natural or human induced hazards and vulnerable conditions.

In UNISDR, risk is conventionally expressed by the notation:

$$\boldsymbol{Risk} = \boldsymbol{Hazards} \times \boldsymbol{Vulnerability}$$

where, in accord with MEDROPLAN definitions, *Hazards* is defined as potentially damaging physical event, phenomenon or human activity that may cause the loss of life or injury, property damage, social and economic disruption or environmental degradation; each hazard is characterised by its location, intensity, frequency and probability, while *Vulnerability* is defined as the conditions determined by physical, social, economic and environmental factors or processes, which increase the susceptibility of a community to the impact of hazards. Beyond expressing a possibility of physical harm, it is crucial to recognize that risks are inherent or can be created or exist within social systems. It is important to consider the social contexts in which risks occur and that people therefore do not necessarily share the same perceptions of risk and their underlying causes.

3.4 Typologies of Water Shortage

The risk of a water shortage in the urban area can derive from different causes and, depending on different aspects it assumes, it can be gathered into three typologies:

- **Water shortage related to water supply infrastructure WSI:** frequent water resource shortage events with impacts that generally generate modest damages of brief duration, particularly tied up to problems of water supply infrastructure reliability; this latter considered as the attitude to carry out its specific functions, when used in its preset conditions, every time that it is requested and for the considered duration; it can derive from water losses that subtract water resource,

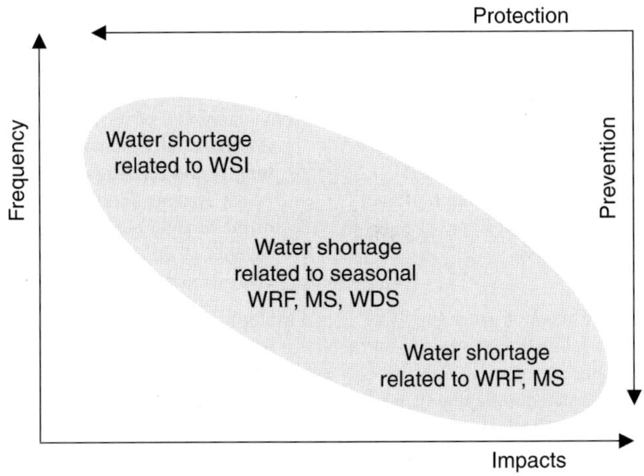

Figure 5. Typologies of water shortage

in short or long periods from breakdowns in the pumping stations, interruption of the electric power, from meaningful variations of water demand with repetitiveness characteristics ascribed to particularly remarkable users;

- **Water shortage related to seasonal hydrologic scenario (WRF, MS) and Water Supply Demand WSD:** frequent or periodic events with impacts giving rise to damages of average intensity; they can derive from the normal seasonal alternation of periods with scarce precipitation and variation of water demand;
- **Water shortage related to hydrologic scenario (WRF, MS):** events with very low frequency and impacts causing serious damages; they are due to anomalous events as drought periods of average and long duration.

The measures, as shown in Figure 5, can be distinguished in prevention interventions, trying to reduce the frequency, and protection interventions, trying to mitigate the consequences.

A qualitative analysis of the potential social, environmental and economic vulnerability can be fulfilled with respect to the main source of the shortage event such as meteorological drought, water balance inconsistency, water demand excess and the level of awareness of the water service manager (Table 2).

4. METHODOLOGY FOR ANALYSIS AND EVALUATION OF THE RISK OF WATER SHORTAGE

4.1 Introduction

The analysis of the risk of water shortage should be a systematic method to appraise the vulnerability of an urban area to situations of water shortage taking into account the level of social, environmental and economic impacts. This analysis should facilitate the decisions on the risk management and on natural calamity declaration.

Table 2. Level of vulnerability of the area

	High-level of vulnerability	Low level of vulnerability
METEOROLOGICAL DROUGHT	Great variability in the precipitations Information lack Passive acceptance of the drought Long duration of the drought period Rapid changes in water supply	Low variability of precipitations Available information on long period Existing drought watch systems Short duration of the drought period Low severity of the deficit Gradual changes in water supply
WATER RESOURCE – WATER DEMAND BALANCE	Single source of water supply or low reliability of water resources Water resources at high risk of contamination Supply of water resources from external systems High population growth or high water demand area	Multiple supply sources or high reliability of water resources Maintenance of water resources Local water resource and locally checked Steady water demand or in diminution
LEVEL OF PREPAREDNESS	Lacking or waiting for the declaration of water supply shortage Decline of responsibilities Lack of agreement and cooperation with external systems Rationing not planned Scarce public awareness	Predetermined and impartial water allocation methods High involvement of the population

The proposed methodology of analysis of the risk keeps on for consecutive phases:

Step 1: Acquisition of historical information;

Step 2: Identification of the principal water shortage risk events that may occur (par. 4.2);

Step 3: Evaluation of the impacts of the events identified on the territorial area including the urban area served through the water supply sector (par. 4.3, 4.4);

Step 4: Identification of the priorities associated to the risk events of water shortage (par. 4.5).

The reconstruction of the past water shortage history and of the events of urban drought is a priority in facing the analysis of the considered area. In order to identify the causes that have determined past urban drought conditions, it is fundamental to characterize the risk typology to which the area is subject and to identify the actions to be implemented. The analysis should be carried out in terms of frequency, spatial distribution and intensity, on the basis of qualitative and quantitative information, considering the changes in land use, in population growth and in its development; this represents an important phase useful to hypothesize what would happen in the actual context if the past meteoclimatic conditions had to occur again. The necessary information also concerns the effects that urban drought have produced on the population in social, environmental and economic terms. To collect this information in a historical file it is very useful to understand the correlation between the gathered meteorological and hydrological variables and the consequences on the ecosystem, so forcing to consider the following steps:

- to better understand, for the urban area, the existing connection among different conditions of water resources shortage, both in terms of water resource and of impact on water demand and drought event;
- to estimate the temporal evolution of the available water resource.

4.2 Identification of Risk Events of Water Shortage

4.2.1 Criterions of identification of risk events

The identification of the water shortage risk events represents a fundamental moment in the management of the risk and in the predisposition of activities aiming to facilitate the emergency management; in fact it is not possible to manage risks that are not recognized and considered. In the identification of risk events, it is important to improve the understanding of the complex interdependences through a general improvement of the information and of the monitoring system, in terms of number of measured variables and formality of acquisition. The identification of risk events of water shortage consists of evaluating the states that actual variables can assume in the three sectors, hydrologic scenario, water supply sector and territorial framework, inside the considered urban area and to which high degree of vulnerability in terms of urban drought corresponds. A risk event of water shortage can be identified by the question: "what happens if..." and its detection can be consequence of a systematic analysis or of observations and experiences; at every revision, the considered risk events must be updated on the basis of the new level of acquired knowledge. This should allow identifying scenarios inside which the management strategies are to be assessed. It is necessary however to consider that the identification of the best option is not clear because of the uncertainty in assessing the impact of the choice itself. If enough information is present, a statistic evaluation of the risk events characterizing a scenario can be of **probabilistic type**; if data are lacking, the events which characterize the scenario can be of **empirical type** in the sense that only the possibility that the event happens is exclusively considered.

The description of a water shortage risk event can be both qualitative and quantitative, through **descriptive variables** or **numerical indicators**. The risk event of water shortage can be associated to duration and the likely duration of the consequent impact should be evaluated too, in order to be able to manage its consequences. The description of the event imposes to identify the source and the nature of the risk. The method considers the following criterions: water deficit, probability, social, economic and environmental impacts, duration associated to an event of risk of water shortage. For every criterion, some classes are individualized on the basis of **Definition**, **Level** and **Equivalent numerical value** connected (Table 4, 5 , 7, 8). The Equivalent numerical values limit for each class has been adopted uniformly for all criterions, except for the probability of water shortage for which the classes have been better detailed. In consideration of the consistence of the available information, it is possible to perform a qualitative evaluation, through the level of water risk; in alternative more precise numerical evaluation methods can be used allowing to assign a quantitative definition to every class and, subsequently, a more objective assessment of each criterion.

4.2.2 Principal risk events of water shortage

The risk events that determine water shortage conditions, with different degree of criticality, are not generally isolated, but more frequently they determine a composed water shortage risk event inside which single events are introduced with different intensity and the urban drought is ascribed to one or more growing causes. Often some events constantly occur in the territorial area including the urban area served through the water supply system. These considerations point out the interaction between the water shortage event and the system components present inside the considered urban area as shown in Figure 4. Among the causes that determine water shortage conditions, those connected to the reduction or to the exhaustion of the water resource due to drought periods are evidently the most dangerous even if not necessarily the most frequent. This is due to the characteristics of unpredictability of the intensity and duration of the drought. Other causes are often consequences of scarce or limited attention to the efficiency of the infrastructures and to the management policies of water demand evolution; in this case even if the effects are apparently analogous, shortage conditions are really predictable and therefore, through adequate measures, they are easier to reduce in frequency and intensity, if not completely eliminable. Besides, many measures connected to water supply infrastructure and to water losses have to be taken into account in the normal planning, while on the contrary drought events, for their random characteristic, require a specific planning. The principal risk events that can occur are pointed out in Table 3; they have to be considered generic and indicative if not contextualized within an urban area and a specific water supply sector. On this subject, the historic reconstruction of the events causing water shortage situations is of fundamental interest.

4.3 Evaluation of The Level of Water Deficit

It is possible to define **water deficit connected to water shortage risk event** as the variation produced in terms of water supply and water consumption; **water deficit connected to scenario** is the unbalance in comparison to the initial situation that allows appraising the unsatisfied water demand connected to a water shortage condition. The analysis of the users considers water needs of each of them in normal conditions. For the industrial users, water needs must be evaluated with reference to the quantity of strictly necessary water to assure production adopting commonly adopted water saving systems. The approaching of water shortage should be pointed out from the monitoring system; it is opportune to determine whether the consequence of the unbalance is due to water supply reduction or water demand increase therefore finding the instigating factor. Particularly as far as water demand increase is the cause of water shortage, situations of endemic water deficit can be

Table 3. Principal risk events

Risk event denomination	Components of territorial area served through water supply sector involved
Reduction or exhaustion of water resource	MS, WR
Pollution of water resource	WR, WSI
Ineffectiveness of potable water treatments	WSI
Mechanical failures of pipelines, pumping stations, reservoirs	WSI
Structural insufficiencies of reservoirs and pipelines that determine functional ineffectiveness	WSI
Operating failure of hydraulic valves	WSI
Insufficient pressure levels	WSI
Deterioration of water quality in water supply infrastructure	WSI
Water losses	WL, WSI
Increase in water demand	WSD

distinguished from situations in which the water deficit derives from a very rapid growth of water demand on a short temporal period, typical of the touristic flows.

The produced impacts are described through:

- area of influence of the event,
- effects on water demand variation ΔWD and on the available water resource that suffers a variation ΔWR,
- the forecast of the duration for the deficit condition.

The evaluation of **water deficit connected to water shortage risk event D_e** is given from:

$$D_e = \Delta WR - \Delta WD \tag{5}$$

The **water deficit connected to scenario D_S,** related to an initial condition (WR_S, WD_S), is valued as:

$$DS = (WR_S + \Delta WR) - (WD_S + \Delta WD) \tag{6}$$

where the difference between the whole available water resource (WR_S+ΔWR) and the overall water demand (WD_S+ΔWD) after that the water shortage risk events included into the scenario are finished or during their manifestation, individuates the relative unbalance which is necessary to provide measures of mitigation and response. The temporal interval of evaluation is related to the cycles of the water supply sector.

The **water deficit coefficient** DC is valued as:

$$DC = (WR_S + \Delta WR)/(WD_S + \Delta WD) \tag{7}$$

Table 4. Level of water deficit

Level	Definition	Equivalent numerical value
severe	$DC \leq 0.50$	1
high	$0.50 < DC \leq 0.70$	Range: $0.65 < \text{value} \leq 1$
moderate	$0.70 \leq DC < 0.90$	Range: $0.35 < \text{value} \leq 0.65$
low	$0.90 \leq DC < 1$	Range: $0 < \text{value} \leq 0.35$
normal	$DC \geq 1$	0

where $DC < 1$ is index of water shortage respect to the normal water use condition. The concept of normal use of water is not easily generalized because it is related to the standards of each specific area considered. Obviously the carrying out of preventive measures allows to reduce the incidence in terms of frequency and intensity of risk events that determine water shortage or to reduce its level when it occurs. Urban drought can be evidently characterized by different degrees of intensity until the worst situation of total suspension of water supply. Table 4 shows the different intensity of water deficit in connection to water deficit coefficient.

The concept of urban drought intensity is tightly connected to the water balance between the input volume and the water demand volume and the provisions will concern the necessary measures to reduce the level of water shortage in the sense to fill the water deficit in comparison to a defined reference level. Drought event effect is perceived by the users with diminution of the pressure, rationing, scarce quality of the water. Water distribution network interposes among water resource and water demand so the structural characteristics and the operational capacities can modify the effects produced by the urban drought. In this way, it is possible to have different water shortage impacts according to the territorial context or to the typology and the vulnerability of the users that insist on the urban area on which the same event appears. Therefore this first important evaluation of water deficit must be followed by the analysis of the events of risk in terms of frequency, impacts, duration.

4.4 Evaluation of the Impacts

The phase of identification of water shortage risk events, and the determination of water deficit, is followed by the phase of impact analysis in relation to multiple criteria. For every risk event or for the connected scenario, the evaluation of the impacts is analysed according to three criteria:

- **probability of every identified risk event of water shortage**;
- **magnitude of the impacts** that it produces on the water supply sector and on the territorial framework;
- **duration of the impacts**;

CRITERION: Probability of every identified risk event of water shortage

The impact of a scarcity event is dependent on initial conditions of the affected area; in this way we should consider the conditioned probability $0 < \text{prob}(A|B) < 1$

Table 5. Level of probability of risk event

Level	Definition	Equivalent numerical value *Probability*
low	Extremely sure not to occur	$0.00 < p \leq 0.05$
low	Almost sure not to occur	$0.05 < p \leq 0.15$
low	Not likely to occur	$0.15 < p \leq 0.25$
low	Not very likely to occur	$0.25 < p \leq 0.35$
average	Somewhat less than an even chance	$0.35 < p \leq 0.45$
average	An even chance to occur	$0.45 < p \leq 0.55$
average	Somewhat greater than an even chance	$0.55 < p \leq 0.65$
high	Likely to occur	$0.65 < p \leq 0.75$
high	Very likely to occur	$0.75 < p \leq 0.85$
high	Almost sure to occur	$0.85 < p \leq 0.95$
high	Extremely sure to occur	$0.95 < p < 1.00$

where A is the event associated risk that has been identified and B the present condition. The level of probability can be appraised as indicated in Table 5.

CRITERION: Magnitude of the impact

For every event of risk, the social, environmental and economic impacts that can be produced on the components of the sectors of which the urban area is composed must be evaluated. The evaluation has the objective to establish a hierarchy of impacts produced by water shortage based on their gravity so to determine impacts that need a more urgent attention. The considerations, useful to appraise the priority of such effects, include costs, extension of the affected urban area, tendencies in the time, public opinion, social equity and ability to react to the impact. This initial assessment identifies the impacts of drought and not the reasons for these impacts. Some of the more common social, environmental and economic impacts consequent to water shortage are synthetically pointed out in Table 6.

The quantification of the impact is complex, so the analysis can be developed to qualitative level defining some classes of impact. These qualitative levels are subsequently classified inside equivalent numerical values or scores and this is repeated for every area of impact.

In Table 7 the level of social, environmental, economic impact is associated to a qualitative level with relative definition and the equivalent numerical value. It is necessary to associate a definition to every impact level for every considered risk event; there will be therefore a chart for every risk event. In order to analyse the impacts associated to risk events, it will be necessary to proceed to an analysis of vulnerability of the users present in the territorial framework from the social, economic and environmental point of view.

The analysis must produce a map of vulnerable users and so a map of the zones more vulnerable inside the urban area. This spatial position, associated to an evaluation of the water demands is important because it allows an overlap with the map of the water supply infrastructure in order to analyze the relative degree of vulnerability. From the social point of view, particular users, as centres

Table 6. Impacts produced by water shortage in urban area

SOCIAL IMPACTS	Physical and mental stress of the population (anxiety, depression, safety loss)
	Hygienic safety problems connected to the scarce flows (diminution of the flow in sewerage, increase of the concentration of the pollutants) and to the use of the water
	Diminution of the ability of struggle against the fires
	Alimentary restrictions (increase of the cost, diminution of the production of some foods)
	Increase of respiratory troubles
	Increase of the political conflicts
	Increase of the conflicts on the management of the water
	Inequity in the distribution of the assistance for the drought
	Reduction or modification of the recreational activities
	Reduction of the quality of life
	Loss of human life
	Loss of aesthetical values
ENVIRONMENTAL IMPACTS	Reduction of the water resources, levels in reservoirs and lakes, flow from springs
	Changes in the boundaries of the salty waters
	Decrease of the levels of groundwater, land subsidence
	Diminution of the flow of the springs
	Effects on the quality of the waters
	Air quality effects
ECONOMIC IMPACTS	Losses in the tourism and in the sectors of recreational activities
	Increase of the cost of the water resources
	Increase of the cost of foods
	Increase of energy demand
	Revenue losses to public administration from reduced tax base
	Cost of increased groundwater depletion (mining), land subsidence
	Cost of new or supplemental water resource development
	General reduction of economic development

Table 7. Level of social, environmental, economic impact associated to a risk event

Level	Equivalent numerical value I
Severe	1
High	Defaults: 0.65, 0.83, 0.95 Range: $0.65 \leq$ allowable value < 1
Moderate	Defaults: 0.35, 0.50, 0.60 Range: $0.35 \leq$ allowable value < 0.65
Low	Defaults: 0.05, 0.18, 0.30 Range: $0 \leq$ allowable value < 0.35
Normal	0

Table 8. Duration of social, environmental, economic impact for each risk event

Level	Equivalent numerical value Duration
Short	Range: $0 \leq$ value < 0.35
Average	Range: $0.35 \leq$ value < 0.65
Long	Range: $0.65 \leq$ value < 1

connected to the health services, schools, etc., and the territorial density of the users, must be considered classifying them on the basis of their need of water. From the economic point of view, the economic – productive turnover must be evaluated, individuating the users of productive, handicraft, zootechnical, tourist, recreational interest classifying them on the basis of their need of water.

Environmental uses of water resources should not be connected with water supply sector that should be addressed to the satisfaction of antropic water needs. These uses should be identified and possibly eliminated or transferred to other water resources. The analysis can be carried out considering the users of water distribution network and directly appraising the contractual specifications and the invoiced volumes.

CRITERION: Duration of impacts

Duration of risk event is the duration of the event itself plus the persistence of the related social, environmental and economic impacts. This second component of the duration of risk event increases notably the uncertainty and the subjectivity in the valuation of the duration, but it allows to make the most dangerous events emerge because the persistence of shortage condition constitutes the element of great trouble for the population and for the economic context. The definition of the risk event duration can be obtained by Table 8 in analogy to the evaluation of the impacts. It is necessary to define a preventive interpretation for each level.

4.5 Individuation of the Priority Associated to Risk Events

To associate a priority to risk events it is necessary to assign an unambiguous score to every event of risk. Different possible techniques are available; in this context, a weighted average model is proposed. Initially it is necessary to appraise an **impact score *IS*** that is the weighted average (in comparison to the weights w_i) of the Equivalent numerical values associated to the impacts (I_i) produced by risk events i:

$$IS = w_1 I_1 + w_2 I_2 + \ldots + w_i I_i \qquad (8)$$

where the weights w_i evidently have to have sum 1.

The **risk score** *RS* associated to a risk event is therefore the weighted average of the scores associated to the considered criterions on the basis of the associated Equivalent numerical value:

- water deficit coefficient *DC*
- probability of every identified risk event of water shortage *Probability*

- score of impact *IS*
- duration of the risk event *Duration*

and it is given by:

$$RS = w_{DC}(1 - DC) + w_{Probability}Probability + w_{IS}IS + w_{Duration}Duration \quad (9)$$

where the weights w_{DC}, $w_{Probability}$, w_{IS} e $w_{Duration}$ evidently have to have sum 1.

On the contrary, the **consequences score** *CS* does not consider the probability that the risk event takes place and therefore it is calculated beginning from the water deficit, from the duration and from the impact independently of its occurring. This parameter is useful in a possible but not probabilistic evaluation of risk events, when the arbitrariness is very high, also for lack of data on the past, in determining the class of probability associated to the event. Its evaluation is given by:

$$CS = w_{DC}/(w_{DC} + w_{IS} + w_{Duration})(1 - DC) + w_{IS}/(w_{DC} + w_{IS} + w_{Duration})IS + \\ + w_{Duration}/(w_{DC} + w_{IS} + w_{Duration})Duration \quad (10)$$

In the above reported brought formulation the weights w_{DC}, w_{IS} e $w_{Duration}$ have to have sum 1 (excluding therefore $w_{Probability}$). For the calculation of the score of risk RS and of the score of the consequences CS it is necessary to valuate the water deficit connected to scenario D_S obtained comparing the connected value of the water resource connected to the considered risk event with the relative water demand; besides it has to be defined a value Δ depending on the safety border that you want to adopt, on the uncertainty on the data used for the preceding analyses and on the threshold of acceptance of water shortage condition. Three cases can occur:

- $DC > 1 + \Delta$ the event does not alter the normal operation of the water sector or water sub-sector, it is set RS = 0, CS = 0 for definition;
- $1 < DC \leq 1 + \Delta$ the event creates a condition of attention for the water sector or water sub-sector, it is set RS = 0, CS = 0;
- $DC \leq 1 + \Delta$ the event determines a state of water shortage, RS or CS are calculated according to their definition.

5. MITIGATION MEASURES

After the characterization of risk events with their probability and the consequent impacts, it is necessary to review possible mitigation measures that can act in two different directions:

- the reduction of system vulnerability acting on the attitude of the system to be affected by water scarcity (this is commonly a medium – long term objective to be achieved through an efficient planning)
- the mitigation of scarcity events damage by the definition of measures and procedures to be activated in case of not expected drought events or events with magnitude higher than those considered in the planning phase (Management of the emergency)

The first group of measures aims to prevent the risk, the second is addressed to tackle the extreme situations in order to guarantee the limitation of the damages and the maintenance of the social structure of the urban area. For sake of simplicity, in the following a simple classification of planning and emergency measures will be given, remanding to the technical literature (Wilhite, 1993a; Wilhite, 1993b; Najarian, 2000; Yevjevich et al., 1983; Rossi, 2000) for details. Wilhite (1993b) divided the mitigative strategies listed by respondents into 9 categories:

- **Water Resource Planning:** planning constitutes the first mitigation measure being able to provide a picture of the resources exploitation and users needs
- **Legislative measures:** such approaches can impose water saving technologies, the increase of the water fees or natural resources preservation
- **Increase in water resources abstraction:** such measures are aimed to the research and the exploitation of new ordinary or alternative water sources
- **Education programs and public involvement:** this category of measures aims to create new consciousness in the customers of water service stimulating water saving and demolishing the common places that fresh water is a free and inexhaustible resource
- **Infrastructures efficiency programs:** such approaches aim to improve the management of water infrastructures reducing losses and water waste
- **Reduction of the water demand:** these measures aim to reduce the urban water demand through the introduction of low consumption technologies both for the industrial and commercial uses and residential ones
- **Resolution of the water conflicts:** this category of measures is aimed to reduce the possible damages of the urban drought event rather than to reduce the vulnerability of the system by avoiding conflicts between the customers that can be generated in cases where the resources are no longer sufficient to cover the demand (for ex. conflicts between agricultural productive uses and industrial use)
- **Emergency plans:** such measures constitute the bridge between the planning and the management of unexpected drought events; in the case of events that fall outside the planning forecasting, the plan defines a clear and unambiguous decisional chain fixing the entities deputed to control the situation and minimise damages; the emergency plan normally sets a temporary concentration of powers that commonly invests the Civil Protection. Table 9 synthesizes the emergency measures to face water shortage in urban area
- **Contingency plans:** such plans are associate to the emergency management plans and they constitute its technical addendum; the contingency plans give a picture of the available resources in a wider range than the one commonly used for planning and they define priority water demands and the procedures of exclusion or reduction of the not essential uses; on the base of the drought magnitude, the Emergency Management Committee picks from the contingency plans the technical solutions for the drastic water consumption reduction or the increase of the water supplies.

Table 9. Emergency measures for water shortage in urban areas

INCREASE OF WATER RESOURCES	Restocking from external systems
	Temporary connections
	Temporary cessation of the water concessions
	Over- exploitation of the groundwater
	Use of additional resources
	Drilling of new wells
	Temporary relaxation of use constraints
	Increase the operational efficiency of water distribution network
	Desalination
	Increase use of non potable water for non potable uses
REDUCTION OF WATER DEMAND	Public information and education campaigns for water saving
	Technologies of water saving for residential and industrial use
	Voluntary limitation in the use of the water
	Adoption of restrictive ordinance for the use of the water
	Management of the pressure
	Tariff politics of limitation of the consumption
MITIGATION OF THE IMPACTS	Information of the population
	Different modalities of management of water resources
	Temporary abilities of accumulation and compensation
	Economic support
	Rationing of water supply
	Implementation of the monitoring system
	Exceptions on the parameters of quality of the waters
	Water transfer among sectors of use
	Reduction or delay of payment deadline

6. CONCLUSIONS

The present chapter presented a simplified approach for analysing the risk of water shortage conditions in urban areas. Water shortage in urban areas can be responsible of high damage on economic and social activities being responsible in the more severe cases of real social threats connected with the research and supply of water resources. The aggregation of several residential and productive uses inside a limited area increases the interest of water asset managers in estimating and forecasting water demand in order to prevent the risk of water scarcity. For this reason the chapter introduces a brief classification of possible mathematical model able to analyse water demand and consumption patterns at different temporal and spatial aggregation scale.

The proposed risk analysis is based on the characteristics of the expected water shortage impacts (duration, magnitude and probability) and not on the drought event because the impacts on antropic activities can be differed and prolonged in time with respect to the generating drought event.

In the last part of the chapter, a short classification of possible mitigation measures has been presented focusing the attention on the differences between planning activities and emergency interventions.

REFERENCES

Alvisi, S., Franchini, M. and Marinelli, A. (2003). A stochastic model for representing drinking water demand at residential level, *Water Resources Management*, 17, 197–222.

American Water Works Association. (1992). Drought Management Planning. Denver, Colo.: American Water Works Association.

American Water Works Association. (1999). AWWA Manual M34, Water Rate Structures and Pricing. Denver, Colo.: American Water Works Association.

Ashu, J., Ashish, K.V., and Umesh, C.J. (2001). Short-term water demand modelling at IIT Kanpur Using Artificial Neural Networks. *Water Resources Management*, 15.

Bender, S. (2002). Development and Use of Natural Hazard Vulnerability-Assessment Techniques in the Americas Natural Hazards Rev., 3(4), 136–138.

Buchberger, S.G., and Wu, L. (1995). A model for instantaneous Residential water demands. *Journal of Hy-draulic Engineering*, ASCE,121(3).

Buchberger, S.G., and Wells, G.J. (1996). Intensity, duration and frequency of residential water demand, *Journal of Water Resources Planning and Management*, 122(1), 11–19.

Cowpertwait, P.S.P., O'Connel, P.E., Metcalf, A.V., and Mawdsley, J.A. (1996). Stochastic point process modelling of rainfall. I. Single-site fitting and validation. *Journal of Hydrology*, 175.

Fontanazza, C.M., Freni, G., La Loggia, G., and Termini, D. (2006). Comparison of different stochastic models for urban water demand forecasting in drought conditions. Proceedings of the 7th International Conference on Hydroinformatics, Nice.

Freni, G., La Loggia, G., Termini, D., and Vitellaro, P. (2004). A water demand model by means of the artificial neural networks method, Proceedings of the 6th International Conference on Hydroinformatics, Singapore, 21–27 June 2004.

Guercio, R., Magini, R., and Pallavicini, I. (2001). Instantaneous residential water demand as stochastic point process, paper presented at Water Resources Management, Halkidiki, Greece, 24–26 September 2001.

Hayes, M.J. Wilhelmi, O.V., and Knutson, C.L. (2004). Reducing Drought Risk: Bridging Theory and Practice Natural Hazards Rev., 5(2), pp. 106–113.

Levenberg, K. (1944). A Method for the Solution of Certain Problems in Least Squares. *Quarterly Applied Mathematics*. Vol. 2, 164–168,.

Margat, J., and Vallée, D. (1999). Water and sustainable development – (Conference on Water security in the Third Millenium. Mediterranean Countries towards a Regional Vision, 12–15 April, Como).

Marquardt, D. (1963). An Algorithm for Least-Squares Estimation of Nonlinear Parameters SIAM. *Journal of Applied Mathematics*. Vol. 11, 432–441.

MEDROPLAN. (2005). Drought Preparedness and Mitigation in the Mediterranean: Analysis of the Organizations and Institutions. Options Méditerranéennes, Série B, No. 51, ISBN 2-85352-320-9.

Najarian, P.A. (2000). An analysis of state drought plans: a model drought plan proposal. Master thesis (Advisor: D.A. Wilhite), University of Nebraska-Lincoln, USA.

Rodriguez-Iturbe, I. (1986). Scale of fluctuation of rainfall models, Water Resources Research, 22(9), 15S–37S.

Rossi, G. (2000). Drought mitigation measures: a comprehensive framework, In: Vogt, J.V. Somma, F. (Eds.), Drought and Drought Mitigation in Europe, Kluwer Academy Publishers, Dordrecht, The Netherlands, pp. 233–246.

Wilhite, D.A. (1993a). An Assessment of Drought Mitigation Technologies in the United States. Final Report to the Soil Conservation Service/USDA. IDIC Technical Report 93-1, University of Nebraska-Lincoln.

Wilhite, D.A. (1993b). Planning for Drought: a Methodology. In: Wilhite, D.A. (Ed.), Chapter 6 (pp. 87–108). Drought Assessment, Management, and Planning: Theory and Case Studies, Kluwer Academic Publishers, Boston, MA.

Yevjevich, V., Da Cunha, L., and Vlachos, E. (Eds.) (1983). Coping with Droughts. Water Resources Publications, Littleton, Colorado.

CHAPTER 19

DROUGHT RISK IN AGRICULTURE IN MEDITERRANEAN REGIONS. CASE STUDY: EASTERN CRETE

G. TSAKIRIS AND D. TIGKAS

National Technical University of Athens, Laboratory of Reclamation Works and Water Resources Management, Athens, Greece

Abstract: The objective of this paper is to present a methodology for the quantification of drought risk in agriculture in Mediterranean regions. The methodology consists of three interrelated components: the estimation of the severity and frequency of droughts, the simulation of water use and crop yield production and the assessment of annualized risk in agriculture. Both rainfed and irrigated agriculture are considered through the concept of vulnerability. The methodology is demonstrated by a case study in Eastern Crete

Keywords: drought risk, crop yield, drought severity, Mediterranean agriculture, Eastern Crete

1. INTRODUCTION

Drought is a regional recurrent meteorological phenomenon influencing most of the economic activities in the area that is affected. Significant water shortage, due to extreme drought events, can affect agriculture, industry, municipalities, tourism, recreation and environmental protection.

Among the various water uses, the most vulnerable ones during drought are those related to environmental protection and agriculture. The difference between these two types of uses is that the impacts on agriculture can be quantified to some extent whereas the impacts on the environment are more obscure and cannot be easily assessed.

The objective of this article is to present a methodology that will be used for risk assessment in agriculture, mainly for the Mediterranean regions. The methodology is based on a minimum data requirements approach and therefore it provides a simple and understandable way to assess risk in agriculture in the Mediterranean. The methodology can also be used in other areas of the world affected by drought, after implementing the necessary modifications.

G. Rossi et al. (eds.), Methods and Tools for Drought Analysis and Management, 399–414.

2. RAINFED AND IRRIGATED AGRICULTURE

Agricultural production is affected by drought, because it is directly dependent on water availability, which is limited under drought. The significance of the impacts of drought on the agricultural sector should be assessed taking into account the severity of the drought (magnitude and duration of the drought episode) and the vulnerability of the agricultural system.

The severity of a drought episode is usually quantified by calculating a set of drought indices which are either general or specific, although this distinction is sometimes obscure. General indices are mostly of the "meteorological" type. The Standardized Precipitation Index (SPI), the Method of Deciles, The Palmer Drought Severity Index (PDSI) and many others have been used extensively in the past to characterize drought severity. The SPI seems to be the most popular among them, due to its simplicity of calculation and the fact that it requires only precipitation series. Quite recently, a similar index, the Reconnaissance Drought Index (RDI) was proposed, including apart from precipitation, the potential evapotranspiration.

Apart from the general indices, several specific indices have been proposed for specific uses such as the agricultural use.

Attempts to relate general indices with the impact in agriculture have failed in most of the cases. Still, there are hopes that some general indices could be used in their initial or modified form. However, additional parameters (e.g. timing, duration, etc.) need to be carefully specified for the better adaptation of the indices to suit the production stages and sensitivities of the crops.

In this case, the RDI could be proposed as a general drought index that can be calculated for the period of the year, which coincides with the entire production cycle of the main crops of the area under study. In this approach irrigated agriculture is considered similar to rainfed agriculture but with an decreased vulnerability.

A further justification for linking the RDI with the impact on agriculture is that it uses the infiltrated precipitation rather than the simple precipitation. The methodology for calculating "infiltrated precipitation" or "effective precipitation" will be presented in section 3.2.

Needless to say that rainfed agriculture is analyzed in a completely different way compared to irrigated agriculture.

3. INPUT PARAMETERS

3.1 Categories of Input Parameters

Obviously numerous factors influence crop production. To avoid the "chaos" created from the big number of parameters involved, the minimum data requirements approach is followed and a large number of parameters with small or medium significance may be assumed as "favourable" or "neutral".

Input parameters may be distinguished into two categories: those related to the phenomenon of drought and those related to the vulnerability of the system. According to a different categorization, the input parameters are distinguished into those related to nature, which are out of farmers' and irrigation managers' control, and those related to agricultural and irrigation practices which can be altered within certain limits. Although the input parameters discussed in this study are mainly studied independently from each other, it would be more realistic if they could be studied in relation to each other. Therefore, considerable simplification is introduced in this analysis by assuming that input parameters are independent.

The following parameters are regarded, as being the most important ones:

- Precipitation
- Potential evapotranspiration
- Soil characteristics
- Water application at the farm level
- Method and technology of water conveyance and distribution
- Fertilizers, cultivation and timing of agricultural activities
- Farmers education, skill and cooperation

3.2 Precipitation and Effective Precipitation

Since a monthly time step is frequently used for calculating drought indices a rather "gross" methodology should be applied to transform monthly precipitation data (P) into effective monthly precipitation data (P_e).

Two widely used approaches will be used for the purpose of this analysis (Stamm, 1967):

(i) Empirical equation

$$P_e = P - (c + P/8), \text{ for } P \geq 7\,\text{mm} \tag{1}$$

where c is between 10 and 20, c=10 for coastal areas and c=20 for mountainous areas, and P and P_e in mm.

(ii) Table 1 is proposed by the U.S. Bureau of Reclamation for calculation based on classes.

Table 1. Estimation of Effective Precipitation

Monthly Precipitation P (mm)	Effective Precipitation (%P)
0.0 - 25.4	90 – 100
25.4 - 50.8	85 – 95
50.8 - 76.2	75 – 90
76.2 - 101.6	50 – 80
101.6 - 127.0	30 – 60
127.0 - 152.4	10 – 40
> 152.4	0 – 10

3.3 Potential Evapotranspiration

The monthly potential evapotranspiration of the reference crop (PET) is obtained using theoretical (e.g. Radiation method, Penman, Penman – Monteith) or empirical methods (e.g. Blaney – Criddle, Thornthwaite). In the case where only mean monthly temperature data are available, the Thornthwaite method is proposed for the calculation of PET. It should be noticed that normally the Thornthwaite method leads to conservative estimates of PET. If analyses from different areas are compared, it is advisable to use the same method for PET estimation. For this purpose empirical relationships may be used in order to compare PET figures estimated by different methods (based on linear regression).

Although it is evident that PET is dependent on rainfall occurrence, if the time step is small, for the sake of simplicity, PET may be considered as an independent variable and can be generated by single-variable models.

3.4 Soil Characteristics

Several soil characteristics play an important role in the water application, redistribution, uptake, consumption, and finally crop yield production processes. The soil characteristics are largely out of the farmers' control although small soil improvements can be achieved by using the appropriate cultivation and irrigation practices. The most important soil characteristics related to water balance and drought analysis are:

(i) Infiltrability
(ii) Water holding capacity
(iii) Characteristic curves $h = h(\theta)$ (water potential as a function of water content) and $K = K(\theta)$ (unsaturated hydraulic conductivity as a function of water content)

The effect on crop yield of every combination of the above parameters can not be easily assessed. Following the minimum data requirement approach only the infiltrability of the soil is used here. Soil infiltrability influences the effective precipitation percentage. Using Table 1, and by integration a useful nomograph is produced (presented in figure 1).

3.5 Water Application at Farm Level, Method and Technology of Water Conveyance and Distribution

Clearly, both the water application at farm level and the method and technology of water conveyance and distribution refer only to irrigated agriculture. 'Water application at farm level' may be substantially improved to be implemented efficiently with smaller quantities of water. This can be realised by either limiting the water losses during and/or after watering the fields, or by improving the effectiveness of useful water. In this context, technological and management approaches may assist in limiting losses due to evaporation and drift, or in limiting runoff and deep percolation at the field level. Alternatively, if the target is not the maximum production

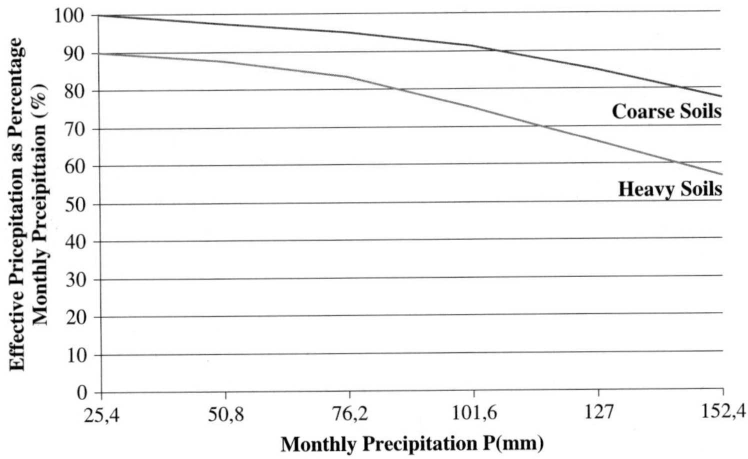

Figure 1. Estimation of Effective Precipitation taking into account the soil characteristics

but the maximum water productivity, significant water quantities can be saved. This happens, because maximum productivity is achieved when less water is applied, as it can be observed in Figure 2.

It is known, that water productivity (WP) is defined as:

$$WP = Y/W, \tag{2}$$

where Y is the yield or economic revenue per unit area (kg/ha) from crop production and W is the water quantity applied (m^3/ha).

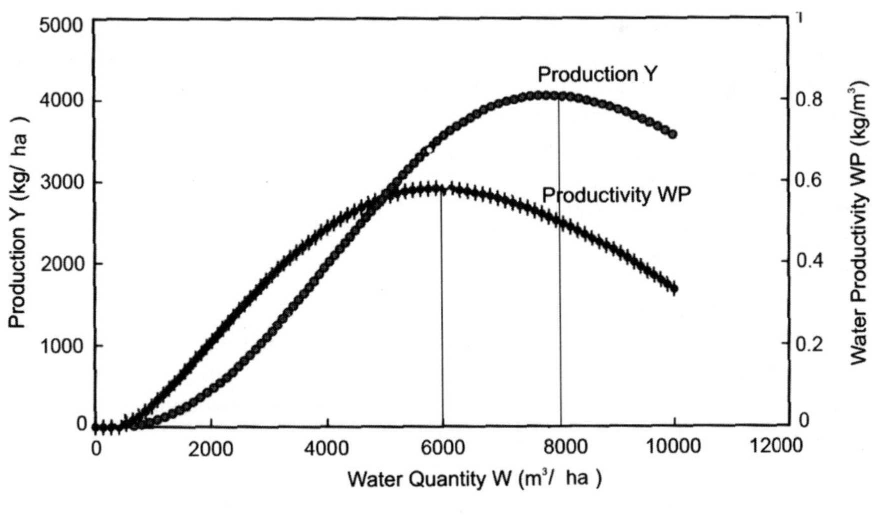

Figure 2. Crop Production and Water Productivity versus water applied (Molden et al., 2003)

"The alternate sets" for sprinkler irrigation, which improves the uniformity of watering, and "the deficit irrigation", which saves water, are promising techniques. These techniques are suitable for increasing the water capacity in the system and consequently for handling effectively the drought conditions.

Finally, another important issue concerning the water requirements at farm level is the transition from water demanding crops to less demanding ones. In this process additional factors, such as local economic and social conditions, may be considered. It can be successful only if the stakeholders are involved positively. It is easily understood, that administrative and legal conditions might favour or impede these changes.

Water conveyance and distribution can be dealt with in a systematic manner to minimize the water losses. As a result, the water productivity will increase. Automation, which can be sensibly introduced, plays a very important role in the effective management of the conveyance and distribution of water (apart from the technological improvements).

3.6 Fertilisers and Cultivation, Farmers' Education, Skill and Cooperation

Fertilisers and cultivation, and the education, skill and cooperation of farmers are important factors for improving the water productivity and for coping with the less quantity of water that is available for irrigation in drought conditions. Particularly the farmers' education may be the most crucial factor for the success of the implementation of any plan aiming at drought mitigation.

4. THE VULNERABILITY OF THE SYSTEM

Vulnerability of a system in reference to a natural hazard is its susceptibility to anticipated damages from the natural hazard. In a simplified form vulnerability may be expressed as a function (from 0 to 1) of variables that reflect the ability of the system to cope with a drought episode of a certain magnitude.

As expected, the vulnerability of an agricultural area affected by drought depends on the following parameters:

- the severity of drought (intensity and duration)
- the time occurrence
- the initial conditions of the system
- the preparedness status
- the skills, education and motivation of people involved
- the institutional set up and the administrative capabilities etc.

In fact only some of these parameters can be modified by human intervention reducing the vulnerability of the agricultural system.

Essentially, irrigated agriculture is less vulnerable to drought than rainfed agriculture. Especially, significant storage facilities (mostly over-year storage facilities) improve substantially the capacity of an agricultural system to withstand drought. Although the vulnerability of irrigated agriculture is a fairly complicated issue, a very simplistic approach was adopted in this text based mainly on advice from agricultural engineers and agronomists. The vulnerability function is devised from expert knowledge as it can be seen in the case study of this paper.

5. ASSESSMENT OF THE CONSEQUENCES

To estimate the damaging effects of a drought episode on an agricultural system a "response system" is needed to act as the "filter" between the input parameters and the final economic output. Evidently the response system may be described at various levels of complexity depending on the desired level of accuracy in the description of the involved processes. Numerous approaches for the study of infiltration processes, the vertical movement of water, water redistribution, water absorption by the roots and crop production can be mentioned. These approaches require a large number of parameters which are rarely available. Furthermore, the variability of these parameters in time and space makes the attempt to simulate the corresponding processes, a rather academic exercise with no practical applicability.

A thorough review of crop models was presented in a volume published by ICID as a result of a specially organized workshop. Some of the models presented in this volume appear on Table 2 and are distinguished in three categories (Pereira et al., 1995).

Such approaches are scientifically very interesting. Nevertheless, their implementation entails some difficulties arising from engineering and management issues. Therefore, simpler approaches have gained universal applicability, improving at the same time the empirical and unreliable yield functions which are based on a single relationship between crop yield and water depth applied. It has been accepted that the ratio of the real evapotranspiration to the potential evapotranspiration (dimensionless water use) is a reliable independent variable used in the prediction of the crop yield. Thus simple conceptual models of deterministic nature may be employed to transform the soil water data in the root zone to dimensionless water use values (Slabbers, 1980).

Using a simple linear function between the dimensionless available soil water (x/x_0) and the ratio of the real to potential evapotranspiration (ET/PET) the dimensionless water use can be estimated for each stage of growth or the entire growing period. Two widely used linear relationships between ET/PET and x/x_0 are shown in Figure 3.

The dimensionless water use $\tilde{w}$ can be estimated by integrating the function describing line (b) in Figure 3. This is a linear function. Therefore, the dimensionless water use can be estimated by subtracting the available soil water of the root zone at the beginning of each stage from its quantity at the end of each stage. If w_0 is the maximum water use then the dimensionless water use is $\tilde{w} = w/w_0$. Detailed description of the simplified crop models of this type in this study can be found in Tsakiris, 1985a and 1985b.

Table 2. Crop simulation and management models (Pereira et al., 1995)

Category I: Soil water balance simulation models for irrigation scheduling	Category II: Water fluxes and crop growth simulation models	Category III: Models for irrigation management at system level
RELREG: a model for real time irrigation scheduling *J.L. Teixeira, R.M Fernando and L.S Pereira*	Design and operation of subsurface irrigation scheme with MUST *P.I.M. de Laat*	USU command area decision support model - CADSM W.R. Walker, S Prajamwong, R.C. Allen and G.P. Merkley
RENANA: a model for irrigation scheduling, employed on a large scale *G. Giannerini*	Application of SWATRE to evaluate drainage of an irrigated field in the Indus Plain, Pakistan *J. Beekma, Th.J. Kelleners, Th.M. Boers and Z.I Raza*	PROREG: a simulation software to design demand in irrigation projects *J.L. Teixeira, M.P Farrajota and L.S. Pereira*
Improved irrigation management under center pivots with SHED *G.W. Buchleiter*	The soyabean model SOYGRO: field calibration and evaluation of irrigation schedules *G. Gerdes, B.E. Allison and L.S. Pereira*	Modeling paddy rice irrigation A.M. Paulo, L.A. Pereira, J.L. Teixeira and L.S. Pereira
Field water balance: BIdriCo 2 F. Danuso, M. Gani and R Giovanardi	Use of the CERES – Millet model for production strategy: analysis in the Sudano-Sahelian Zone *B.E. Allison, J. Fechter, A. Leucht and MVK Sivakumar*	Crop water modeling with an operational management perspective *R.J. Verhaeghe and W.N.M van der Krogt*
An irrigation scheduling model combining slow mobile water changes *M. Parkes, R Bailey, D. Williams and Y. Li*	SWARD: a model of grass growth and the economic utilization of grassland *A.C. Armstrong, D.A. Castle and K.C. Tyson*	HYDRA: a decision support model for irrigation water management G. Jacucci, P.Kabat, P.J. Verrier, J.L. Teixeira, P. Steduto, G. Bertanzon, G. Giannerini, J. Huygen, R.M Fernando, A.A Hooijer, W. Simons, G. Toller, G. Tziallas, C. Uhric, B.J. Van den Broek, J. Vera Munoz and P. Yovchev
A soil-water-crop model for large and small applications *F Mannocchi and P Mecarelli*	A finite element model for simulation of soil moisture balance during plant growth *V.N. Sharda and S.R. Singh* Simulation of crop water balance with OPUS *R.E. Smith*	

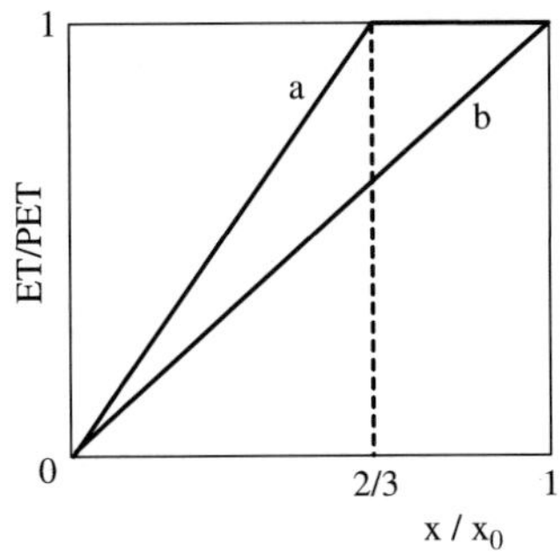

Figure 3. Linear relationships between dimensionless soil water (x/x_0) and dimensionless evapotranspiration (ET / PET)

To simplify the computations of soil water simulation and the anticipated yield losses, x/x_o is considered proportional to ET/PET and could be taken as equal to $\tilde{w}$.

Yield functions were employed to correlate water use and crop yield. The most commonly used crop yield function is the following linear relationship:

$$\tilde{y} = a_1 \tilde{w} + (1 - a_1), \tag{3}$$

where $\tilde{y}$ and $\tilde{w}$ are dimensionless yield and water use during the growing season and a_1 is the slope of the linear yield function expressing the average sensitivity of the crop to water shortage during the growing season.

For illustration purposes, the values of slope a_1 for a number of crops are presented in Table 3. These values were obtained by regression analysis.

For simplicity purposes it is assumed that $\tilde{w}$ may be directly estimated from annual precipitation ratio $P/\bar{P}$ with which it is related linearly.

Aside from the limiting assumptions, the main disadvantage of the approach presented above is the fact that no differential response of the crop to water deficit during the growing season is taken into account. This response can be accounted for, by adopting a multiplicative or an additive dated production function. There are crops which conform to the geometric principle (multiplicative function) and others to the arithmetic principle (additive function) (Tsakiris, 1982, 1985a and 1985b).

Table 3. Slope a_1 of the linear yield function for various crops

Crop	*Slope* a_1	*Source*
Wheat	1.400	Doorenbos & Kassam, 1979
Corn	1.452	Martin et al., 1984
Sorghum	1.100	Garrity, 1980
Soybeans	1.090	Manam, 1974

Multiplicative yield functions have been used by many researchers in the past for studies related to water shortage (Hall and Butcher, 1968; Hanks, 1974; Minhas et al., 1974, Doorenbos & Kassam, 1979). A typical multiplicative dated production function is the one presented by Jensen (1968)

$$\tilde{y} = \prod_{j=1}^{j=m} \tilde{w}_j^{\lambda_j}, \tag{4}$$

where m is the number of stages in the growing cycle, $\tilde{w}_j$ is the dimensionless water use and λ_j is the sensitivity index of the crop, during the *j-th* stage of growth.

6. DROUGHT RISK ESTIMATION

Risk may be defined as a real threat to a system, given its existing vulnerability. For natural phenomena, the risk of the system can be calculated as the product of natural hazard and the system's vulnerability.

$$\{R\} = \{H\} \times \{V\}\,. \tag{5}$$

More correctly the multiplication sign could be replaced by a functional sign describing the combination between hazard and vulnerability. Some agencies also use other expressions incorporating quantities which reflect the coping capacity of the system.

In mathematical terms, the average (annualized) risk can be calculated as follows:

$$R(D) = \int_0^\infty x \cdot V(x) \cdot f_D(x)\, dx, \tag{6}$$

where x is the potential consequence caused by the phenomenon of corresponding magnitude, its probability density function of which is $f_D(x)$ and $V(x)$ is the vulnerability of the system towards the corresponding magnitude of the phenomenon. An assumption is made that vulnerability may be represented by a dimensionless figure between 0 and 1.

It is obvious that if vulnerability is equal to one, the average risk calculated by Eq. 6 coincides with the average hazard.

By analyzing a number of years, the frequency of drought occurrences with a certain severity could be identified. For each level (class) of severity, following the procedure presented above, an estimation of the anticipated losses from the affected agricultural area in monetary units can be obtained.

In this way, a three column table may be produced. Using SPI as the drought index of severity and analyzing 65 years of data, an illustrative example is presented in Table 4. Assuming that the system's vulnerability is 1 the annualized risk or hazard is calculated as follows:

$$R(D) = H(D) = 20 \cdot 1/3 + 150 \cdot 1/7 + 400 \cdot 1/12 + 900 \cdot 1/25 = 97.16\, K€/yr \tag{7}$$

Table 4. Drought severity, probability of occurrence and the anticipated losses from the affected agricultural area

Severity of annual drought	*Probability of occurrence*	*Anticipated losses (K€)*
0> SPI >-1	1:3	20
-1> SPI >-1.5	1:7	150
-1.5 > SPI > -2	1:12	400
SPI < -2	1:25	900

Let us assume that the vulnerability of the system with respect to drought can be reduced according to the following scenario (Table 5).

Table 5. The reduced vulnerability associated with the severity of drought

Severity of drought	*Vulnerability*
0> SPI >-1	0.000
-1> SPI >-1.5	0.667
-1.5 > SPI > -2	0.750
SPI < -2	0.944

The improved annualized risk (R') is now calculated, based on the above vulnerability, as follows:

$$R'(D) = 20 \cdot 0 \cdot 1.3 + 150 \cdot 0.667 \cdot 1/7 + 400 \cdot 0.75 \cdot 1/12 + 900 \cdot 0.944 \cdot 1/25$$
$$= 73.1K€/yr \tag{8}$$

In the example presented here the risk was reduced from 97.16 to 73.1K€/yr. This is a 25% reduction.

7. APPLICATION IN EASTERN CRETE

The methodologies presented above, concerning the estimation of risk due to drought in agricultural areas and specifically the estimation of drought consequences in the crop production, were implemented in a case study at eastern Crete.

Eastern Crete comprises of two prefectures: Heraklion and Lassithi. To apply the methodologies to rather homogeneous areas, each prefecture was divided into two areas with more or less similar characteristics (Figure 4). For simplicity sake, three categories were considered in the analysis: olives, grapes and a third category comprising of non-arable land and scattered crops. The areas covered by each category appear in the Table 6.

The data concerning potential yield vary widely and therefore average values were used in the calculations. For instance the average quantity of grapes produced

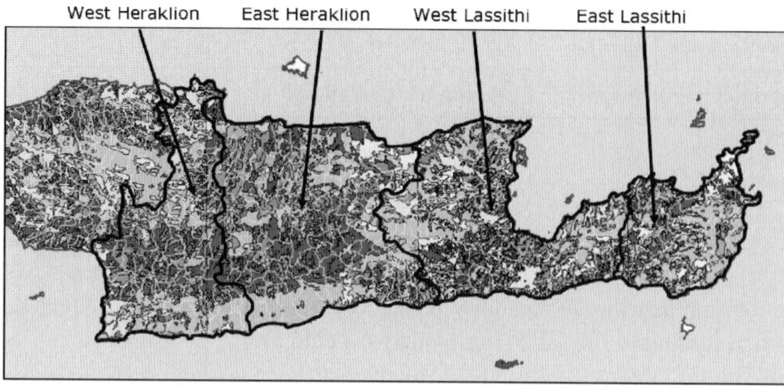

Figure 4. Eastern Crete divided into four areas

per year is 16000 kg/ha and the average price was estimated to be equal to 0.5 €/kg. The corresponding value for olives was 1500 kg/ha of oil with average price 3 €/kg.

Simple lumped crop yield functions were used similar to the one described in Eq. 3 with $a_1 = 1.50$ for grapes and $a_1 = 1$ for olives.

For the computation of the anticipated crop yield losses the following algorithm should be employed:

$$\text{Precipitation} \rightarrow \text{Effective precipitation} \rightarrow \text{water use} \rightarrow \text{crop yield} \rightarrow \text{marketable yield}$$

For simplicity reasons, as already mentioned, the dimensioneless water use was estimated directly from the ratio $P/\bar{P}$, which in turn was used as an input in the single linear crop yield function (Eq. 3).

By analyzing drought in the four homogeneous areas severity indices such as the SPI were estimated for the historical record (1973/74 to 2003/04). The results of this analysis in annual values are presented in graphical form in Figure 5.

A rather gross assumption was made concerning the vulnerability of the agricultural systems. It was assumed that 50% of the crops were rainfed and 50% were

Table 6. Areas covered by cultivated crops in eastern Crete

		Olives		*Grapes*		*Residual Land Use (Bare soil, Forests, Cereal crops, etc)*	
	Total Area (ha)	Area (ha)	Perc. (%)	Area (ha)	Perc. (%)	Area (ha)	Perc. (%)
West Heraklion	89,897	15,268	17.0	4,632	5.2	69,996	77.9
East Heraklion	187,098	45,721	24.4	22,530	12.0	118,848	63.5
West Lassithi	114,706	17,716	15.4	118	0.1	96,872	84.5
East Lassithi	59,123	5,772	9.8	741	1.3	52,609	89.0
Total	450,824	69,210		23,389		358,225	

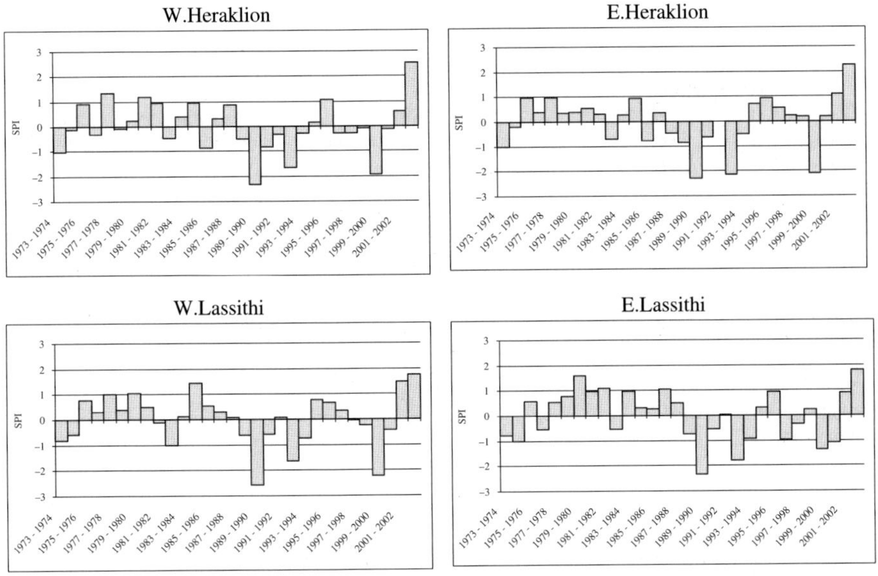

Figure 5. The annual SPI for each hydrological year (October – September) for the four areas of Eastern Crete for the period 1973/74 – 2003/04

receiving supplemental irrigation. As a result, the values of the vulnerability function were set to 1 for the rainfed crops. Also, the values presented in Table 7 were adopted for the irrigated crops for each class of drought severity.

It should be noted that the vulnerability function of irrigated agriculture was determined after long discussions with agricultural engineers and agronomists from the area of eastern Crete.

Based on the above gross assumption the average hazard and risk were estimated for all four areas studied for the two main crops: grapes and olives. Table 8 shows the results for the average annualized hazard and the risk per hectare and Table 9 presents the same quantities in monetary values.

It can be noticed that with 50% of the land supported by irrigation, the annualized risk is substantially reduced. For the example of W. Heraklion, the initial estimated hazard was equal to 734€/ha. Due to the positive influence of supplemental irrigation in half of the area the risk was reduced to 621€/ha. Table 8 shows the

Table 7. Vulnerability of irrigated agricultural lands

Severity of drought	*Vulnerability*
0> SPI >-1	0.5
-1> SPI >-1.5	0.7
-1.5 > SPI > -2	0.9
SPI < -2	1

Table 8. Annualized Hazard and Risk per unit area

		Hazard (€/ha · yr)		
	W. Heraklion	E. Heraklion	W. Lassithi	E. Lassithi
Grapes	734	543	559	609
Olives	281	217	219	255
		Risk (€/ha · yr)		
	W. Heraklion	E. Heraklion	W. Lassithi	E. Lassithi
		Grapes		
Rainfed	367	271	280	305
Suppl. Irrig.	254	218	213	217
Total	621	489	493	522
		Olives		
Rainfed	141	109	110	128
Suppl. Irrig.	98	87	83	90
Total	239	196	193	218

comparisons and provides data for sensitivity analysis. As expected, by reducing the vulnerability through irrigation or any other protection system, the risk will be reduced. By comparing the additional costs for improvements, such as irrigation projects, (expressed per hectare) with the monetary values of risk (per hectare) the rational level of improvements may be proposed.

From the Tables 8 and 9 several important conclusions can be derived. It is apparent that rainfed agriculture faces a more severe threat due to drought occurrence compared to the agricultural areas supported by supplemental irrigation. It is also shown that grapes are more vulnerable to drought than olives both as rainfed and irrigated crops, respectively.

Table 9. Total annualized Hazard and Risk for each area

		Hazard (M€/yr)		
	W. Heraklion	E. Heraklion	W. Lassithi	E. Lassithi
Grapes	3.4	12.2	0.3	0.5
Olives	4.3	9.9	3.9	1.5
		Risk (M€/yr)		
	W. Heraklion	E. Heraklion	W. Lassithi	E. Lassithi
		Grapes		
Rainfed	1.7	6.1	0.1	0.2
Suppl. Irrig.	1.2	4.9	0.1	0.2
Total	2.9	11.0	0.2	0.4
		Olives		
Rainfed	2.2	5.0	1.9	0.7
Suppl. Irrig.	1.5	4.0	1.5	0.5
Total	3.7	9.0	3.4	1.2

From the same values a great variability of results can be observed when comparing the four homogeneous areas. Regarding the total values it is observed that hazard and risk are more important in the two areas of Heraklion as compared to the areas of Lassithi.

8. CONCLUDING REMARKS

The methodology and the analysis presented in this text is the first step for devising a rational procedure to assess hazard and risk in an agricultural area. Based on this methodology which should be further refined, the necessary information can be produced for implementing preparedness plans to face droughts and water shortages. Very important values concerning risk can be used for setting priorities / regions in a descending order of urgency for improvement, in order to reduce the vulnerability in agricultural systems due to drought.

ACKNOWLEDGEMENTS

The work presented in this paper was supported by the project SEDEMED II (Interreg Community Initiative framework). The constructive comments of the anonymous referees are kindly acknowledged.

REFERENCES

Doorenbos, J. and Kassam, A.H. (1979). Yield response to water, FAO, *Irr. and Drain.* Paper 33, Rome.

Garrity, D.P. (1980). *Moisture deficits and gain Sorghum performance,* PhD thesis, University of Nebraska, Lincoln.

Hall, W.A. and Butcher, W.S. (1968). Optimal timing of irrigation, *J. Irr. and Drain.* Div. 94 (IR2): 267–275.

Hanks, R.J. (1974). Model for predicting plant yield as influenced by water use, *Agronomy Journal*, 66(5): 660–664.

Jensen, M.E. (1968). Water consumption by agricultural plants, *Water deficits and plant growth*, Ed. T.T. Kozlowski, Vol. 2: 1–22, Academic Press, NY.

Manam, R. (1974). *Physiological and Agronomic Studies in Soil - Plant - Water Relations of Soybeans*, PhD thesis, Kansas State University, Manhattan.

Martin, D.L., Watts, D.G. and Gilley, J.R. (1984). Model and production function for irrigation management, *J. Irr. and Drain. Eng.* 110(2): 149–164.

Minhas, B., Parkh, K. and Srinivasan, T.N. (1974). Toward the structure of a production function for wheat yield with dated inputs of irrigation water, *Water Resour. Res.* 10(3): 383–393.

Molden, D., Murray-Rust, H., Sakthivadivel, R. and Makin, I. (2003). A water productivity framework for understanding and action. In J. Kijne, R. Barker and D. Molden (Eds) *Water Productivity in Agriculture - Limits and Opportunities for Improvement*, Cabi Publishing, UK, pp. 1–18.

Pereira, L.S., van den Broek, B., Kabat, P. and Allen R. (Ed.) (1995). *Crop-Water-Simulation Models in Practice*, ICID, Wageningen Press, The Netherlands.

Slabbers, R.J. (1980). Practical prediction of actual evapotranspiration, *Irr. Sci.*1: 185–196.

Stamm, G.G. (1967). Problems and procedures in determining water supply requirements for irrigation projects. In R. Hagan *et al.* (Eds) *Irrigation of agricultural lands*, Am. Soc. Agron. Madison, WI., pp. 771–784.

Tsakiris, G. (1982). A method for applying crop sensitivity factors in irrigation scheduling, *Agric. Water Manage.* 5: 335–343.

Tsakiris, G. (1985a). Evaluating the effect of non-uniform and deficient irrigation (I), *Adv. Water Resour.* 8: 77–81.

Tsakiris, G. (1985b). Evaluating the effect of non-uniform and deficient irrigation (II), *Adv. Water Resour.* 8: 82–85.

SUBJECT INDEX

Water Science and Technology Library

1. A.S. Eikum and R.W. Seabloom (eds.): *Alternative Wastewater Treatment*. Low-Cost Small Systems, Research and Development. Proceedings of the Conference held in Oslo, Norway (7–10 September 1981). 1982 ISBN 90-277-1430-4
2. W. Brutsaert and G.H. Jirka (eds.): *Gas Transfer at Water Surfaces*. 1984 ISBN 90-277-1697-8
3. D.A. Kraijenhoff and J.R. Moll (eds.): *River Flow Modelling and Forecasting*. 1986 ISBN 90-277-2082-7
4. World Meteorological Organization (ed.): *Microprocessors in Operational Hydrology*. Proceedings of a Conference held in Geneva (4–5 September 1984). 1986 ISBN 90-277-2156-4
5. J. Němec: *Hydrological Forecasting*. Design and Operation of Hydrological Forecasting Systems. 1986 ISBN 90-277-2259-5
6. V.K. Gupta, I. Rodríguez-Iturbe and E.F. Wood (eds.): *Scale Problems in Hydrology*. Runoff Generation and Basin Response. 1986 ISBN 90-277-2258-7
7. D.C. Major and H.E. Schwarz: *Large-Scale Regional Water Resources Planning*. The North Atlantic Regional Study. 1990 ISBN 0-7923-0711-9
8. W.H. Hager: *Energy Dissipators and Hydraulic Jump*. 1992 ISBN 0-7923-1508-1
9. V.P. Singh and M. Fiorentino (eds.): *Entropy and Energy Dissipation in Water Resources*. 1992 ISBN 0-7923-1696-7
10. K.W. Hipel (ed.): *Stochastic and Statistical Methods in Hydrology and Environmental Engineering*. A Four Volume Work Resulting from the International Conference in Honour of Professor T.E. Unny (21–23 June 1993). 1994
    10/1: Extreme values: floods and droughts ISBN 0-7923-2756-X
    10/2: Stochastic and statistical modelling with groundwater and surface water applications ISBN 0-7923-2757-8
    10/3: Time series analysis in hydrology and environmental engineering ISBN 0-7923-2758-6
    10/4: Effective environmental management for sustainable development ISBN 0-7923-2759-4
    Set 10/1–10/4: ISBN 0-7923-2760-8
11. S.N. Rodionov: *Global and Regional Climate Interaction: The Caspian Sea Experience*. 1994 ISBN 0-7923-2784-5
12. A. Peters, G. Wittum, B. Herrling, U. Meissner, C.A. Brebbia, W.G. Gray and G.F. Pinder (eds.): *Computational Methods in Water Resources X*. 1994 Set 12/1–12/2: ISBN 0-7923-2937-6
13. C.B. Vreugdenhil: *Numerical Methods for Shallow-Water Flow*. 1994 ISBN 0-7923-3164-8
14. E. Cabrera and A.F. Vela (eds.): *Improving Efficiency and Reliability in Water Distribution Systems*. 1995 ISBN 0-7923-3536-8
15. V.P. Singh (ed.): *Environmental Hydrology*. 1995 ISBN 0-7923-3549-X
16. V.P. Singh and B. Kumar (eds.): *Proceedings of the International Conference on Hydrology and Water Resources* (New Delhi, 1993). 1996
    16/1: Surface-water hydrology ISBN 0-7923-3650-X
    16/2: Subsurface-water hydrology ISBN 0-7923-3651-8

16/3: Water-quality hydrology ISBN 0-7923-3652-6
16/4: Water resources planning and management ISBN 0-7923-3653-4
Set 16/1–16/4 ISBN 0-7923-3654-2

17. V.P. Singh: *Dam Breach Modeling Technology*. 1996 ISBN 0-7923-3925-8
18. Z. Kaczmarek, K.M. Strzepek, L. Somlyódy and V. Priazhinskaya (eds.): *Water Resources Management in the Face of Climatic/Hydrologic Uncertainties*. 1996 ISBN 0-7923-3927-4
19. V.P. Singh and W.H. Hager (eds.): *Environmental Hydraulics*. 1996 ISBN 0-7923-3983-5
20. G.B. Engelen and F.H. Kloosterman: *Hydrological Systems Analysis*. Methods and Applications. 1996 ISBN 0-7923-3986-X
21. A.S. Issar and S.D. Resnick (eds.): *Runoff, Infiltration and Subsurface Flow of Water in Arid and Semi-Arid Regions*. 1996 ISBN 0-7923-4034-5
22. M.B. Abbott and J.C. Refsgaard (eds.): *Distributed Hydrological Modelling*. 1996 ISBN 0-7923-4042-6
23. J. Gottlieb and P. DuChateau (eds.): *Parameter Identification and Inverse Problems in Hydrology, Geology and Ecology*. 1996 ISBN 0-7923-4089-2
24. V.P. Singh (ed.): *Hydrology of Disasters*. 1996 ISBN 0-7923-4092-2
25. A. Gianguzza, E. Pelizzetti and S. Sammartano (eds.): *Marine Chemistry*. An Environmental Analytical Chemistry Approach. 1997 ISBN 0-7923-4622-X
26. V.P. Singh and M. Fiorentino (eds.): *Geographical Information Systems in Hydrology*. 1996 ISBN 0-7923-4226-7
27. N.B. Harmancioglu, V.P. Singh and M.N. Alpaslan (eds.): *Environmental Data Management*. 1998 ISBN 0-7923-4857-5
28. G. Gambolati (ed.): *CENAS. Coastline Evolution of the Upper Adriatic Sea Due to Sea Level Rise and Natural and Anthropogenic Land Subsidence*. 1998 ISBN 0-7923-5119-3
29. D. Stephenson: *Water Supply Management*. 1998 ISBN 0-7923-5136-3
30. V.P. Singh: *Entropy-Based Parameter Estimation in Hydrology*. 1998 ISBN 0-7923-5224-6
31. A.S. Issar and N. Brown (eds.): *Water, Environment and Society in Times of Climatic Change*. 1998 ISBN 0-7923-5282-3
32. E. Cabrera and J. García-Serra (eds.): *Drought Management Planning in Water Supply Systems*. 1999 ISBN 0-7923-5294-7
33. N.B. Harmancioglu, O. Fistikoglu, S.D. Ozkul, V.P. Singh and M.N. Alpaslan: *Water Quality Monitoring Network Design*. 1999 ISBN 0-7923-5506-7
34. I. Stober and K. Bucher (eds): *Hydrogeology of Crystalline Rocks*. 2000 ISBN 0-7923-6082-6
35. J.S. Whitmore: *Drought Management on Farmland*. 2000 ISBN 0-7923-5998-4
36. R.S. Govindaraju and A. Ramachandra Rao (eds.): *Artificial Neural Networks in Hydrology*. 2000 ISBN 0-7923-6226-8
37. P. Singh and V.P. Singh: *Snow and Glacier Hydrology*. 2001 ISBN 0-7923-6767-7
38. B.E. Vieux: *Distributed Hydrologic Modeling Using GIS*. 2001 ISBN 0-7923-7002-3
39. I.V. Nagy, K. Asante-Duah and I. Zsuffa: *Hydrological Dimensioning and Operation of Reservoirs*. Practical Design Concepts and Principles. 2002 ISBN 1-4020-0438-9

Water Science and Technology Library

40. I. Stober and K. Bucher (eds.): *Water-Rock Interaction*. 2002 ISBN 1-4020-0497-4
41. M. Shahin: *Hydrology and Water Resources of Africa*. 2002 ISBN 1-4020-0866-X
42. S.K. Mishra and V.P. Singh: *Soil Conservation Service Curve Number (SCS-CN) Methodology*. 2003 ISBN 1-4020-1132-6
43. C. Ray, G. Melin and R.B. Linsky (eds.): *Riverbank Filtration*. Improving Source-Water Quality. 2003 ISBN 1-4020-1133-4
44. G. Rossi, A. Cancelliere, L.S. Pereira, T. Oweis, M. Shatanawi and A. Zairi (eds.): *Tools for Drought Mitigation in Mediterranean Regions*. 2003 ISBN 1-4020-1140-7
45. A. Ramachandra Rao, K.H. Hamed and H.-L. Chen: *Nonstationarities in Hydrologic and Environmental Time Series*. 2003 ISBN 1-4020-1297-7
46. D.E. Agthe, R.B. Billings and N. Buras (eds.): *Managing Urban Water Supply*. 2003 ISBN 1-4020-1720-0
47. V.P. Singh, N. Sharma and C.S.P. Ojha (eds.): *The Brahmaputra Basin Water Resources*. 2004 ISBN 1-4020-1737-5
48. B.E. Vieux: *Distributed Hydrologic Modeling Using GIS. Second Edition*. 2004 ISBN 1-4020-2459-2
49. M. Monirul Qader Mirza (ed.): *The Ganges Water Diversion: Environmental Effects and Implications*. 2004 ISBN 1-4020-2479-7
50. Y. Rubin and S.S. Hubbard (eds.): *Hydrogeophysics*. 2005 ISBN 1-4020-3101-7
51. K.H. Johannesson (ed.): *Rare Earth Elements in Groundwater Flow Systems*. 2005 ISBN 1-4020-3233-1
52. R.S. Harmon (ed.): *The Río Chagres, Panama*. A Multidisciplinary Profile of a Tropical Watershed. 2005 ISBN 1-4020-3298-6
53. To be published.
54. V. Badescu, R.S. Cathcart and R.D. Schuiling (eds): Macro-Engineering: A Challenge for the Future. 2006 ISBN 1-4020-3739-2
55. To be published.
56. G. Salvadori, C. De Michele, N.T. Kottegoda and R. Renzo: *Extremes in Nature*. An Approach Using Copulas. 2007 ISBN 1-4020-4414-3
57. S.K. Jain, R.K. Agarwal and V.P. Singh: *Hydrology and Water Resources of India*. 2007 ISBN 1-4020-4414-4
58. To be published.
59. M. Shahin: *Water Resources and Hydrometeorology of the Arab Region*. 2007 ISBN 1-4020-4577-8
60. To be published.
61. R.S. Govindaraju and B.S. Das: *Moment Analysis for Subsurface Hydrologic Applications*. 2007 ISBN 978-1-4020-5751-9
62. Giuseppe Rossi, Teodoro Vega and Brunella Bonaccorso (eds.): *Methods and Tools for Drought Analysis and Management*. 2007 ISBN 978-1-4020-5923-0